XXX IUPAP Conference on Computational Physics (CCP2018)

Journal of Physics: Conference Series Volume 1290

Davis, California, USA
29 July - 2 August 2018

Editors:

Barry M. Klein
Warren E. Pickett

ISBN: 978-1-5108-9948-3
ISSN: 1742-6588

Printed from e-media with permission by:

Curran Associates, Inc.
57 Morehouse Lane
Red Hook, NY 12571

Some format issues inherent in the e-media version may also appear in this print version.

Institute of Physics
Dirac House, Temple Back
Bristol BS1 6BE UK

Phone: 44 1 17 929 7481
Fax: 44 1 17 920 0979

techtracking@iop.org

Additional copies of this publication are available from:

Curran Associates, Inc.
57 Morehouse Lane
Red Hook, NY 12571 USA
Phone: 845-758-0400
Fax: 845-758-2633
Email: curran@proceedings.com
Web: www.proceedings.com

TABLE OF CONTENTS

IOP Conf. Series: Journal of Physics: Conf. Series **1290** (2019) 011001 doi:10.1088/1742-6596/1290/1/011001

XXX IUPAP Conference on Computational Physics (CCP2018)

Barry M Klein[1], Warren E Pickett[2] and Richard T Scalettar[2]

Department of Physics, University of California, Davis CA 95616, -USA
[1]Conference Chair and Proceedings Co-Editor
[2]Conference Vice Chair and Proceedings Co-Editor

E-mail: bmklein@ucdavis.edu, wepickett.ucdavis.edu, rtscalettar@ucdavis.edu

Abstract. This paper represents a preface to the Proceedings of the XXX IUPAP Conference on Computational Physics held at the University of California, Davis, 29 July – 2 August 2018. Background information and the organizational structure of the meeting, and acknowledgements of the contributions of the many people who made the conference a success are presented.

1. Background

The XXX Conference on Computational Physics (CCP2018) took place July 29 – August 2, 2018 at the University of California, Davis. CCP is a series of conferences held annually under the auspices of the International Union of Pure and Applied Physics (IUPAP) on the basis of endorsement by its Commission on Computational Physics (C20). The purpose of the conference series is to bring together computational scientists working in physics and closely related areas to exchange the latest developments in computational techniques and their applications.

There were more than 200 attendees from 23 countries at CCP2018. The conference featured plenary, invited and contributed talks as well as poster presentations in a wide number of subject areas of active interest to the computational physics community. -The full program is available at http://ccp2018.physics.ucdavis.edu/ where details of the eight plenary sessions, 24 parallel sessions, usually headed by one or more invited presentations, and the poster session, are presented, including the associated abstracts.

The multiple focus areas of CCP2018 included presentations on: materials physics; statistical mechanics and complex systems; soft matter; biological physics; fluid dynamics, turbulence, nanofluidics, and magnetohydrodynamics; quantum many-body physics; quantum computation; lattice field theory in particle and nuclear physics; astrophysics, space-plasma physics, gravitation, and cosmology; computational physics education; novel hardware and software paradigms; energy, environment and climate modeling; machine learning and applications; and algorithm development.

Professor Noa Marom of Carnegie Mellon University was awarded the *2018 IUPAP Young Scientist Prize in Computational Physics* for her research on developing and implementing approaches to computationally design materials with desired properties for target applications. She gave a plenary lecture at CCP2018 describing her research. We offer our congratulations and best wishes to Professor Marom for continued success.

The conference participants and their guests were also treated to a stimulating lecture at the conference banquet by Tony DeRose of *PIXAR Animation Studios* on "Computational Physics at PIXAR." We extend our thanks to Dr. DeRose.

2. Acknowledgements

We gratefully acknowledge the many people who were key to the success of CCP2018, including the International Advisory Board and the Program and Local Committee whose membership is listed below. Their recommendations included the selection of the plenary and invited speakers at CCP2018.

Special thanks to Cindy Choi, Giacomo Resta and Teresa Alameda for excellent and gratefully appreciated conference support, and Professors Ching Fong and Bernd Hamann of UC Davis for their work on the conference and its program.

We also gratefully acknowledge and thank the organizations that sponsored and supported CCP2018: The International Union of Pure and Applied Physics, The National Science Foundation, the U.S. Department of Energy, the Division of Computational Physics of the American Physical Society, the Los Alamos National Laboratory, the UC Davis Department of Physics, the UC Davis Office of Research, the UC Davis Colleges of Letters and Science, Engineering, and Biological Sciences, and UC Davis Global Affairs.

3. CCP2018 Organization

Barry M. Klein, Conference Chair
Warren E. Pickett, Conference Vice Chair
Richard T. Scalettar, Conference Vice Chair

International Advisory Board

Joan Adler (Technion, Israel)
Constantia Alexandrou (U of Cyprus, Cyprus)
Marc Etienne Brachet (ENS, France)
Nithaya Chetty (U of Pretoria, South Africa)
Marvin Cohen (UC Berkeley, USA)
Glen Brian Evenbly (U of Sherbrooke, Canada)
Hans Fangohr (U of Southampton, UK)
James Freericks (Georgetown U, USA)
Sharon Glotzer (U of Michigan, USA)
James Gubernatis (Los Alamos Nat'l Lab)
Hans Hagen (U of Kaiserslautern, Germany) Suklyun Hong (Sejong U, Korea)
Barbara Jones (IBM San Jose, USA)
Georg Kresse (U of Vienna, Austria)
David Landau (U of Georgia, USA)
Hai-Qing Lin (Beijing CSRC, China)
Richard Liska (Czech Tech'l U, Czech Republic)
Priya Mahadevan (SNBNCBS, India)
Richard Martin (U of Illinois Urbana Champaign)
Rossitza Pentcheva (U Duisburg-Essen, Germany)
Marcos Rigol (Pennsylvania State U, USA)
Mathieu Salanne (Sorbonne U, France)
Anders Sandvik (Boston U, USA)
Sitangshu Bikas Santra (Indian Institute of Technology Guwahati, India)
Purusattam Ray (Institute of Mathematical Sciences, Chennai, India)
Friederike Schmidt (U of Mainz, Germany)
Barry Schneider (NIST, USA)
Lev Shchur (Landau Institute for Theoretical Physics, Russia)

IOP Conf. Series: Journal of Physics: Conf. Series **1290** (2019) 011001 doi:10.1088/1742-6596/1290/1/011001

Program and Local Committee
Jerome Buerki (California State U, Sacramento)
Steve Carlip (UC Davis)
Lincoln Carr (Colorado School of Mines)
Hai-Ping Cheng (U of Florida - Gainesville)
Aurora Clark (Washington State U)
Valentino Cooper (Oak Ridge Nat'l Lab)
Daniel Cox (UC Davis)
Raissa D'Souza (UC Davis)
Fereydoon Family (Emory U)
Marivi Fernández-Serra (Stony Brook U)
Ching Fong (UC Davis)
Giulia Galli (U of Chicago)
Nir Goldman (Lawrence Livermore Nat'l Lab)
Jim Gubernatis (Los Alamos Nat'l Lab)
Francois Gygi (UC Davis)
Bernd Hamann (UC Davis)
Michele Johannes (U.S. Naval Research Lab)
Paul Kent (Oak Ridge Nat'l Lab)
Ying Wai Li (Oak Ridge Nat'l Lab)
Michael Mehl (U.S. Naval Academy)
Peter Nugent (Lawrence Berkeley Nat'l Lab)
Dimitri Papaconstantopoulos (George Mason U)
Warren Pickett (UC Davis)
Sergey Savrasov (UC Davis)
Richard Scalettar (UC Davis)
Rajiv Singh (UC Davis)
Mark Stevens (Sandia Nat'l Lab)
Jean-Luc Vay (Lawrence Berkeley Nat'l Lab)
Andrew Wetzel (UC Davis)
Gergely Zimanyi (UC Davis)

IOP Conf. Series: Journal of Physics: Conf. Series **1290** (2019) 011002 doi:10.1088/1742-6596/1290/1/011002

Peer review statement

All papers published in this volume of *Journal of Physics: Conference Series* have been peer reviewed through processes administered by the proceedings Editors. Reviews were conducted by expert referees to the professional and scientific standards expected of a proceedings journal published by IOP Publishing.

IOP Conf. Series: Journal of Physics: Conf. Series **1290** (2019) 012001 doi:10.1088/1742-6596/1290/1/012001

Cosmology data analysis challenges and opportunities in the LSST sky survey

J Anthony Tyson

Department of Physics
Univerity of California, Davis, USA

tyson@physics.ucdavis.edu

Abstract. Fueled by advances in software, microelectronics, and large optics fabrication, a new type of sky survey will soon begin. In a relentless campaign of 15 second exposures with a 3 gigapixel camera, the Large Synoptic Survey Telescope will cover the sky deeply every week for ten years. LSST will chart billions of remote galaxies, providing multiple probes of the mysterious Dark Matter and Dark Energy. Multiple probes of the effects of dark energy over an unprecedented volume of the universe will allow us to measure how dark energy behaves over time to high precision. Hundreds of petabytes of high dimensional complex data will be mined and compared with Exascale simulations. After reviewing the LSST project, I will describe some of the computational challenges and opportunities.

1. Introduction

It is remarkable that fully 96% of our universe remains a mystery, evidence for new physics. Moreover there are two "dark" components. Dark matter is responsible for the gravitational assembly of galaxies, and while we can produce images of dark matter utilizing the gravitational mirage it produces on the background galaxies, we have no idea of its physical nature. Similarly we as yet have no understanding of the physics of the late-time acceleration of the expansion of the universe, "dark energy." Dynamically, these two dark components oppose each other: dark matter causes galaxies to cluster, but dark energy expands space-time at recent times.

To explore the nature of this, we need to measure the way the expansion of our universe and the dark matter mass structure changes with cosmic time. Since the expansion of the universe has been accelerating, the development of mass structures via ordinary gravitational in-fall will be impeded. Measuring how dark matter structures and ratios of distances grow with cosmic time – by observing the cosmic mirage of background galaxies produced by foreground mass -- will provide clues to the nature of dark energy. A key strength of the next generation surveys is the ability to survey huge volumes of the universe. Such a probe will be a natural part of the all-sky imaging survey: billions of distant galaxies will have their shapes and colors measured. Sufficient color data will be obtained for an estimate of the distance to each galaxy by using redshifts derived from the multi-band photometry. By combining with dynamical data from spectroscopy, we can test whether the "dark energy" is due to a breakdown of General Relativity on large scales. These diverse goals require the construction of a new astronomical facility capable of surveying billions of galaxies – measuring their positions, shapes, and distances – over the full visible sky and back in cosmic time.

IOP Conf. Series: Journal of Physics: Conf. Series **1290** (2019) 012001 doi:10.1088/1742-6596/1290/1/012001

1.1 Exploration in astronomy

The various dimensions of exploration in astronomy (wavelength, angular resolution, depth, sky coverage, and time resolution) normally are impossible to achieve simultaneously with the same facility. The rate at which we can survey the sky to a given faintness is proportional to the product of telescope light collecting area and the size of the focal plane imaging array. This is also the data rate. The technologies of microelectronics, computation and software, and large optics fabrication drive the new sky. Aided by rapid progress in these areas, CCD camera sky surveys are changing the way we view and study the universe. Historically there has been complementary progress in detectors (imaging arrays) and computation.

Figure 1 charts this trend in optical sky surveys over 40 years. The effect of technology is clear. While the total light collecting area of telescopes (blue curve) has remained comparatively constant, the information content of sky surveys (using the number of galaxies measured per year to a given signal-to-noise ratio as a proxy) has risen exponentially. This is mainly due to high efficiency imaging arrays growing to fill the available telescope focal plane, and to the increase in processing power to handle the resulting data volume. Development of software to analyze these digital data in new ways has resulted in a corresponding increase in science output. Photographic surveys had the advantage of large focal plane area early on, but have been eclipsed by more sensitive CCD surveys, driven by the exponential rise in pixel count and computer processing capability – both enabled by the microelectronics "Moore's Law". Plotted versus time is the cumulative sum of all CCD pixels in sky survey cameras, as well as the number of transistors in a typical CPU. Processing capability keeps up with the data rate. Also plotted is one result of CCD surveys – the number of galaxies measured per unit time – ranging from a survey using a single 164 Kpixel CCD on a 4m telescope to a 3.2 Gpixel camera on an 8.4m telescope: the Large Synoptic Survey Telescope (LSST) sky survey.

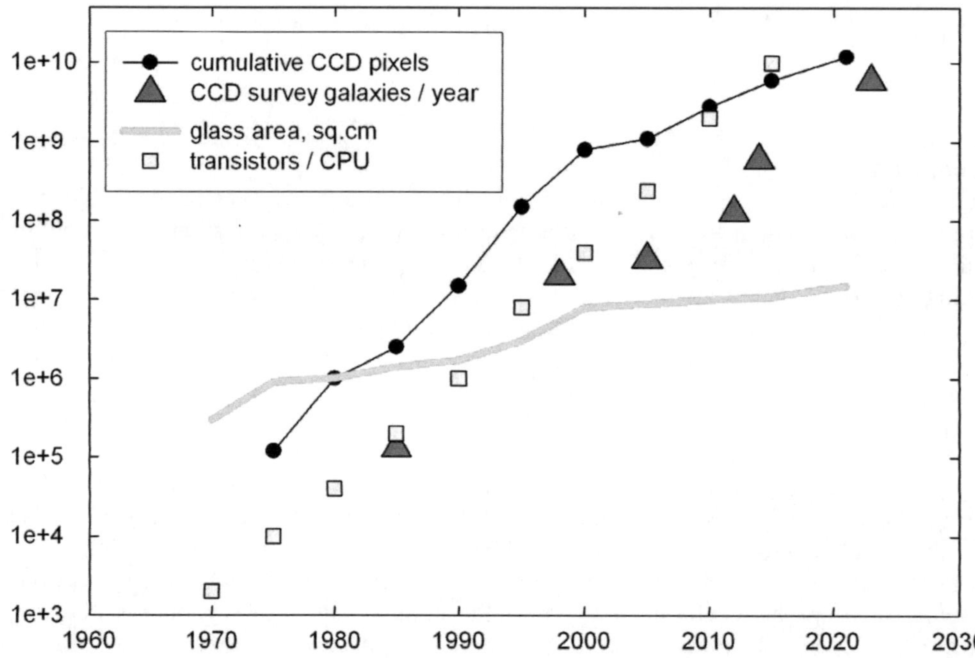

Figure 1. Data trends in optical surveys of the sky. Data rate is proportional to the product of telescope light collecting area and the size of the focal plane CCD array. Information content (galaxies surveyed per unit time) in digital sky surveys roughly follows Moore's law, and processing capability has kept up with pixel count. The LSST wide-fast-deep sky survey 2022-2032 will open new windows on the universe.

IOP Conf. Series: Journal of Physics: Conf. Series **1290** (2019) 012001 doi:10.1088/1742-6596/1290/1/012001

2. LSST

LSST combines a large light collecting capability with a giant wide-field digital camera: an 8.4 meter diameter mirror and a 3.2 gigapixel camera covering ten square degrees per exposure (an area on the sky equivalent to 40 full moons). The camera focal plane is tiled with 189 4Kx4K CCDs of special design which enables good spectral response from 300nm to 1100nm. Each CCD is segmented into 16 sections, each with its own output amplifier. Due to this massively parallel architecture, the entire focal plane is read out synchronously within 2 seconds. The LSST has the capability of surveying the deep universe in a novel way. This system will have unprecedented optical throughput and can image about 10,000 square degrees of sky in three clear nights using 30 second "visits" per each sky patch 2-3 times per night. Beginning in 2022 the LSST will begin tiling the entire visible southern sky with 1000 exposures per sky patch. This will continue every night for ten years, creating a catalog from its 30 trillion measurements of 37 billion objects. Figure 2 shows LSST at its location on Cerro Pachon in northern Chile. By October 2018 the facility was nearly complete. A live web cam at Cerro Pachon on the gallery page at http://lsst.org shows the current status.

The sky survey capability of LSST is unprecedented. Because each 30 second exposure goes so faint, rapid tiling of the sky enables a new kind of exploration of the time domain in astronomy. Unexplored regions of luminosity-timescale for explosive events will be explored. The huge volume of high-quality data enables a wide range of science opportunities, opening a new era of precision cosmology with unprecedented statistical power [1].

Figure 2. The LSST Observatory: artist's rendering of the dome enclosure with the attached summit support building on Cerro Pachon in northern Chile. [LSST Project/NSF/AURA]

2.1 The New Sky
Since LSST will cover the entire visible sky deeply and rapidly, unexplored regimes become accessible. The unprecedented data from the LSST sky survey will open new windows of understanding on cosmology, the assembly of our Milky Way galaxy, and near-Earth asteroids. New types of objects will likely be found, whose luminosity changes rapidly. In cosmology the combination of a survey of a deep volume of the universe (a wide range of cosmic time) with a wide sky area will bring novel probes of the physics of dark matter and dark energy. Surveying a wide range of cosmic time (wide redshift range) is important both for optimizing the gravitational lens mirage and for detection of the evolution of cosmic dark matter structure.

2.1.1. Multiple probes of cosmology. The mean density of dark matter in the universe is only one piece of the puzzle. More information is needed to decipher the physics of dark matter. The late-time cosmic acceleration [dark energy] is a profound challenge to our understanding of the universe. So far investigations have been focused on two classes of models: a new mass-energy having effectively negative pressure and modified gravity. There are also models that have no new mass-energy components or physics but attribute the acceleration to a breakdown of the cosmological principle or an oversimplification of General Relativity effects in the real universe. These models are less popular, but they will be tested with the multiple probes together with measurements of the isotropy of dark energy.

The data from the LSST sky survey may be used to probe the physics of dark energy and dark matter in multiple complementary ways. Measurements of galaxy clustering, dark matter distribution vs cosmic time, the evolution of geometry, and any anisotropy all give clues to the nature of these mysteries. For a recent review of LSST cosmological probes, see [2].

3. LSST Data
During operations the LSST facility will produce 20 TB of on-sky imaging data per night for ten years, in addition to daily calibration images. These data will result in a relational database including 20 billion galaxies and a similar number of stars, and will serve the majority of the primary science programs. While this data volume is not challenging either for storage or processing, the high dimensionality of the data is challenging for analysis. For example, each of the approximately 20 billion galaxies imaged and cataloged by LSST have hundreds of effective dimensions. Data collected by the LSST telescope and camera will be processed to data products -- catalogs, alerts, and reduced images -- by the LSST Data Management system. The science community will work directly with these data products, without having to work directly with the raw pixels.

The data processing from images to catalogs requires 1.8 PFLOPS continuous. This includes pixel processing, calibration, transient object detection, co-addition of all images of the same part of the sky, and sky catalog generation. Already by 2018 a million lines of code have been written. All the LSST code has been designed and implemented following software engineering best practices, including modularity, clear definition of interfaces, continuous integration, unit testing, and a single set of documentation and coding standards [3]. The primary implementation language is Python and, where necessary for performance reasons, C++. Most of the LSST software is licensed under the terms of the GNU General Public License (GPL), Version 3, and can be found at https://github.com/lsst The documentation for the LSST Science Pipelines is available at https://pipelines.lsst.io .

The data management system will span four key facilities on three continents: the Summit Facility on Cerro Pachon in Chile; the Base Facility in La Serena, Chile (which will serve as a retransmission for data uploads to North America, as well as the Data Access Center for the Chilean community); the Data Processing and Archiving Facility at the National Center for Supercomputing Applications (NCSA) in Champaign-Urbana, IL and the Satellite Processing Facility at CC-IN2P3 in Lyon, France. The data will be transported between the centers over redundant existing and new high-speed optical fiber links from South America to the U.S.

IOP Conf. Series: Journal of Physics: Conf. Series **1290** (2019) 012001 doi:10.1088/1742-6596/1290/1/012001

3.1 Data Products

After the pixel processing, the data processing branches into a prompt alert processing pipeline and the "static sky" deep survey pipeline. In each case catalogs are created. The large data volume, the fast detection and alert aspect, and the complexity of processing involved makes it impractical to rely on the end users for the data reduction. Instead, the data collected by the LSST system will be automatically reduced to scientifically useful catalogs and images. Over the ten years of LSST operations and 11 data releases, this processing will result in cumulative processed data of about 500 PB for imaging, and over 50 PB for the catalog databases. The final data release catalog database alone is expected to be approximately 20 PB in size.

The data from the LSST camera will be automatically processed to data products: catalogs, alerts, and reduced images. These products enable the majority of LSST science cases, without the need to work directly with the raw pixels. Details of the LSST data products may be found in the LSST Data Products Definition Document [https://ls.st/dpdd], which is periodically updated. These data will be served via a relational database. The scale of the LSST data release catalogs -- together with the rapid needed response -- present some engineering challenges. The LSST project has been developing *Qserv*, a shared-nothing massively parallel processing database system, to meet these needs. Catalog data within *Qserv* is spatially partitioned, and hosted on servers running on dedicated hardware within the LSST Data Facility. The servers locally leverage conventional relational database management system technologies, running behind custom front-end codes which handle query analysis, rewrite, distribution, and result aggregation. The *Qserv* servers also provide cross-user synchronization of full-table scans in order to provide predictable query response times when serving many users concurrently. More details about *Qserv* can be found in the LSST document LDM-135 [4].

Two major categories of data products will be produced and delivered by the LSST: Prompt and Data Release. Prompt products are time domain data products which support the discovery, characterization, and rapid follow-up of time-dependent phenomena. Data Release products are the deep universe catalogs obtained from co-added multi-band images and are most relevant to cosmology: they are designed to enable systematics and flux-limited science, and will be made available in annual Data Releases. The LSST data products will be made available to the U.S. and Chilean scientific communities and to international partners with no proprietary period. The ultimate goal is to make LSST data products including the relational database available to the public and scientists around the world.

3.2 Data Enabled Discovery

Jim Gray commented that with the new generation of experiments we are entering a 4th paradigm: data driven discovery. As an example of this new paradigm, LSST is a kind of astronomy genome project. Rather than undertake separate new sky surveys to test new idea as we have done in the past, a single comprehensive sky survey will gather sufficient data to serve the entire astrophysics community. In effect the sky will be transformed onto spinning disks with high-dimensional catalogs of features for 37 billion astronomical objects. This Exascale data enables many separate "experiments". It also enables automated searches for the unexpected.

The LSST science community has organized into several science collaborations to carry out specific experiments centered around supernovae, active galaxies, stars, etc. These organized teams are working together to optimize science analyses and assess the importance of systematic uncertainties on the derived results. For example, the LSST Dark Energy Science Collaboration (DESC) has members with interests in the study of dark energy and related topics in fundamental physics with LSST data. The LSST science community will perform its analyses using LSST computational facilities at NCSA, and the system has been sized accordingly.

IOP Conf. Series: Journal of Physics: Conf. Series **1290** (2019) 012001 doi:10.1088/1742-6596/1290/1/012001

4. Curse of Dimensionality: Analysis Computing

The LSST sky survey will produce large uniform data sets with high quality for multiple cosmological probes. These probes are affected by sample selection and various systematic errors and are sensitive to cosmology in different ways. Joint analyses of multiple probes will improve parameter constraints, and more importantly, also enable cross-calibrations of known systematics via breaking single-probe degeneracies, and finally detection and characterization of unknown systematics. It is important to have a consistent treatment of shared physics (there is only one universe with one dark matter distribution) and systematics across all probes.

This enables robust tests on cosmological models and exploration of the fundamental physics of the universe beyond current understanding. Automated data quality assessment is key; there is too much data to be looked at by humans. The challenges of dimensionality become apparent when we consider the computational effort for the LSST 3-probe investigation of dark energy. The data vector will have 6500 elements, the covariance matrix will have 43 million elements, and the cosmology likelihood analysis is expected to take 1M CPU-hours per run. There is also the challenge of optimization and sampling in high-dimensional parameter spaces. As described below however, there is a far larger effort required in forward simulating the observations themselves.

4.1 Cosmology simulations

We must undertake simulations of the universe for many different cosmological models in order to solve the inverse problem of determining cosmological parameters, investigating systematics in the survey observations, as well as astrophysical systematics which can mimic physics effects of the dark sector. These cosmological simulations are also needed for calibration of errors in the process of cosmological parameter likelihood estimation via precision covariance estimates, and for testing, optimizing, and validating observational strategies with synthetic catalogs. Data richness (high dimensionality) is an issue, requiring suites of very expensive cosmological simulations. These simulations must be sufficiently large in volume to cover the observed universe and of sufficient mass and force resolution to resolve the smallest relevant objects. The simulations must also span the model parameter space, and reach the precision enabled by the LSST data, and be thoroughly validated. Typically each of these simulations of cosmic structure formation in large volumes take weeks on supercomputers [5].

These n-body simulated catalogs of dark matter structure formation must also supplemented with realistic distributions of astrophysical objects, including galaxies which form gravitationally at the peaks of the dark matter. Their number, size, brightness, colors, and shape distributions must match what is known from deeper and complementary surveys. These galaxies are assigned self-consistently to corresponding peaks in the n-body simulated dark matter 3D field. It is critical to validate the consistency of these catalogs of the simulated astrophysical sky [6].

4.2 Simulations of observations

Ultimately observations must be compared with simulated observations based on the astrophysical sky. The cosmological simulated catalogs discussed above are used as input to simulate mock observed images and mock observational catalogs. It is important to include the selection effects due to observation parameters as well as the effects of image analysis and catalog generation software. Because the photometric pipeline processing from images to catalogs is nonlinear and complex, we must first simulate mock observed images, and then process those in exactly the same way that the LSST images are processed. There are two reasons to simulate the observations: (1) this permits unit testing of analysis algorithms, including cosmological parameter estimation (see below), and (2) production of realistic covariance matrices. These simulations of LSST images take months on supercomputers. It is useful to benchmark the computation required ultimately for the simulation of the survey. Using an LSST image simulator which follows each photon from the cosmos through the detector, and then extrapolating to the full 18,000 sq.deg implies we would need 1.3 billion CPU-hours! Better use of GPUs will help, but

IOP Conf. Series: Journal of Physics: Conf. Series **1290** (2019) 012001 doi:10.1088/1742-6596/1290/1/012001

there are also possible simulation algorithm efficiency improvements now being developed. Finally, there is the question of whether we need to simulate the full survey images, rather than a representative subset. Obviously the forward simulation of the imaging data is expensive. The computational burden of simulating these representative images is far larger than any other task, including all the data management required for processing the images to catalogs.

4.3 Cosmological parameter estimation

We wish to compare the theoretical prediction to the observed measurements. In a Bayesian analysis, a likelihood function is developed to estimate the probability of the observational data given a proposed theory (and its parameters). Modular methods have been developed which make it efficient to incorporate the observational selection and systematics in a self-consistent way for any combination of correlation functions for the observational probes [7]. The goal is to determine the posterior distribution of cosmological parameters given the simulation output and the observational data. In parallel with the observations, simulations of many candidate cosmologies are carried out, and converted to catalogs with the addition of galaxies (and their properties) and known astrophysical systematics. This parallel chain is shown in figure 3. Large simulations of the full observations yield a covariance matrix. After incorporating the known observational sample selection and systematics, the model is compared with observations via a likelihood analysis incorporating errors from the full covariance matrix. This must be repeated for each candidate cosmology model.

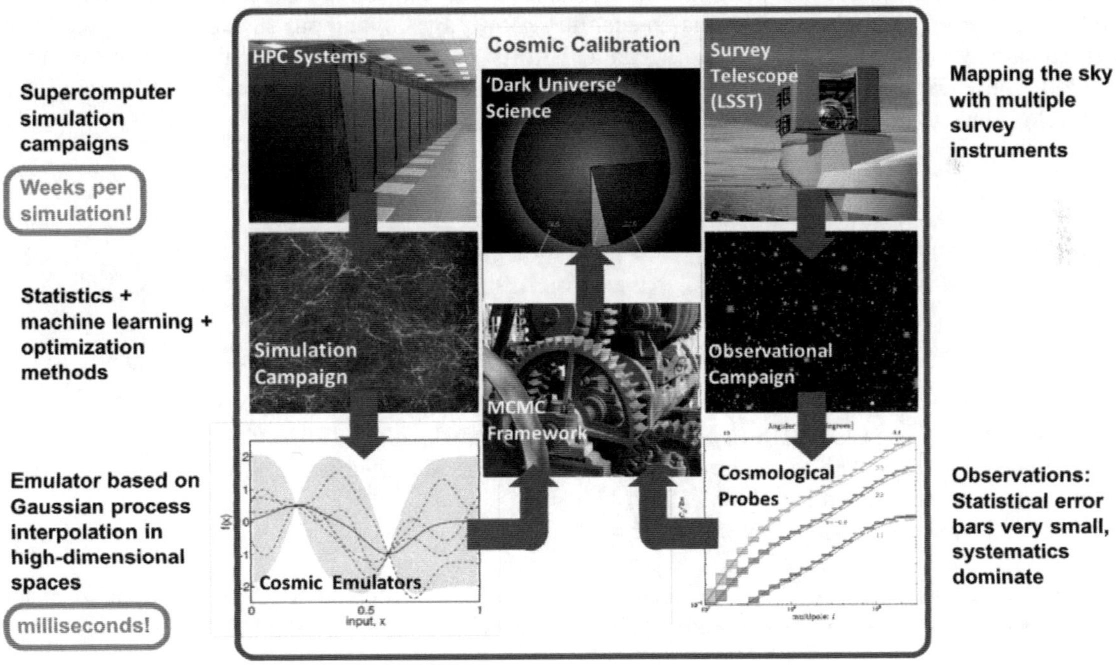

Figure 3. The Bayesian process of cosmological parameter estimation begins with two parallel campaigns: astronomical observations and simulated universes. After incorporating the known observational sample selection and systematics, the model is compared with observations via a likelihood analysis incorporating errors from a full covariance matrix. [S. Habib]

IOP Conf. Series: Journal of Physics: Conf. Series **1290** (2019) 012001 doi:10.1088/1742-6596/1290/1/012001

The level of sophistication of this whole process has developed over the years, tracking computation technology. This increase in capability has led to several generations of frontiers in complexity-volume space as applied to cosmology. Terascale: first generation surveys and single cosmological probe simulations. Petascale: second generation surveys, two-probe simulations, intermediate accuracy cosmological parameter estimation. Exascale: next generation surveys (like LSST), end-to-end multi-probe full survey scale simulations, and multiple cross-calibrated probes. Although the cosmic inference MCMC is not parallelizable leading to weeks of supercomputer time, by far the largest computation is the simulation of realistic observed images.

4.4 Opportunity for discovery of the unexpected

While the normal approach of hypothesis testing, i.e. comparing the data with various cosmological models will certainly be done, the richness of the data enables more adventurous explorations; we can in principle let the data speak for itself. Correlations over the high-dimensional database may discover unexpected cosmological trends. Non-parametric machine learning approaches – trained on the set simulations for of candidate cosmologies – may reveal outliers in feature space. There has been some initial progress, with ML applied to galaxy distance estimation.

The LSST holds great promise for breakthrough discoveries in cosmology, and it comes with challenges. While the LSST project completes the systems engineering and construction work, scientists have much to prepare for. Since the galaxy sample size is so large, the statistical noise is sub-dominant: the LSST data will be limited by systematic errors arising from instruments, observations, data reduction, analysis methods, and even deficiencies in theories. The good news is that the LSST survey will bring unprecedented statistical precision enabling knowledge beyond our current horizon, requiring extraordinary effort and novel use of the data. Efforts on detecting and controlling systematics, especially those not yet known, are crucial for the LSST to reach for its full potential. An imperative is the development of tools for new modes of data exploration.

References

[1] Abell P A *et al.* (The LSST Collaboration) 2009 LSST Science Book arXiv:0912.0201
[2] Zhan H and Tyson J A 2018 *Rep. Prog. Phys.* **81** 066901 (arXiv:1707.06948)
[3] Jenness T (LSST Data Management Team) 2017 *ASPC* **512** 297 (arXiv:1511.06790)
[4] Becla J *et al.* 2017 https://ls.st/LDM-135
[5] Habib S *et al.* 2016 *New Astronomy* **42** 49-65
[6] Mao Y *et al.* 2018 *ApJS* **234** 36
[7] Zuntz J, Paterno M, Jennings E, Rudd E, Manzotti A, Dodelson S, Bridle S, Sehrish S and Kowalkowski J 2015 *Astronomy & Computing* **12** 45 (arXiv:1409.3409)

IOP Conf. Series: Journal of Physics: Conf. Series **1290** (2019) 012002 doi:10.1088/1742-6596/1290/1/012002

Towards efficient density functional theory calculations without self-interaction: The Fermi-Löwdin orbital self-interaction correction

K. A. Jackson, J. E. Peralta, R. P. Joshi, K. P. Withanage, K. Trepte, K. Sharkas and A. I. Johnson

Physics Department and Science of Advanced Materials Ph.D. Program, Central Michigan University, Mt. Pleasant, MI USA

E-mail: `jacks1ka@cmich.edu`, `peral1j@cmich.edu`

Abstract. The Fermi-Löwdin orbital (FLO) approach to the Perdew-Zunger self-interaction correction (PZ-SIC) to density functional theory (DFT) is described and an improved approach to the problem of optimizing the Fermi-orbitals in order to minimize the DFT-SIC total energy is introduced. To illustrate the use of the FLO-SIC method, results are given for several applications involving problems where self-interaction errors are pronounced.

1. Introduction

Density functional theory (DFT) is the workhorse of computational condensed matter physics. Nowhere is this more evident than in a materials genome project where DFT calculations, typically using the Perdew-Burke-Ernzerhof (PBE) [1] generalized gradient approximation (GGA) to the exchange-correlation functional, are performed for thousands of materials to create a large database of calculated results that can be mined to discover property trends. A number of these projects were discussed at CCP2018. PBE is used for these calculations because it combines reliability with efficiency, *i.e.* accurate predictions of materials properties can be produced at relatively low computational cost. A subtle factor in the success of this enterprise is that the calculations are done for atomic configurations at or near equilibrium. In such situations PBE and similar functionals generally perform well; however, the reliability of these methods deteriorates when the atoms are far from equilibrium. For example, chemical transition states typically involve strongly stretched bonds and PBE and other functionals significantly underestimate the corresponding reaction barriers. The culprit behind this breakdown is self-interaction error (SIE), which is present in all approximate semilocal density functionals due to an inexact cancellation of self-Coulomb and self-exchange-correlation energies. In addition to too-low reaction barriers, SIE causes a raft of other problems in DFT calculations, from orbital energies that are too high, to problems in the description of charge transfer.

The Perdew-Zunger self-interaction correction (PZ-SIC) [2] was introduced in the 1980's to remove electron self-interaction from DFT. Early tests indicated that many SIE-related problems were indeed fixed by PZ-SIC, but the method was not widely adopted for two reasons. First, the method is computationally demanding relative to uncorrected DFT, and second, while the method can improve the success of DFT in situations where SIE is important, it degrades the performance of PBE and related functionals in settings where DFT already works well [3, 4].

The Fermi-Löwdin orbital self-interaction correction (FLO-SIC) [5–7] is a recent implementation of PZ-SIC that offers a potential computational advantage over earlier approaches. A long-term goal of our research is to use FLO-SIC to explore new approaches of combining SIC with DFT to remove the effects of SIE without diminishing the otherwise good performance of DFT. A first step in this direction is to more fully assess the method for problems where SIE is known to be important. In this paper we present a brief overview of the FLO-SIC formalism, including a discussion of recent methodological developments that improve its efficiency and functionality. We then present illustrative examples to demonstrate how FLO-SIC successfully addresses the effects of electron self-interaction. We conclude with a reflection on remaining challenges to implementing FLO-SIC more broadly.

2. Theoretical background

In DFT, the electrostatic Coulomb energy of the electrons E_{H} includes the interaction of each electron with itself. In exact DFT, this self-Coulomb energy is exactly cancelled by the exchange-correlation (XC) energy E_{XC}, but residual self-interaction remains when an approximate XC functional is used instead. In PZ-SIC, this residual energy is subtracted out on an orbital-by-orbital basis [2]

$$E^{\text{DFT-SIC}} = E^{\text{DFT}}[n^{\alpha}, n^{\beta}] - \sum_{i\sigma}(E_{\mathrm{XC}}[n_i^{\sigma}, 0] + E_{\mathrm{H}}[n_i^{\sigma}]), \tag{1}$$

where n_i^{σ} are single orbital densities of spin σ, E^{DFT} is the DFT energy functional, E_{XC} is the exchange-correlation functional used and E_{H} is the Hartree, or Coulomb, energy

$$E_{\mathrm{H}} = \frac{1}{2}\int \frac{n(\mathbf{r})n(\mathbf{r}')}{|\mathbf{r}-\mathbf{r}'|}\mathrm{d}^3r\mathrm{d}^3r'. \tag{2}$$

Schrödinger-like DFT-SIC equations are derived analogously to the Kohn-Sham (KS) equations using a standard variational method, yielding

$$\left(\hat{H}_{\sigma}^{\mathrm{DFT}} + V_{i\sigma}^{\mathrm{SIC}}\right)\phi_{i\sigma} = \hat{H}_{i\sigma}^{\text{DFT-SIC}}\phi_{i\sigma} = \sum_{j=1}^{M_{\sigma}} \lambda_{ji}^{\sigma}\phi_{j\sigma}. \tag{3}$$

Here $\hat{H}_{\sigma}^{\mathrm{DFT}}$ is the DFT Hamiltonian, $V_{i\sigma}^{\mathrm{SIC}}$ is the SIC potential of orbital i, and $\hat{H}_{i}^{\text{DFT-SIC}}$ is the corresponding SIC Hamiltonian. The λ_{ji}^{σ} are the Lagrange multipliers introduced in the variational procedure to ensure orthogonality between the M_{σ} occupied spin orbitals.

A challenging feature of PZ-SIC is that the total energy $E^{\text{DFT-SIC}}$ and the DFT-SIC equations are orbital-dependent. This means that individual orbital densities must be optimized in order to obtain the lowest $E^{\text{DFT-SIC}}$. The key idea behind FLO-SIC is to use localized Fermi orbitals (FO) in the PZ energy functional [5–7]. A FO is defined as

$$F_{i\sigma} = \sum_{j=1}^{M_{\sigma}} \frac{\psi_{j\sigma}^{*}(\mathbf{a}_{i\sigma})\psi_{j\sigma}(\mathbf{r})}{\sqrt{\sum_{j=1}^{M_{\sigma}}|\psi_{j\sigma}(\mathbf{a}_{i\sigma})|^2}}, \tag{4}$$

where $\psi_{j\sigma}(\mathbf{r})$ is any set of orbitals that spans the occupied orbital space (*i.e.* $\sum_{j=1}^{M_{\sigma}}|\psi_{j\sigma}(\mathbf{r})|^2 = n_{\sigma}(\mathbf{r})$). The $\mathbf{a}_{i\sigma}$ are called Fermi-orbital descriptors (FODs). They are M_{σ} distinct points in space that characterize the transformation. The FO are normalized, but not orthogonal, so the Löwdin method [8] is applied to obtain orthonormal Fermi-Löwdin orbitals $\phi_{i\sigma}$ (FLOs).

Minimizing $E^{\text{DFT-SIC}}$ requires that both the total density and the FLOs are optimized. The latter is accomplished by rearranging the FOD positions. As discussed further below, the FOD positions can be optimized by using the gradients of the total energy with respect to the $\mathbf{a}_i$, which we refer to as FOD forces. A flowchart summarizing a FLO-SIC energy minimization is given in Fig. 1.

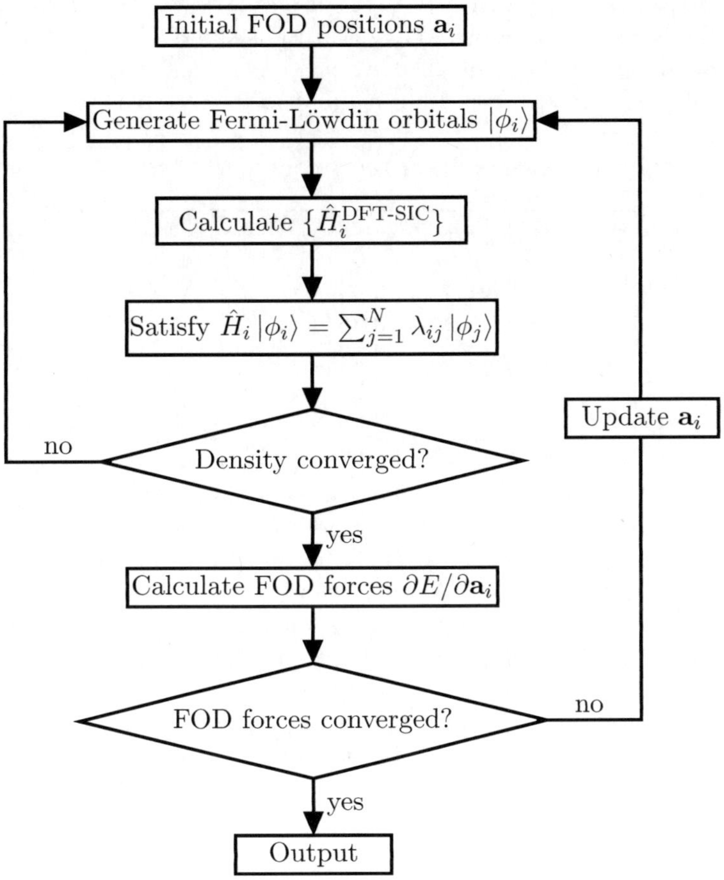

Figure 1: General outline of the FLO-SIC method.

3. Improving FOD optimization

In FLO-SIC, the total energy is minimized by optimizing the FOD positions. Pederson [6] derived an expression for the gradients $\mathbf{g}_i = \partial E^{\text{DFT-SIC}}/\partial \mathbf{a}_i$ and used these in a conjugate gradient algorithm to optimize the FODs for some closed-shell atoms. The method is akin to that used to optimize the arrangements of atoms in a molecule. Using a similar approach, we determined minimum-energy FODs for all the atoms Li to Kr [9]. In practice, the conjugate gradient algorithm may take hundreds of steps to determine optimal FOD positions, and the number of steps required grows as the number of FODs increases. It is clearly important to make the FOD optimization as efficient as possible.

IOP Conf. Series: Journal of Physics: Conf. Series **1290** (2019) 012002 doi:10.1088/1742-6596/1290/1/012002

To gain insight into the nature of the FOD energy surface, we computed second derivatives of the energy with respect to the FOD positions at the minimum energy FOD arrangement for several closed-shell atoms. For a particular FOD **a**, the second derivative $h_{kl} = \partial^2 E^{\text{DFT-SIC}}/\partial a_k \partial a_l$ was calculated numerically via finite differences of the corresponding gradients

$$h_{kl} = \frac{(g_k|_{+\Delta x_l} - g_k|_{-\Delta x_l})}{2\Delta x_l}, \tag{5}$$

where g_k is the gradient in the k direction ($k = x$, y or z) and $\pm\Delta x_l$ is an FOD displacement in the l direction. We used displacements of 0.001 Bohr to compute the derivatives. The average of the diagonal element ($h_D = \frac{1}{3}\sum_k h_{kk}$) is a measure of how steeply curved the energy surface is for the given FOD. For a typical atom, the curvature is similar for all FODs corresponding to local orbitals in the same electronic shell. The averaged, per-shell values of h_D for the closed-shell atoms Ne, Mg, Ar, Ca, Zn, Kr are presented in table 1.

Table 1: Averaged diagonal elements h_D of the second derivative matrix for each shell (a.u).

	Z	1s	2s2p	3s3p3d	4s4p
Ne	10	0.52	0.011	—	—
Mg	12	1.3	0.044	—	—
Ar	18	6.2	0.36	0.012	—
Ca	20	12.	0.92	0.026	—
Zn	30	42.	7.8	0.30	—
Kr	36	81.	18.	0.81	0.0022

h_D for the 1s FOD is at least an order of magnitude larger than those for valence FODs for all atoms. For example, for Kr the 1s second derivative (81 a.u.) is four orders of magnitude larger than that of the 4s4p FODs (0.0022 a.u.). This implies that for similar displacements of the FODs from equilibrium, the 1s FOD gradient would be vastly larger than a 4s4p gradient. This is a signature of an ill-conditioned optimization problem that can result in slow convergence to a solution using a straightforward application of gradient-based algorithms.

To accelerate the convergence, we used the averaged diagonal elements h_D of the second derivative matrix to scale the gradients (**g**) and the FOD coordinates (**a**) as follows

$$\mathbf{g}' = h_D^{-1/2}\mathbf{g} \tag{6}$$

$$\mathbf{a}' = h_D^{1/2}\mathbf{a}, \tag{7}$$

prior to using the coordinates and gradients in the optimization algorithm. The effect is to make the scaled second derivatives $\partial g_k'/\partial a_l'$ approximately the same size for all FODs. To carry out this preconditioning step for atoms not listed in table 1, we fit the values of h_D for each shell vs atomic number Z. For each shell, a cubic polynomial gives an excellent fit to the data in the table. Values for any atom can be extracted from these fits.

The scaled gradients and coordinates can be used in conjunction with any gradient-based optimization algorithm. We used them with the conjugate gradient (CG) and L-BFGS algorithms. To compare the performance of these methods with and without preconditioning, we considered the Kr and Cr atoms. We used the same starting set of FODs for all methods. To create the starting sets, we displaced each FOD from its ideal position given in Ref. [9], while maintaining the same approximate distance to the nucleus. For the scaled cases, we simply applied the scaling transformations to the gradient and coordinate vectors immediately before the call to the optimization subroutine and reversed the scaling for the updated coordinate vector returned from the subroutine. Each FOD was then moved using the unscaled update.

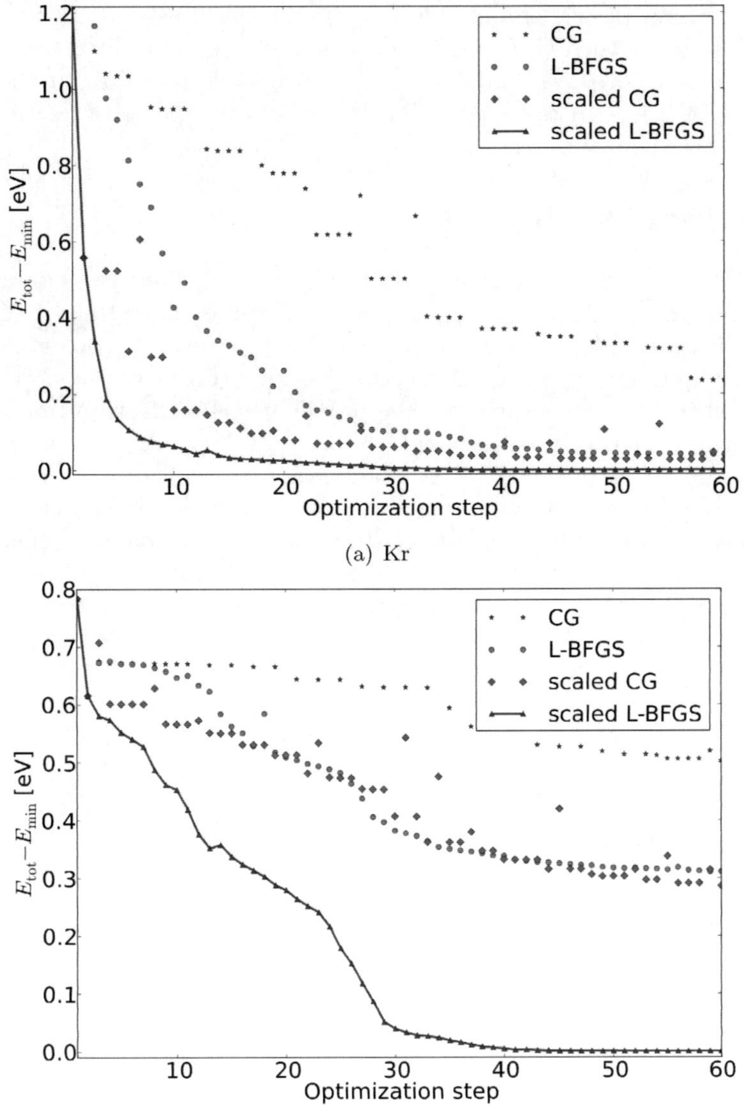

(a) Kr

(b) Cr

Figure 2: Total energy over number of FOD optimization steps using different algorithms (CG, scaled CG, L-BFGS, scaled L-BFGS) for (a) Kr and (b) Cr atom.

The total energy versus the optimization step for each method is shown in Fig. 2. It is clear that scaling improves both the CG and L-BFGS algorithms. The scaled L-BFGS performs the best of the methods tested. For both Kr and Cr, the minimum energy is reached after approximately 40 steps. None of the other methods converges to the lowest energy within the 60 steps shown. The unscaled CG method has the slowest convergence for both Kr and Cr. The scaled CG and unscaled L-BFGS show similar performance after 60 steps, although the scaled CG leads to a

IOP Conf. Series: Journal of Physics: Conf. Series **1290** (2019) 012002 doi:10.1088/1742-6596/1290/1/012002

faster initial drop in energy.

To implement the scaled L-BFGS method for molecules, we simply use the h_D value for the appropriate shell of the nearest atom to the FOD position. This is clear for the core orbitals, since the core FODs have positions quite close to the nuclear position. For valence states and particularly bonding states, the FODs can lie between atoms, potentially complicating the choice of scaling parameters; however, we find that simply using the h_D value for the valence shell of the nearest atom is effective. In tests for molecules, we obtain speedups for the scaled L-BFGS method similar to those shown in Fig. 2.

4. Results of applying FLO-SIC

4.1. Atomic orbital energies

Because the potential an electron "sees" in a DFT calculation includes an interaction with itself, at a large distance r from a neutral atom the potential goes exponentially to zero, rather than exhibiting the physically correct $-1/r$ behavior. This leads to several problems in the description of atoms and other finite systems. In this section we focus on orbital energies, which are too high in DFT calculations, making them poor approximations of electron removal energies. For example, the 1s orbital in the H atom should be at -13.6 eV. As seen in Fig. 3, the actual orbital energy in a spin-polarized DFT-LDA calculation is -7.3 eV. On the other hand, the FLO-SIC treatment of H (again spin-polarized), which contains only one electron, is exact and the 1s orbital energy in FLO-SIC-LDA equals the experimental removal energy. FLO-SIC is not exact for He, but the He 1s energy level is clearly brought into much better agreement with experiment in FLO-SIC-LDA than in the uncorrected theory.

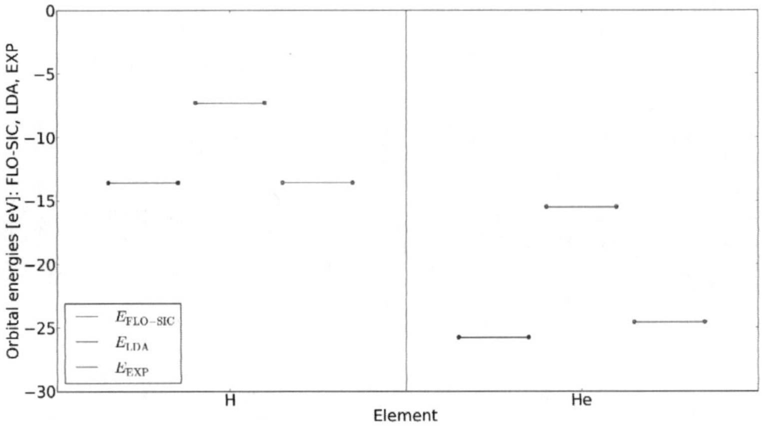

Figure 3: Atomic orbital energies for the 1s orbitals of H and He. The FLO-SIC values for H are exact while for He they are in much better agreement to experiment than the parent LDA functional.

It is interesting to look at other elements and at different energy levels (valence, semi-core, core) to determine whether all energy levels are corrected in a FLO-SIC calculation. This is done in Fig. 4, where the orbital energies of the 2s2p as well as 3s3p shells for the atoms from Na-Ar are shown. Even these deeper energy levels are strongly corrected towards the experimental removal energies when comparing DFT-LDA and FLO-SIC-LDA. (All calculations involving open shell atoms are done spin-polarized. The splitting of the majority and minority spin levels is not visible on the scale of Fig. 4, though all levels are plotted.)

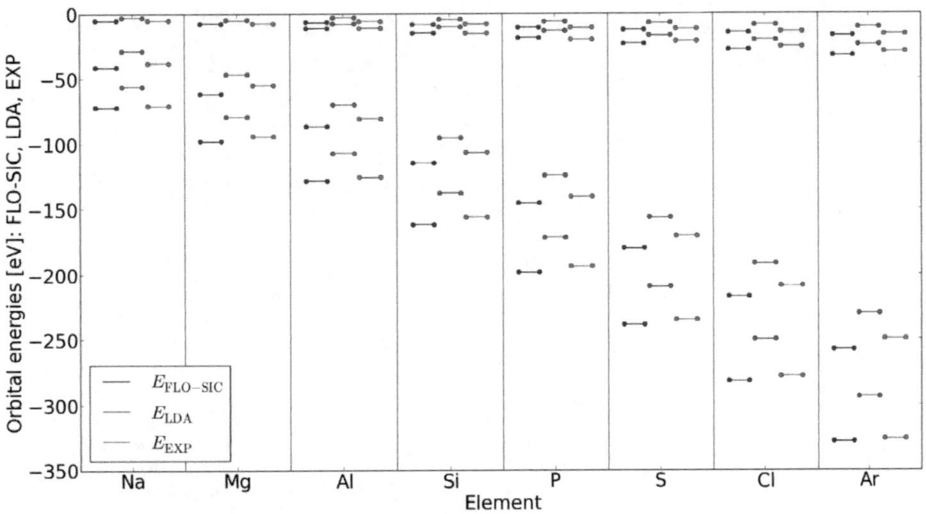

Figure 4: Atomic orbital energies for the 2s2p and 3s3p shells from LDA, FLO-SIC-LDA and experiment for Na-Ar. The FLO-SIC values are in much better agreement with experiment than those of the parent LDA functional.

Taking all the orbital levels for all elements from H to Zn and comparing them to experiment, we find the errors given in table 2. It can be seen that not only the value of the highest occupied orbital is corrected [9, 10], but all other orbital levels are as well, down to the deepest core levels. Accordingly, FLO-SIC restores to all orbital energies the physical significance of experimental removal energies.

Table 2: Mean absolute error (MAE) and mean absolute percentage error (MAPE) of the atomic orbital energies for H-Zn in comparison to experimental values. The notation 'all' means that all orbital energies were taken into account (from valence to core) while 'HOO' refers to the highest occupied orbital and '1s' to the lowest occupied orbital. The agreement of FLO-SIC with experimental data is significantly better than that of the parent LDA functional and close to Hartree-Fock [11].

Orbitals	Error	LDA	FLO-SIC-LDA	Hartree-Fock
all	MAE [eV]	34.3	7.2	9.1
	MAPE [%]	32.8	5.8	6.5
HOO	MAE [eV]	4.2	0.7	0.7
	MAPE [%]	70.6	6.2	9.1
1s	MAE [eV]	104.9	19.8	18.9
	MAPE [%]	9.9	1.9	1.8

4.2. Dissociation Curves

To investigate the performance of FLO-SIC in stretched bond situations, we computed dissociation curves by systematically removing a single H from LiH, BeH_2, BH_3, CH_4, NH_3, H_2O, and HF, while holding the remaining atoms fixed. An example of such a curve for the CH_4 molecule is shown in Fig. 5, where a comparison between LDA, FLO-SIC-LDA and reference CASPT2 calculations is shown. The LDA and FLO-SIC-LDA calculations were done without spin polarization. Energies are plotted with respect to the separated limit of an isolated H atom and an isolated molecular fragment.

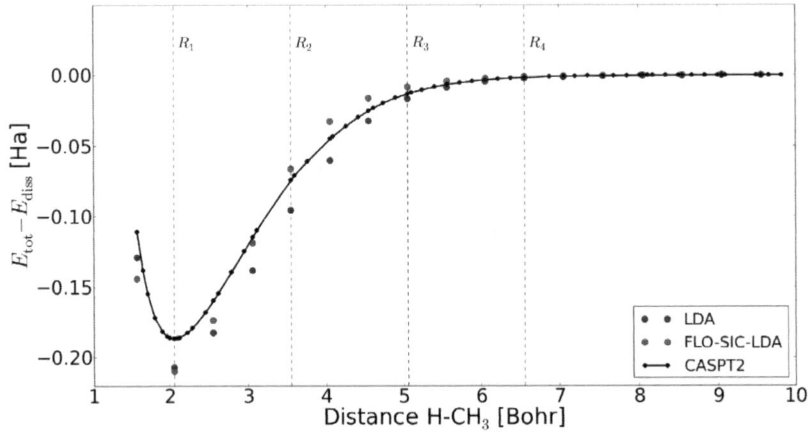

Figure 5: Energy with respect to the dissociated limit E_{diss} for moving a single H away from CH_4 calculated with LDA, FLO-SIC-LDA and CASPT2. The distances used in the error calculation are indicated by dashed lines.

For a more quantitative analysis of the results for all seven molecules, the error (calculation−reference) at four equidistant points along the energy curves is determined for LDA, FLO-SIC-LDA, PBE, and FLO-SIC-PBE. The first of these points is the equilibrium X-H distance where the energy has the value $E = E_{\text{b}}$, the binding energy of the bond. The second is the separation at which $E = E_{\text{b}}/2$. The third and fourth points are then determined by making two additional steps of the same length outward from the second point. The mean error and the mean absolute error with respect to the reference values for all seven molecules is given in Fig. 6. It can be seen that FLO-SIC-LDA always improves the LDA energy, while FLO-SIC-PBE tends to over-correct the already good description of PBE. For the largest separations, FLO-SIC reduces the mean absolute error in both LDA and PBE, for LDA dramatically.

4.3. Magnetic Exchange Couplings

As shown in Section 4.1, SIE strongly affects orbital energy levels and hence it is also expected to impact molecular properties. One property of particular interest for possible nanotechnology applications are magnetic exchange couplings (J) of transition metal systems. The couplings gauge the nature and strength of spin interaction among metal centers [12, 13]. The effect of SIE on J has been analyzed in the past for transition metal complexes empirically using several widely available hybrid functionals with different amount of exact Hartree-Fock exchange [14]. Here we show the effect of explicitly removing SIE on the magnetic exchange couplings of the transition metal complex $[Fe_2OCl_6]^{2-}$ using the FLO-SIC method.

The computational burden associated with optimizing hundreds of FODs in a large system can be reduced by using effective core potentials (ECPs), where only the valence electrons (and hence valence FODs) are treated explicitly. This simple strategy enables calculations on large systems by substantially reducing the computational time. The use of ECPs has been shown to work well for a number of organic radicals as well as for another transition metal complex ($[Cu_2Cl_6]^{2-}$) [15]. Here, we used the Stuttgart effective core potentials [16] for Fe and Cl atoms with the corresponding uncontracted basis sets, whereas the O atom was treated in an all-electron manner. We calculate the magnetic exchange couplings (J) using the spin-projected

IOP Conf. Series: Journal of Physics: Conf. Series **1290** (2019) 012002 doi:10.1088/1742-6596/1290/1/012002

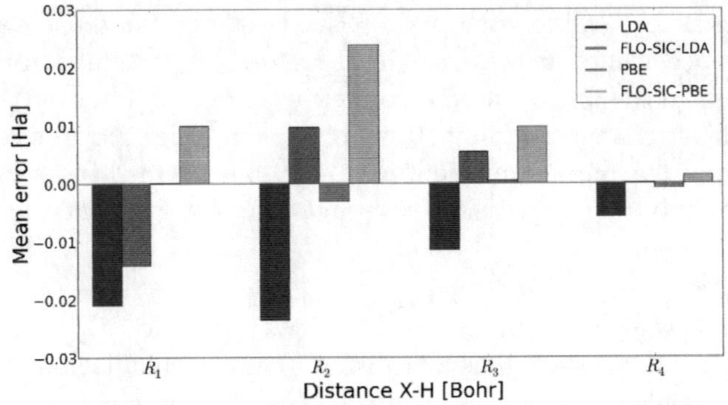

(a) Mean error

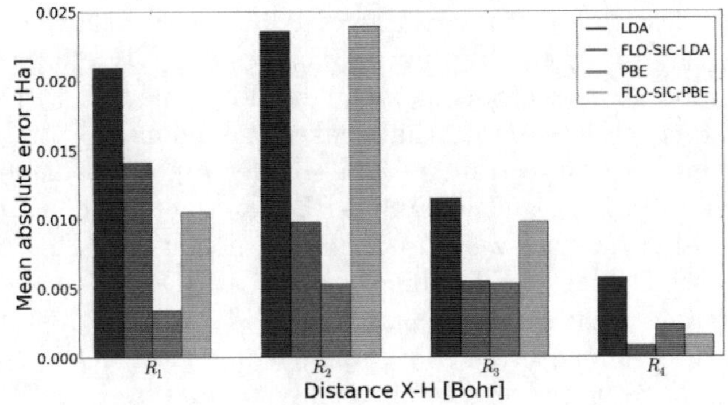

(b) Mean absolute error

Figure 6: Errors of LDA, FLO-SIC-LDA, PBE and FLO-SIC-PBE respect to the CASPT2 reference values at four different points along the X-H dissociation curve (see figure 5) for LiH, BeH_2, BH_3, CH_4, NH_3, H_2O, and HF. See the text for more details.

approach of Noodleman [17],

$$J = \frac{E_{\mathrm{BS}} - E_{\mathrm{HS}}}{2 S_A S_B}, \tag{8}$$

where E_{HS} and E_{BS} are the energies for the high spin state (HS) and a broken-symmetry spin state (BS), respectively, and S_A and S_B are the nominal spins on the two centers.

Table 3: Magnetic exchange couplings (cm^{-1}) for $[Fe_2OCl_6]^{2-}$ using different methods.

Methods	J
LDA	−495
FLO-SIC	−97
PBEh	−377
B3LYP	−443
M062X	−284
ωB97XD	−384
Experiment [18]	−112

In table 3, we present the J values obtained from plain LDA and FLO-SIC-LDA (dubbed FLO-SIC in the table), along with the ones from multiple hybrid functionals. The experimental value is listed as a reference. The J value with plain LDA is overestimated by a large amount when compared to the reference value. However, with FLO-SIC it is significantly corrected and the resulting J value is close to the one obtained from experiment. All hybrid functionals considered in this work correct the exchange coupling toward the experimental reference similar to FLO-SIC and lie between the LDA and experimental values. The hybrid functionals partially correct for self-interaction by including a fraction of exact Hartree-Fock exchange in the exchange-correlation energy.

4.4. Charge transfer and self-interaction

SIE also impairs the description of charge transfer in DFT calculations. For example, it has been shown that heteronuclear dimers dissociate to fractionally charged atoms in DFT, due to unphysical positions of the atomic energy levels caused by self-interaction [19]. A related, but more subtle effect is the description of the charge distribution in polar molecules at equilibrium bond lengths. For molecules near equilibrium, PBE gives a good description of properties like binding energies, yet properties like dipole moments do not agree well with experimental values. This is a result of electron self-interaction that makes the anions relatively less stable in DFT calculations, resulting in less charge transfer and smaller dipole moments. This can be seen in table 4 where results from PBE and FLO-SIC-PBE are compared with reference CCSD(T) values [20]. Experimental geometries were used for all calculations. PBE results in dipole moments that are systematically too small, while FLO-SIC-PBE produces values in much closer agreement with the reference values. For the six molecules shown in the table, the average PBE result underestimates the reference values by 5.2%, while the FLO-SIC-PBE values have an average error of only 1.8%, an improvement of roughly 3 times over PBE.

We also show calculated Mulliken charges for the anions in table 4. The increase in the dipole moments in FLO-SIC-PBE is clearly due to an increase in the charge transferred to the anion. For the six molecules shown, the charge on the anion is greater on average by 0.07 electrons.

Table 4: Dipole moments (in Debye) and Mulliken charge (Q, in electrons) on the anion computed for ionic molecules in PBE and FLO-SIC-PBE. Reference values for the dipole moments were obtained using the CCSD(T) method [20].

	Dipole (Debye)			Q (e)	
	PBE	FLO-SIC-PBE	CCSD(T)	PBE	FLO-SIC-PBE
LiH	5.570	5.943	5.829	0.477	0.585
LiF	6.129	6.347	6.288	0.833	0.889
LiCl	6.890	7.281	7.096	0.726	0.773
NaH	5.751	6.689	6.397	0.413	0.497
NaF	7.698	8.191	8.134	0.814	0.896
NaCl	8.475	8.984	9.007	0.721	0.757

5. Summary and future directions

The applications of the FLO-SIC method presented above illustrate the success of FLO-SIC and the PZ-SIC formalism in removing the effects of SIE from commonly used exchange-correlation functionals in situations where these errors are important. Similar improvements over uncorrected functionals have been found in other benchmarking applications carried out by our group [21] and others [4]. Challenges remain to making FLO-SIC more efficient for practical calculations. We are currently working on creating improved starting points for FOD positions, based on optimized results for the free atoms and accumulating experience gained in describing molecules. Good starting points can greatly reduce the number of optimization steps needed to find energy-minimizing FLOs. Further improvements to the optimization schemes are also possible, with the goal of accelerating convergence still further. We are also pursuing algorithmic improvements to make the solution of the DFT-SIC equations more efficient. Also needed is the implementation of periodic boundary conditions, to make FLO-SIC calculations possible for crystaline materials. A more fundamental goal is to improve the performance of FLO-SIC for problems where the underlying DFT functional already works well. As illustrated by the data for R_1 in Fig. 6, FLO-SIC-PBE yields worse binding energies than PBE for molecules near their equilibrium geometries. More work is needed to determine how to best pair SIC with functionals such as PBE and the promising new SCAN functional [22] in order to remove SIE when it is prominent, without diminishing their performance when it is not. This, combined with more efficient computational tools, can make self-interaction-free DFT calculations a reality.

6. Acknowledgments

We gratefully acknowledge the financial support of the U.S. Department of Energy, Office of Science, Office of Basic Energy Sciences as part of the Computational Chemical Sciences Program under award #DE-SC0018331.

References

[1] Perdew J P, Burke K and Ernzerhof M 1996 *Physical Review Letters* **77** 3865
[2] Perdew J P and Zunger A 1981 *Physical Review B* **23** 5048
[3] Vydrov O A and Scuseria G E 2004 *Journal of Chemical Physics* **121** 8187
[4] Schwalbe S, Hahn T, Liebing S, Trepte K and Kortus J 2018 *Journal of Computational Chemistry* **39** 2463–2471
[5] Pederson M R, Ruzsinszky A and Perdew J P 2014 *Journal of Chemical Physics* **140** 121103
[6] Pederson M R 2015 *The Journal of Chemical Physics* **142** 064112
[7] Yang Z h, Pederson M R and Perdew J P 2017 *Physical Review A* **95**(5) 052505
[8] Löwdin P O 1950 *Journal of Chemical Physics* **18** 365–375
[9] Kao D y, Withanage K, Hahn T, Batool J, Kortus J and Jackson K 2017 *The Journal of Chemical Physics* **147** 164107
[10] Withanage K P K, Trepte K, Peralta J E, Baruah T, Zope R and Jackson K A 2018 *Journal of Chemical Theory and Computation* **14** 4122–4128
[11] Mann J B 1968 *Report LA-3690 (Los Alamos National Laboratory)*
[12] Abate B A, Joshi R P and Peralta J E 2017 *Journal of Chemical Theory and Computation* **13** 6101–6107
[13] Joshi R P, Phillips J J and Peralta J E 2016 *Journal of Chemical Theory and Computation* **12** 1728–1734
[14] Ruiz E, Alvarez S, Cano J and Polo V 2005 *Journal of Chemical Physics* **123** 164110
[15] Joshi R P, Trepte K, Withanage K P K, Sharkas K, Yamamoto Y, Basurto L, Zope R R, Baruah T, Jackson K A and Peralta J E 2018 *Journal of Chemical Physics* **149**(16) 164101
[16] Feller D 1996 *Journal of Computational Chemistry* **17** 1571–1586
[17] Noodleman L 1981 *Journal of Chemical Physics* **74** 5737–5743
[18] Haselhorst G, Wieghardt K, Keller S and Schrader B 1993 *Inorganic Chemistry* **32** 520–525

[19] Ruzsinszky A, Perdew J P, Csonka G I, Vydrov O A and Scuseria G E 2006 *Journal of Chemical Physics* **125** 194112

[20] Hait D and Head-Gordon M 2018 *Journal of Chemical Theory and Computation* **14** 1969–1981

[21] Sharkas K, Li L, Trepte K, Withanage K P K, Joshi R P, Zope R R, Baruah T, Johnson J K, Jackson K A and Peralta J E 2018 *Journal of Physical Chemistry A* **122** 9307–9315

[22] Sun J, Ruzsinszky A and Perdew J P 2015 *Physical Review Letters* **115** 036402

IOP Conf. Series: Journal of Physics: Conf. Series **1290** (2019) 012003 doi:10.1088/1742-6596/1290/1/012003

Near-term Quantum Algorithms for Quantum Many-body Systems

Mark B. Ritter

IBM T.J. Watson Research Center, Yorktown Heights, NY 10598

E-mail: mritter@us.ibm.com

Abstract. There has been a good deal of work on algorithms to simulate quantum many-body systems with fault-tolerant quantum computers- those with full error correction. Fault-tolerant quantum computers of scale requisite to achieve computational advantage for these problems are likely over a decade away. Moreover, devices that we can build in the near term, called Noisy Intermediate Scale Quantum computers (NISQ), have too much noise to implement the long circuits required by these algorithms. We review heuristic, short-depth quantum algorithms more suited to NISQ computers; specifically, their scaling properties when applied to electronic and nuclear structure calculations, including Hamiltonian complexity with particle number, ansatz state preparation, convergence, and noise. We will present examples of actual quantum structure calculations with NISQ computers, as well as a newly-developed error mitigation technique that significantly improves accuracy. We end with an outlook for "advantage" – when NISQ systems might excel conventional HPC approaches for comparable problems.

1. Introduction

Feynman's early insisght [1] was this: researchers knew that quantum many-body calculations required to describe all matter require conventional computers to grow exponentially with the problem size. However, he speculated, if a computer could be built with quantum devices, the quantum resources required would scale linearly with problem size, and could solve far larger problems that would be forever intractable for conventional computers. That early insight first inspired theorists to explore algorithms, such as Shor's factoring algorithm or Grover's search algorithm [2] [3]. Naturally, since little was actually known about actual quantum devices, theorists started with the assumption that quantum calculations could be performed perfectly, with no errors occurring during the manipulation or storage of the quantum states. However, we now realize that the fragility of quantum devices, qubits, will require complex coding with great hardware overhead to realize fault-tolerant quantum computers - quantum computers that perform calculations free from errors. Therefore, our team is exploring algorithms that can be performed with a quantum computer that has little or no error correction - a computational paradigm that John Preskill has named "Noisy Intermediate-Scale Quantum" computing, or NISQ [4].

2. Back to Feynman's Insight

Quantum many-body problems seem more appropriate for NISQ systems than the mathematical tasks like Shor or Grover– for there is no such thing as an approximate prime or approximate match! However, quamtum chemistry or materials science can benefit from an approximate computation of bond length or angle if the computation can yield a sufficiently accurate result on problems intractable to conventional approximate methods. NISQ algorithms for strongly-interacting fermionic many-body problems have also been an area of theoretical work [10, 6], and only relatively recent demonstrations of one and two-qubit calculations on actual hardware [9]. In this paper we summarize the steps required to: a) transform a many-body Hamiltonian into the qubit basis, b) solve this problem on a quantum computer, and c) mitigate noise on the quantum hardware to improve the accuracy of the result. We give a summary of our new work [7] that extend computations to six qubits to model the triatomic molecule BeH_2 .

2.1. Simplifying the Hamiltonian and Transforming the Basis

To make the discussion more concrete we will describe the computational flow used to solve the electronic structure problem, though the same methods can be used to solve other many-body problems like nuclear states or fermionic-Hubbard models. In essence, these quantum algorithms use the quantum computer to represent and explore the exponentially-large Hilbert space of fermionic particles to find the eigenstate wavefunctions, and use these wavefunctions to estimate energy eigenvalues. Analysis of the ground-state eigenvalue of a chemical bond as a function of bond length then gives information about the equilibrium bond length, bond angle, and dissociation energy. The flow (Figure 1) begins with writing the first-quantization Hamiltonian using the Born-Oppenheimer approximation.

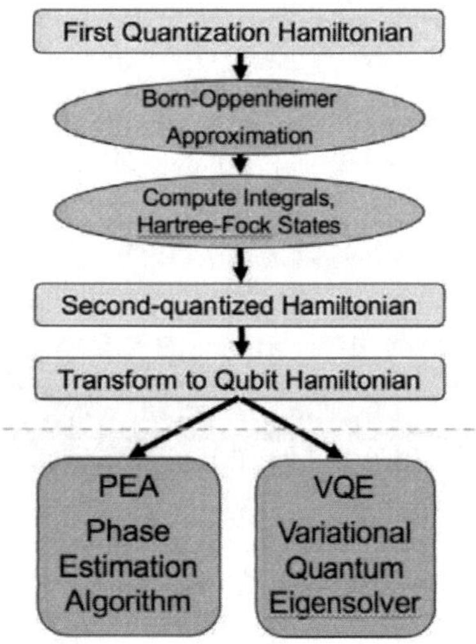

Figure 1. Flow of electronic structure computation for NISQ quantum computation. The first-quantized Hamiltonian is simplified as much as possible before computing the integrals to generate the second-quantized Hamiltonian. There are several transformations to the qubit basis states (text), allowing the transformed problem to be solved with quantum-classical algorithms (green boxes).

We do not need to make this approximation to solve the problem– we are only trying to simplify the problem to use as few qubits as possible because they are currently so precious. (Note: if the reader wishes, they can follow the computational process in this paper using IBM's Open Source qiskit.org software and Qiskit Aqua for quantum chemistry.) Standard quantum chemistry packages can be used to compute the integrals in a chosen basis (STO3G, STO6G, etc.) for a particular molecular configuration of bond lengths and angles. With this preparation we calculate the second-quantized Hamiltonian for the problem at hand.

The next step is a basis transformation of the problem to encode the fermionic degrees of freedom using qubits [13]. Encoding efficiency is important to scaling: we want to use the least number of qubits possible and the least number of terms in the qubit Hamiltonian. There are a number of tranformation rubrics; the Jordan-Wigner, Parity, and Binary Tree encodings. All three formulations require N-qubits to represent N-modes; however, the locality of the operators in the Hamiltonian scales as $\mathcal{O}(N)$ for the Parity and Jordan-Wigner hamiltonians while the locality in the Binary Tree scales as $\mathcal{O}(log(N))$. Further, if the symmetry of the Hamiltonian is analyzed, at least two qubits can be eliminated and even more based on further symmetries [13].

The result of the transformation takes the second quantized fermionic Hamiltonian and produces a qubit Hamiltonian which is a weighted sum of Pauli operators. Each term in the qubit Hamiltonian has a coefficient which depends on bond length, angle, and other desired features of the many-body problem. For example, the qubit Hamiltonian for the hydrogen atom using the Binary Tree encoding has five terms which can be expressed using only two qubits to represent the four spin orbitals [7].

For quantum chemistry, the number of terms in the mapped Hamiltonian grows as $\mathcal{O}(n^4)$, where n is the number of spin orbitals.

There are two classes of algorithms to arrive at the eigenstate of a qubit Hamiltonian: the Phase Estimation Algorithm algorithm (PEA), and a Variational Quantum Eigensolver algorithm (VQE). We next describe and contrast these two solution methods.

2.2. Phase Estimation Algorithm

The PEA starts with an approximation to the groundstate, usually chosen as the (approximate) Hartree-Fock (HF) ground state expressed in qubit states, $\widetilde{\psi(0)}$ and evolves this in time using the Hamiltonian in the qubit basis (see equation 1) to arrive at the actual groundstate of the Hamiltonian $|\psi(t)\rangle$. This method works as long as the HF groundstate has sufficient overlap with the actual groundstate. The propagator in equation [5] can be expressed as a sum of projectors onto eigenstates of the Hamiltonian, each with different phase, ϕ_n, with phases related to eigenstate energies $E_n = \phi_n t$. [10, 11, 12].

$$|\psi(t)\rangle = e^{iHt}|\widetilde{\psi(0)}\rangle = \left(\sum_n e^{i\phi_n t}|\psi_n\rangle\langle\psi_n|\right)|\widetilde{\psi(0)}\rangle \tag{1}$$

The PEA algorithm allows one to measure the phase ϕ_n on a quantum computer. The probability for measuring the phase ϕ_0 is given by $|\langle\psi_0|\widetilde{\psi(0)}\rangle|^2$, from which the ground state energy is easily determined.

Implementing this algorithm requires Trotterization of the exponential, which for most molecules will require $\mathcal{O}(N^5)$ terms and a circuit depth which has been improved [8] to be $\mathcal{O}(N^7)$, where N is the number of spin orbitals. The depth of circuit required is beyond present NISQ capabilities except for the simplest molecules H_2 and HeH^+. The bond distance dependence of the ground state energy has been computed using PEA with photonic and superconducting qubits [9, 10], but the results show more error and higher noise than the VQE algorithm.

2.3. Variational Quantum Eigensolver (VQE) Algorithm

The variational quantum eigensolver (VQE) approach consists in using a qubit processor to approximate the ground state eigenvector, $|\psi\rangle$, with

$$H_q|\psi\rangle = E_q|\psi\rangle, \tag{2}$$

in terms of a classically-intractable trial state. A quantum computer is used to represent the trial state $|\psi(\theta)\rangle$ in its physical memory consisting of qubits. On the quantum computer, a tentative variational eigenstate, usually called the Ansatz or guess eigenstate, $|\psi(\theta)\rangle$ is prepared using a sequence of parameterized quantum gates labeled by control variables $\overrightarrow{\theta}$. One approach to create an ansatz uses the UCCSD (Unitary Coupled Cluster Singlet and Doublet excitation) approximation which must be implemented using Trotterization and is not hardware efficient–

just like the PEA approach discussed above. Shown below is an example of an "Hardware-efficient" ansatz[7], that is constructed as an interleaved sequence of parametrized single qubit rotations and entangling blocks U_{ENT}. As long as this heuristic ansatz state overlaps sufficiently with the true ground state, optimization by varying the control variables $\overrightarrow{\theta}$ is possible.

$$|\psi(\overrightarrow{\theta_k})\rangle = \prod_{q=1}^{M}\left(U^{q,0}(\overrightarrow{\theta_k})\right)\times U_{ENT}\times\prod_{q=1}^{M}\left(U^{q,1}(\overrightarrow{\theta_k})\right)\times\ldots\times U_{ENT}\times\prod_{q=1}^{M}\left(U^{q,d}(\overrightarrow{\theta_k})\right)|0,0\ldots0\rangle \quad (3)$$

Estimates of the average energy of the Hamiltonian, $E = \langle\Psi(\overrightarrow{\theta_k})|H_q|\Psi(\overrightarrow{\theta_k})\rangle$, are obtained by repeated quantum measurements on the qubits prepared in the state $|\Psi(\overrightarrow{\theta})\rangle$. This requires projective measurement of the prepared trial state to compute the expectation value for the $\mathcal{O}(N^4)$ terms in the Hamiltonian. The control variables $\overrightarrow{\theta_k}$ are then changed, for instance, with a gradient descent feedback mechanism on the estimated energies, until the lowest value for E_q is found. This corresponds to the set of controls θ_k^* that create the state with lowest energy–the groundstate E_g.

Simulations of this method for generating an ansatz show that a fixed, partial entanglement U_{ENT} (not a full CNOT) and a depth of approximately six entanglers is necessary to represent the groundstate of LiH with chemical accuracy (1 mHartree or better). Because Ansatz generation starts with all qubits in the groundstate, the first set of Z rotations of the operator $U^{q,0}(\overrightarrow{\theta})$ are not implemented. Therefore, for d entanglers the total number of independent parameters to generate the Ansatz is $p = N(3d + 2)$.

3. VQE Results for Three Molecules

The results of the VQE algorithm for the simulation of the Hamiltonians for H_2 using four qubits, LiH using four, and BeH_2 using six qubits are shown in Figure 2 a, b, and c. The figure includes exact groundstate energy (solid line), the VQE algorithm result averaged over repeated experiments, or shots, (solid circles) and a classical simulation of the VQE algorithm that accounts for decoherence and the effect of finite sampling (fuzzy regions). Here the molecular

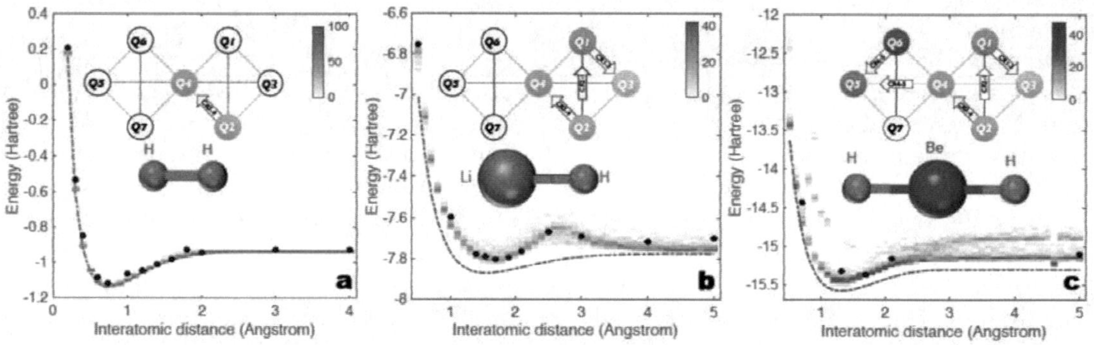

Figure 2. Experimental results (black circles), exact energy surfaces (dotted lines) and density plots (fuzzy) of outcomes from numerical simulations of three molecules. These results are for one entangler circuits (d=1). Error bars on the experimental data are much smaller than the circles. Figure reproduced from [7]

.

energies in units of Hartree as a function of interatomic distance; in the case of BeH_2 it is assumed that both bonds have the same length. For all three molecules, only one entangler was used; otherwise the effect of decoherence overpowers the benefit of using larger depths. The energy discrepancies between the exact and quantum VQE algorithm agree qualitatively with a classical simulation including amplitude damping and dephasing noise, showing how decoherence affects the quantum computation.

4. Error Mitigation

Quantum information is fragile in all known qubit technologies; this is a consequence of the small energy difference between the ground and first excited states of the qubit, which are the two states comprising the computational basis for any qubit. Topological parity codes have been invented to protect quantum information by spreading a single qubit state across $\mathcal{O}(d^2)$ qubits, where d is the code distance [14]. However, the overhead of any known code which protects against all errors is very large– to implement many-body physics algorithms would require hundreds to thousands of physical qubits per protected logical qubit, and hundreds to thousands of these logical qubits to perform computations that are out of reach for conventional computers [6]. We are a long way away from the hundreds of thousands to millions of physical qubits necessary to perform fault-tolerant (error free) computations.

However, we have been exploring error mitigation techniques to reduce error without the large overhead required to correct all errors [15] which can be applied to NISQ hardware. We have implemented one of the ideas, Richardson extrapolation, to reduce the error of the VQE algorithm. The idea is elegant: if we could have an experimental parameter that we could vary to increase noise in the calculation, then we could measure energy as a function of that noise parameter and extrapolate to the noiseless value (theoretically). This approach would be valid as long as the source(s) of decoherence are time invariant over the period of the extrapolation experiments and as long as the noise itself is small compared to the average energy of the Hamiltonian being computed.

We have found a way to implement this error mitigation technique, given that the above assumptions hold. If we take a longer time to generate $|\psi(\overrightarrow{\theta_k})\rangle$, the noise will be roughly proportional to the time taken. This is an easy "knob" for experimentalists since the time it takes to perform a rotation or entanglement operation on qubits is roughly proportional to the amplitude of the RF pulses applied. By stretching out the time of a pulse and reducing amplitude we can perform the same operation slower and slower.

In [16], error mitigation was applied to a variational quantum eigensolver, addressing the electronic structure problem for H_2 and LiH. The ability to mitigate for the effect of decoherence enabled access to longer circuit depths and achieved significant improvements in the estimates of the ground state energy for both molecules.

5. Conclusions

We have attempted to summarize progress simulating many-body systems with NISQ quantum computers. These algorithms are limited by errors in quantum gates and coherence; the PEA algorithm is more sensitive to hardware quality as it requires a polynomial circuit depth, while the VQE algorithm, which is linear in depth, is more robust. Newly-proposed error mitigation techniques have proven able to reduce VQE algorithm error, and this progress is likely to inspire more work on mitigation approaches with less overhead than required for full error correction. It is difficult to predict with certainty when or if NISQ hardware and these heuristic algorithms will improve to the point that we can solve problems intractable for conventional computers. Nevertheless, progress has been rapid and shows promise for these near-term algorithms solving many-body problems we can't now even attempt. This promise is being realized primarily because qubits can represent the highly-entangled states of many-body systems much more efficiently and natively than any conventional hardware.

Acknowledgments
Acknowledgments The author has summarized work of IBM colleagues Abhinav Kandala, Jay Gambetta, Kristan Temme, Sergey Bravyi, Antonio Mezzacapo, Antonio Corcoles, Maika Takita, Markus Brink and Jerry Chow.

6. References

[1] Richard P. Feynman, "Simulating physics with computers," *International Journal of Theoretical Physics* **21**, 467 (1982).
[2] Peter W. Shor, "Why Haven't More Quantum Algorithms Been Found?," *Journal of the ACM* **50**, 87, (2003).
[3] Lov K. Grover, "A fast quantum mechanical algorithm for database search," Available from: arXiv:quant-ph/9605043 (1996).
[4] John Preskill, "Quantum Computing in the NISQ era and beyond," Available from: arXiv:1801.00862 (2018).
[5] Ryan Babbush, Jarrod McClean, Dave Wecker, Alan Aspuru-Guzik, Nathan Wiebe, "Chemical Basis of Trotter-Suzuki Errors in Quantum Chemistry Simulation," *Phys. Rev. A* **91**, 022311 (2015).
[6] Markus Rieher, Nathan Wiebe, Krysta M. Svore, Dave Wecker and Matthias Troyer, "Elucidating reaction mechanisms on quantum computers," *Proc Natl Acad Sci U S A* **114**, 7555 (2017).
[7] Abhinav Kandala, Antonio Mezzacapo, Kristan Temme, Maika Takita, Markus Brink, Jerry Chow, and Jay M. Gambetta, "Hardware-efficient variational quantum eigensolver for small molecules and quantum magnets," *Nature* **549**, 242 (2017).
[8] Matthew B. Hastings, Dave Wecker, Bela Bauer, Matthias Troyer, "Improving quantum algorithms for quantum chemistry," *Quantum Information Computation* **15**, 1 (2015).
[9] Peter J.J. O'Malley, Ryan Babbush, Ian D. Kivlichan, J. Romero, Jarrod R. McClean, Rami Barends, Julian Kelly, Pedram Roushan, Andrew Tranter, Nan Ding, B. Campbell, Yu Chen, Zijun Chen, Ben Chiaro, Andrew Dunsworth, Austin G. Fowler, Evan Jeffrey, Erik Lucero, Anthony Megrant, Josh Y. Mutus, M. Neely, Charles Neill, Chris Qintana, Daniel Sank, Amit Vainsencher, Jim Wenner, Ted C. White, Peter V. Coveney, Peter J. Love, Hartmut Neven, Alain Aspuru-Guzik, and John M. Martinis, "Scalable Quantum Simualtion of Molecular Energies," *Phys. Rev. X* **6**, 031007 (2016).
[10] Alain Aspuru-Guzik, Anthony D. Dutoi, Peter J. Love, Martin Head-Gordon, "Simulated quantum computation of molecular energies" *Science* **309**, 1704 (2005).
[11] Matthew B. Hastings, Dave Wecker, Bela Bauer, and Matthias Troyer, "Improving Quantum Algorithms for Quantum Chemistry" Available from: arXiv:1403.1539v2 (2014).
[12] Dave Wecker, Matthew B. Hastings, Matthias Troyer, "Towards Practical Quantum Variational Algorithms" *Phys. Rev. A* **92**, 042303 (2015).
[13] Sergey Bravyi, Jay M. Gambetta, Antonio Mezzacapo, and Kristan Temme, "Tapering off qubits to simulate fermionic Hamiltonians," Available from: arXiv:1701.08213 (2017).
[14] Sergey B. Bravyi, Alexei Yu. Kitaev, "Quantum codes on a lattice with boundary," arXiv: quant-ph/9811052 (1998).
[15] Kristan Temme, Sergey Bravyi, and Jay M. Gambetta, "Error mitigation for short depth quantum circuits," *Phys. Rev. Lett.* **119**, 180509 (2017).
[16] Abhinav Kandala, Kristan Temme, Antonio D. Corcoles, Antonio Mezzacapo, Jerry M. Chow, Jay M. Gambetta, "Extending the computational reach of a noisy superconducting quantum processor" Available from: arXiv:1805.04492v1 (2018).

IOP Conf. Series: Journal of Physics: Conf. Series **1290** (2019) 012004 doi:10.1088/1742-6596/1290/1/012004

Quantum Monte Carlo with the Langevin Equation: Coupled Bose-Fermi Systems

G. George Batrouni

Université Côte d'Azur, INPHYNI, CNRS, 0600 Nice, France
MajuLab, CNRS-UCA-SU-NUS-NTU International Joint Research Unit, 117542 Singapore
Centre for Quantum Technologies, National University of Singapore, 117542 Singapore
Beijing Computational Science Research Center, Beijing 100193, China

Richard T. Scalettar

Department of Physics, University of California, Davis, California 95616, USA

Abstract. We describe the algorithmic details and a performance evaluation of a Langevin approach to a strongly interacting electron-phonon system, and show it has a near linear scaling with lattice size N_s. Many of the limitations of previous attempts to employ such methods to condensed matter lattice Hamiltonians are absent. In particular, the iterative linear algebra solution remains well behaved at strong coupling and low temperatures. The use of Fourier Acceleration is crucial for efficiency, and its use makes the method competitive with the widely-used local update methods, which scale as N_s^3 for on-site interactions and N_s^4 for long range electron-phonon coupling, even on rather small lattice sizes.

1. Introduction

Monte Carlo methods for interacting classical and quantum systems are generally most easily formulated with local moves which update a single degree of freedom at a time. This has the advantage of simplicity of coding, ease of ensuring detailed balance is satisfied, and low computational cost in evaluating the change in energy required for the acceptance/rejection step. There is, however, a major drawback. Changing the field at a single coordinate, or a collection thereof, yields a new configuration which is highly correlated with its predecessors. Many such moves are required before an independent configuration is attained. This problem becomes increasingly challenging with larger lattice sizes and in the vicinity of second order phase transitions, where the correlation length diverges.

Algorithms which change many degrees of freedom at a time will be effective in solving this problem only if the cost to evaluate the energy change is not exorbitant, and also if the acceptance rate for the large proposed change is reasonably high. In certain situations, highly efficient 'cluster moves' have been formulated. The most famous examples are the Wolff and Swendsen-Wang approaches to the Ising model, which combine all the desiderata of simplicity, high acceptance rates, and inexpensive accept-reject decision [1, 2, 3]. "Loop" and "worm" methods have extended large scale moves to Quantum Monte Carlo (QMC) for bosons and quantum spins within the context of world line, stochastic series expansion, and related approaches [4, 5, 6, 7].

IOP Conf. Series: Journal of Physics: Conf. Series **1290** (2019) 012004 doi:10.1088/1742-6596/1290/1/012004

Simulations of quantum *fermionic* systems which move many degrees of freedom simultaneously have achieved great success in the lattice gauge theory community, and are used there almost universally. Their application to condensed matter has been more problematic. One reason is the relatively high efficiency of local-move based approaches. The cost of evaluation the change in the fermionic action scales only as the square of the spatial system size N_s for local moves, as opposed to the naive $\mathcal{O}(N_s^3)$ expectation for the evaluation of the fermion determinant. Global moves which require explicit $\mathcal{O}(N_s^3)$ evaluation of the determinant must overcome the factor of N_s increase in computational cost. The second reason is that the matrices used in molecular dynamics and Langevin algorithms become ill-conditioned. As a consequence, the solution of the linear algebra kernel gets very expensive, or simply fails [8, 9, 10, 11].

The advent of "machine learning" has lent many new perspectives to the field, including more efficient updates, recognition of phases, and even the sign problem [12, 13, 14, 15]. In particular, self-learning Monte Carlo (SLMC) [16, 17, 18, 19] has been shown to generate moves which provide large changes to the configuration at low computation cost and with high acceptance rates. Originally formulated for classical models, SLMC has now been developed for fermionic systems [20]. The key concept is leveraging inexpensive simulations on small lattices which use the full determinant to generate a determinant-free effective model with an energy which can be very rapidly evaluated, but which is close enough to the original model that moves are accepted with high probability. Somewhat remarkably, simulations on these small lattices are found to produce models which remain accurate even when large lattices are studied.

2. Model and algorithm

In this paper we return to the lattice gauge theory (LGT) idea of using Langevin dynamics to generate and sample equilibrium configurations of a quantum system [21, 22, 23, 24]. We show that there is a class of Hamiltonians for which such approaches are very promising. These involve models of electrons, with creation (destruction) operators $\hat{c}^\dagger_{j\sigma}(\hat{c}_{j\sigma})$ on site j and with spin σ, coupled to phonon degrees of freedom $\hat{a}^\dagger_j(\hat{a}_j)$:

$$\begin{aligned}\hat{H} &= -t\sum_{\langle ij\rangle\sigma}\left(\hat{c}^\dagger_{i\sigma}\hat{c}_{j\sigma}+\hat{c}^\dagger_{j\sigma}\hat{c}_{i\sigma}\right)-\mu\sum_{i\sigma}\hat{n}_{i\sigma}, \\ &+\omega_0\sum_i(\hat{a}^\dagger_i\hat{a}_i+\frac{1}{2})+g\sum_{i,r,\sigma}f(r)(\hat{a}^\dagger_i+\hat{a}_i)\,\hat{c}^\dagger_{i+r,\sigma}\,\hat{c}_{i+r,\sigma}\end{aligned} \tag{1}$$

Here t is the kinetic energy scale of the fermions, which we set to $t=1$ as our unit of energy, and $\langle ...\rangle$ denotes near neighbors. ω_0 is the bare boson energy scale, and $gf(r)$ is the coupling of bosons on site i with fermions on site $i+r$. We choose a specific form $f(r)=e^{-r/\xi}\,/\,(1+r^2)^{3/2}$ derived by Fröhlich [25] which is appropriate to electron-phonon interactions. The phonon operators are related to the nuclear position and momentum by

$$\hat{x}_j=\frac{1}{\sqrt{2\omega_0}}\left(\hat{a}^\dagger_j+\hat{a}_j\right),\quad \hat{p}_j=\frac{1}{i}\sqrt{\frac{\omega_0}{2}}\left(\hat{a}_j-\hat{a}^\dagger_j\right). \tag{2}$$

The partition function of the system is given by $\mathcal{Z}=\mathrm{Tr}\,e^{-\beta\hat{H}}=\mathrm{Tr}e^{-\Delta\tau\hat{H}}e^{-\Delta\tau\hat{H}}\ldots e^{-\Delta\tau\hat{H}}$, where we have discretized the imaginary time direction (*i.e.* β) such that $\beta=\Delta\tau L_\tau$ with L_τ the number of discrete steps. The partition function is expressed as a path integral over nuclear positions by inserting complete sets of coordinate and momentum states between each pair of exponentials and integrating out the momenta. Since the Hamiltonian is quadratic in

IOP Conf. Series: Journal of Physics: Conf. Series **1290** (2019) 012004 doi:10.1088/1742-6596/1290/1/012004

the fermion operators, the trace can be performed exactly leading to [26],

$$\mathcal{Z} = \int \mathcal{D}x(i,\tau)\; e^{-S_{\rm bose}} \left(\det M(\{x(i,\tau)\})\right)^2 \tag{3}$$
$$S_{\rm bose} = \frac{1}{2}\Delta\tau\omega_0^2 \sum_{i,\tau} x(i,\tau)^2 + \frac{1}{2}\Delta\tau \sum_{i,\tau} \Big(\frac{x(i,\tau+1)-x(i,\tau)}{\Delta\tau}\Big)^2.$$

The fermion matrix, $M(\{x(i,\tau)\})$ depends on the phonon field, $\{x(i,\tau)\}$, which is space and imaginary-time dependent. We note that the square of the determinant can be written in the form $\det\left(M^T M\right)$ which will be used in what follows. As is well known [27], the determinant can be written as $\det M = \det\left[I + B_{L_\tau} \ldots B_3 B_2 B_1\right]$ where B_ℓ is the $N_s \times N_s$ matrix, $B_\ell = {\rm e}^{-{\rm d}\tau k}{\rm e}^{-{\rm d}\tau v_\ell}$. N_s is the number of sites, and, for a one-d chain with near-neighbor hopping,

$$k = -\begin{bmatrix} \mu & t & 0 & 0 & \ldots & 0 & t \\ t & \mu & t & 0 & \ldots & 0 & 0 \\ 0 & t & \mu & t & \ldots & 0 & 0 \\ \ldots & \ldots & \ldots & \ldots & \ldots & \ldots & \ldots \\ t & 0 & 0 & 0 & \ldots & t & \mu \end{bmatrix}, \tag{4}$$

and

$$v_\ell = -\begin{bmatrix} x_g(1,\ell) & 0 & 0 & 0 & \ldots & 0 \\ 0 & x_g(2,\ell) & 0 & 0 & \ldots & 0 \\ 0 & 0 & x_g(3,\ell) & 0 & \ldots & 0 \\ \ldots & \ldots & \ldots & \ldots & \ldots & \ldots \\ 0 & 0 & 0 & 0 & \ldots & x_g(N_s,\ell) \end{bmatrix}. \tag{5}$$

The term $x_g(j,\ell)$ in the interaction matrix v_ℓ is defined by

$$x_g(j,\ell) = g\sum_{k=1}^{N_s} f(j+k)x(k,\ell), \tag{6}$$

which includes the interaction of site j with all other sites according to Eq.(1).

Instead of performing a single-site Monte Carlo update of the phonon field, we first write Eq.(3) as,

$$\mathcal{Z} = \int \mathcal{D}x(i,\tau)\; e^{-S}, \tag{7}$$
$$S = S_{\rm bose} - \ln\left(\det(M^T M)\right), \tag{8}$$

and we define fictitious evolution dynamics governed by the Langevin equation,

$$\frac{{\rm d}x(j,\tau,t)}{{\rm d}t} = -\frac{\partial S}{\partial x(j,\tau,t)} + \sqrt{2}\,\eta(j,\tau,t), \tag{9}$$

with the stochastic variable η satisfying

$$\langle \eta(j,\tau,t)\rangle = 0 \qquad \langle \eta(j,\tau,t)\eta(r,\tau',t')\rangle = \delta_{j,r}\delta_{\tau,\tau'}\delta(t-t'). \tag{10}$$

j labels the spatial coordinate of a site, τ its imaginary time and t the Langevin time. In the stationary limit, after transients have died out, the configurations generated by this equation are indeed governed by the Boltzmann weight ${\rm e}^{-S}$ [21, 28].

We discretize the Langevin time in order to integrate Eq.(9) numerically,

$$x(j,\tau,t+\mathrm{d}t) = x(j,\tau,t) - \mathrm{d}t\frac{\partial S}{\partial x(j,\tau,t)} + \sqrt{2\mathrm{d}t}\,\eta(j,\tau,t), \tag{11}$$

with $\langle \eta(j,\tau,t)\rangle = 0$ and $\langle \eta(j,\tau,t)\eta(r,\tau',t')\rangle = \delta_{j,r'}\delta_{\tau,\tau'}\delta_{t,t'}$, which can be modeled with Gaussian random numbers, the discrete analogs of Eq. 10. Here we used the simplest (Euler) discretization, and note the square root of the Langevin time step, $\mathrm{d}t$, in Eq. (11) which comes from replacing the Dirac δ-function in Eq. (10) by the discrete Kronecker form, $\delta(t-t') \to \delta_{t,t'}/\mathrm{d}t$. This results in the linear systematic error $\mathcal{O}(\mathrm{d}t)$ as opposed to $\mathcal{O}(\mathrm{d}t^2)$ for deterministic differential equations. One can, instead, use Runge-Kutta discretization [21] to reduce the errors to $\mathcal{O}(\mathrm{d}t^2)$ here too, but we have found the Euler discretization to be sufficiently accurate (see below).

The derivative of the action in Eq.(11) can be easily calculated,

$$\frac{\partial S}{\partial x(j,\tau,t)} = \frac{\partial S_{\rm bose}}{\partial x(j,\ell,t)} - 2\,\mathrm{Tr}\left(\frac{\partial M}{\partial x(j,\ell,t)}M^{-1}\right). \tag{12}$$

To avoid the costly computation of the trace term, we note that, given a vector of Gaussian random numbers, $\vec{g}$, we have $\langle \vec{g}^T A \vec{g}\rangle = \mathrm{Tr}A$, where the average is taken over the Gaussian distribution. We, therefore, replace the trace term by such a stochastic estimator and write,

$$\frac{\partial S}{\partial x(j,\tau,t)} = \frac{\partial S_{\rm bose}}{\partial x(j,\tau,t)} - 2\vec{g}^T\left(\frac{\partial M}{\partial x(j,\ell,t)}M^{-1}\right)\vec{g}. \tag{13}$$

At first sight it may appear that we have not gained much because we still need to perform the costly calculation of M^{-1}. However, this is not so since we only need $M^{-1}\vec{g}$ which is calculated by solving $M\vec{x} = \vec{g}$ using the bi-Conjugate Gradient (bCG) algorithm. The number of bCG iterations is set by the tolerance level (typically we take $10^{-4} \sim 10^{-5}$), and also by the other physical parameters of the problem (discussed below). Note that in this algorithm, the entire phonon field, $x(j,\tau)$, is updated in a single Langevin step. All the operations are simple sparse matrix-vector multiplies which can be easily optimized.

A well known problem facing simulations of electron-phonon systems is the very long autocorrelation times. To mitigate this difficulty, we introduce Fourier Acceleration (FA) of the Langevin dynamics. The idea is that different length scales can evolve at different rates. Our goal is to force them to evolve at, more or less, the same rate. To implement this, we note that for an arbitrary positive definite matrix, Q, the stationary state configurations produced by the generalized Langevin equation,

$$\frac{\mathrm{d}\vec{x}(t)}{\mathrm{d}t} = -Q\frac{\mathrm{d}S}{\mathrm{d}\vec{x}(t)} + \sqrt{2Q}\,\vec{\eta}(t), \tag{14}$$

are guaranteed to have the correct Boltzmann weigh, e^{-S}. Our aim, then, is to find a Q which is inexpensive to compute and which reduces the autocorrelation times. The form we choose is inspired by the noninteracting limit, $g = 0$ where we have:

$$\frac{\mathrm{d}S}{\mathrm{d}x(i,\tau)} = \Delta\tau\,\omega_0^2 x(i,\tau) + \frac{[x(i,\tau+1)+x(i,\tau-1)-2x(i,\tau)]}{\Delta\tau}, \tag{15}$$

which becomes, after Fourier transforming along imaginary time,

$$\frac{\mathrm{d}\tilde{S}}{\mathrm{d}\tilde{x}(i,k_\tau)} = \left(\Delta\tau\,\omega_0^2 + [2-2\cos(2\pi k_\tau/L_\tau)]/\Delta\tau\right)\tilde{x}(i,k_\tau), \tag{16}$$

with $-L_\tau/2+1 \le k_\tau \le L_\tau/2$. The ratio of the slowest to fastest mode is $(\Delta\tau\,\omega_0)^2/\left(4+(\Delta\tau\,\omega_0)^2\right) \ll 1$, exposing the critical slowing down of the phonons in the imaginary time direction, which gets worse for smaller $\Delta\tau$ and ω_0. To compensate for this, we choose the matrix Q to be diagonal in imaginary time Fourier space and given by,

$$\tilde{Q}(k_\tau) = \frac{\Delta\tau\omega_0^2 + 4/\Delta\tau}{\Delta\tau\omega_0^2 + (2 - 2\cos(2\pi k_\tau/L_\tau))/\Delta\tau}, \tag{17}$$

which is normalized so that $\tilde{Q}(L_\tau/2) = 1$. This choice shows clearly that the fastest mode is not accelerated, it does not need to be, and the slower the mode the stronger the acceleration gets. In the noninteracting limit, this choice will totally eliminate critical slowing down. This is not true when $g \neq 0$. Nonetheless we find that this form, motivated by the noninteracting limit, is inexpensive to calculate, works very well and helps convergence even in the strongly interacting case. We, therefore, use Eq. (17) in all the follows.

The Fourier accelerated Langevin equation then becomes,

$$\begin{aligned}\frac{\mathrm{d}\vec{x}(t)}{\mathrm{d}t} &= -\hat{\mathbf{F}}^{-1}\tilde{Q}(k_\tau)\hat{\mathbf{F}}\frac{\mathrm{d}S}{\mathrm{d}\vec{x}(t)} + \hat{\mathbf{F}}^{-1}\sqrt{2\tilde{Q}(k_\tau)}\,\hat{\mathbf{F}}\vec{\eta}(t)\\ &= -\hat{\mathbf{F}}^{-1}\left[\tilde{Q}(k_\tau)\hat{\mathbf{F}}\frac{\mathrm{d}S}{\mathrm{d}\vec{x}(t)} + \sqrt{2\tilde{Q}(k_\tau)}\,\hat{\mathbf{F}}\vec{\eta}(t)\right],\end{aligned} \tag{18}$$

where $\hat{\mathbf{F}}$ is an FFT operator, $\tilde{Q}(k_\tau)$ is given by Eq. (17), and $\mathrm{d}S/\mathrm{d}\vec{x}$ is given by Eq. (13). In what follows, we use this equation with the simple Euler discretization of the Langevin time.

Physical phonon quantities can be easily calculated since the QMC evolves the phonon field directly. All fermionic quantities can be calculated once the Green function is obtained by,

$$G(i,j) = \left\langle (M[\{x\}])^{-1}_{i,j} \right\rangle, \tag{19}$$

where the sites i and j can be at equal or unequal imaginary time. Indeed, an additional advantage of the Langevin approach is that one does not need separate (and computationally costly) routines to evaluate the unequal time Green function. As we did in the update steps, we avoid evaluating M^{-1} by using a stochastic estimator, to calculate the Green function

$$G(i,j) = \left\langle \gamma_i \left(M[\{x\}]^{-1}\vec{\gamma}\right)_j \right\rangle, \tag{20}$$

where γ_j is a Gaussian random number. $M^{-1}\vec{\gamma}$ is calculated with the bCG algorithm. Once $G(i,j)$ is available, all fermionic quantities, *e.g.* kinetic energy, density correlations, structure factor etc, can be obtained.

3. Benchmarking the Algorithm

We now study the performance of the algorithm and assess how well it works compared with determinant quantum Monte Carlo (DQMC). We will examine the systematic errors due to the discretization of imaginary time and Langevin time in addition to the benefits of Fourier acceleration and the scaling of computation time as a function of the system size. We focus on the charge density wave structure factor,

$$S(\pi,\pi) = \frac{1}{N_s}\sum_{i,j}\langle n(i)n(j)\rangle\,(-1)^{i+j}\ , \tag{21}$$

which sums the density correlations over the entire spatial lattice with a phase $(-1)^{i+j}$ which is +1 if the sites are on the same sublattice, and -1 if they are on different sublattices, of the

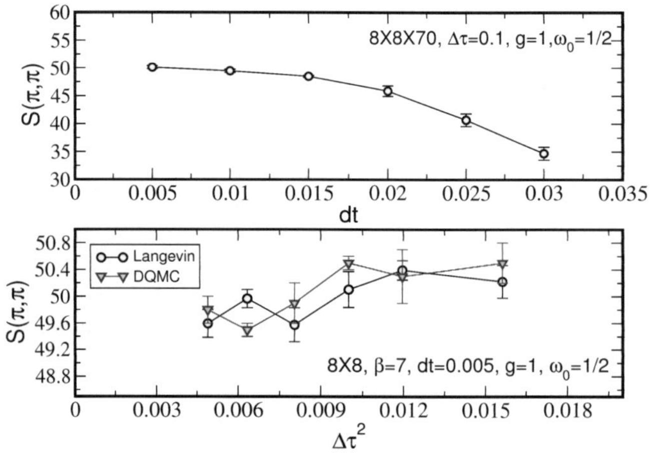

Figure 1. (Color online) Top: The dependence of the structure factor, $S(\pi,\pi)$, on the Langevin time step, dt, in the ordered phase. For d$t \leq 0.015$ the discretization error becomes linear in dt. Bottom: The dependence of $S(\pi,\pi)$ on the imaginary time step, $\Delta\tau$, for both DQMC and Langevin. The Trotter-Suzuki errors are comparable in the two algorithms. d$t = 0.005$ used to integrate the Langevin equation gives excellent agreement with DQMC. The data are for the Holstein limit ($\xi \to 0$).

bipartite square geometry. $S(\pi,\pi)$ signals the transition into a charge density wave (CDW) phase at low temperatures and half-filling. Because it is non-local, $S(\pi,\pi)$ poses a stringent test on equilibration and autocorrelation times.

We start with the Langevin and imaginary time discretization. The top panel of Fig. (1) shows the effect of the Langevin time step, dt, on the value of the structure factor, $S(\pi,\pi)$, in the CDW ordered phase. We see that for d$t \leq 0.015$, the error becomes linear in dt. For $0.015 \leq$ d$t \lesssim 0.03$, the error is beyond the linear regime and increases rapidly and, eventually, for d$t > 0.03$, the numerical integration is unstable. The stability limit, the largest allowed dt, is set by the largest eigenvalue of the iterative process [21], while the critical slowing down is controlled by the ratio of the smallest to largest eigenvalues. The lower panel of Fig. 1 shows the effect of $\Delta\tau$, the imaginary time step, in comparison with DQMC, again in the CDW phase. We see that the two algorithms have comparable dependence on $\Delta\tau$, which is not surprising since this parameter appears in the same way in both. This also demonstrates that the Langevin algorithm gives the correct physical results. We note that these results were obtained with the Fourier accelerated algorithm (more below).

Having shown in Fig. 1 that the Langevin method gives results which are in agreement with DQMC, we now examine the approach to equilibrium. The left panel of Fig. 2 shows the evolution of $S(\pi,\pi)$ as a function of Langevin time for a system in the CDW phase. We see that, starting from a random phonon field, the unaccelerated evolution fails to find the CDW phase even long after the accelerated dynamics have converged. We also see that when the parameter choice is such that the system is in a CDW phase, convergence is helped by choosing an initial phonon field which has the CDW motif. If the parameters put the system in a disordered phase but the initial configuration is CDW, the checkerboard pattern will quickly melt away and the proper phase found. For these reasons, in what follows we will always use Fourier accelerated

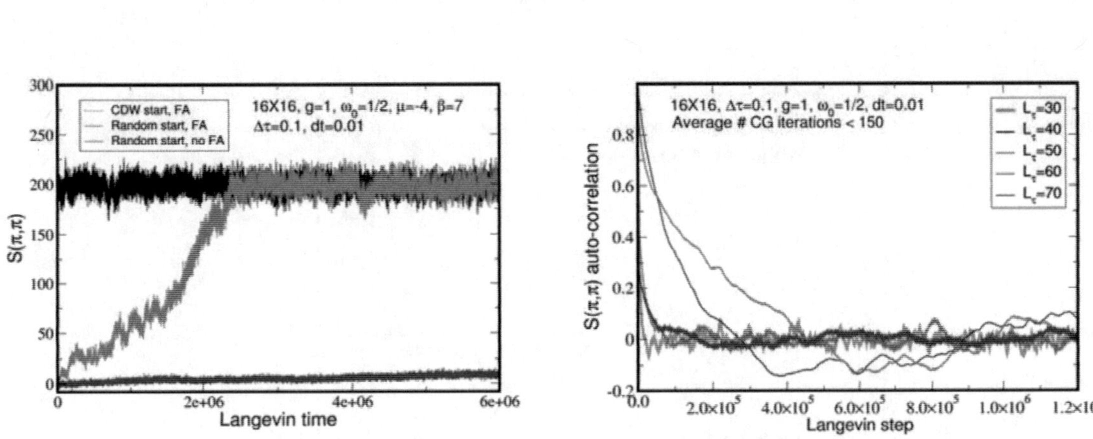

Figure 2. (Color online) Left: $S(\pi, \pi)$ as a function of Langevin time comparing Fourier acceleration (FA) with unaccelerated evolution. The red (FA) and blue (no FA) curves have the same random initial configuration of the phonon field. It is clear that for large systems FA is crucial for equilibration. The initial configuration for the black curve is CDW: The phonon field has an initial checkerboard configuration which triggers the CDW phase for the electrons. Right: The autocorrelation function of accelerated Langevin dynamics of the Holstein model for several β values. The relaxation time is longest for $L_\tau = 40, 50$ ($\beta = 4, 5$) near the critical temperature, $\beta \approx 4.5$.

dynamics.

The approach to equilibrium of the accelerated dynamics is illustrated by studying the autocorrelation time. We show this in the right panel of Fig. 2 for a 16×16 system at five values of β ranging from smaller than $\beta_c \approx 4.5$ to larger. We see that the longest relaxation times are $L_\tau = 40, 50$ ($\beta = 4, 5$) which are closest to β_c. The relaxation times for $\beta = 3, 6, 7$ are very short.

A very important question is how the numerical complexity scales with system size for fixed parameters. As mentioned above, DQMC scales as N_s^3 for contact interaction and as N_s^4 for long range interactions. Figure 3 shows the scaling of CPU time with system size for DQMC and Langevin for the case of contact interaction. For DQMC, the scaling is cubic, as expected, while for Langevin it is essentially linear for the two values of β shown. When the interaction is long ranged, we cut off the range of the interaction by keeping only interactions larger than 10^{-7} (a cutoff which can be adjusted to ensure physical observables are unaffected) which leads, again, to linear scaling for Langevin while DQMC scales as N_s^4. This means that as the system size increases, for fixed parameters, the number of bCG iterations necessary to calculate $M^{-1}\vec{g}$ remains essentially constant. This is very different from the behavior of bCG for other model. For example, for the Hubbard model, where the electron-electron interaction is decoupled via a Hubbard-Stratonovitch (HS) transformation, the number of iterations increases very rapidly with system size. We believe the difference in behavior for these two models is due to the fact that in the Hubbard model, the HS field does not have any intrinsic dynamics, it is not a physical field, and the properties of the matrix depend on the essentially random HS field. On the other hand, in the case of the Holstein model, the phonon field is physical and has its own dynamics, and in particular, has a second order derivative in the imaginary time direction (see Eq.(15)) which smooths out the phonon field. The eigenvalues of the matrix are, therefore, physical and depend on the parameters rather than the size of the system.

Two other interesting aspects of the number of bCG iterations needed are their dependence

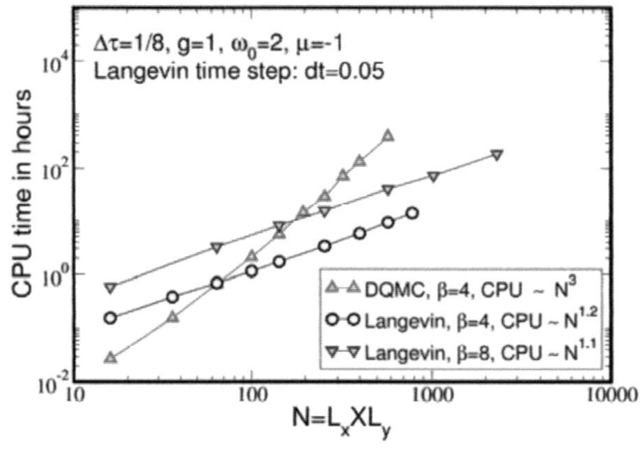

Figure 3. (Color online) CPU time versus system size in the Holstein limit ($\xi \to 0$) for DQMC and Langevin algorithms. DQMC scales as N_s^3 while Langevin scales essentially linearly for both values of β.

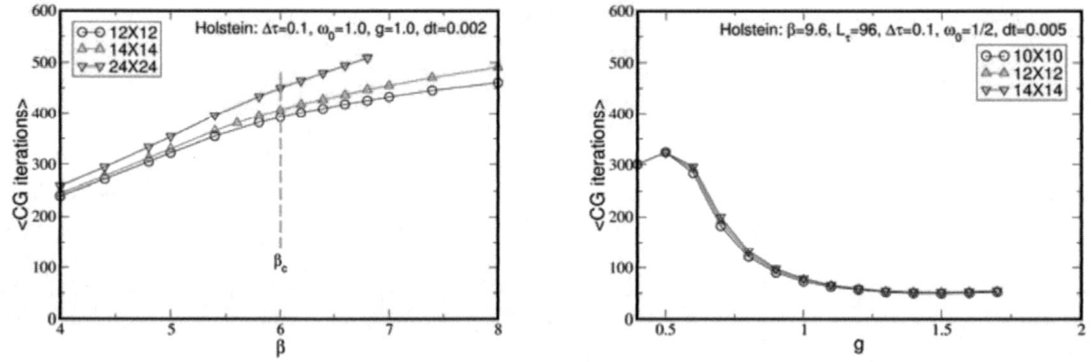

Figure 4. (Color online) Left: Dependence of the number of bCG iterations on the inverse temperature. Right: Dependence of the number of bCG iterations on the electron-phonon coupling constant, g: The number of iterations decreases with increasing g. Notice also the insensitivity to system size.

on β and on g. The left panel of Fig. 4 shows how this number scales with β while all other parameters are fixed. We see that as β_c is approached from the disordered side (small β) the number of iterations increases linearly and continues to do so for $\beta > \beta_c$ but with a smaller slope. The right panel of Fig. 4 shows the dependence of the number of iterations on the electron-phonon coupling constant, g. That the number of iterations decreases with increasing g leads to faster simulations and relaxation times since this behavior is connected with the eigenvalues of the fermion matrix. This behavior is in marked contrast with the Hubbard model, where increasing U leads to very rapid increase in the number of iterations. As discussed above, the difference in behavior is due to the fact that for the Hubbard model, the Hubbard-Stratonovitch field has no dynamics of its own whereas in the phonon model, the phonon momentum couples the field in the imaginary time direction, leading to milder fluctuations. In addition to these considerations, one must choose a stopping criterion for the bCG iterations by choosing a tolerance for the solution. We have found that a tolerance of 10^{-4} to 10^{-5} ensures that errors are much smaller than stochastic and statistical errors and therefore do not affect the evaluation of physical quantities.

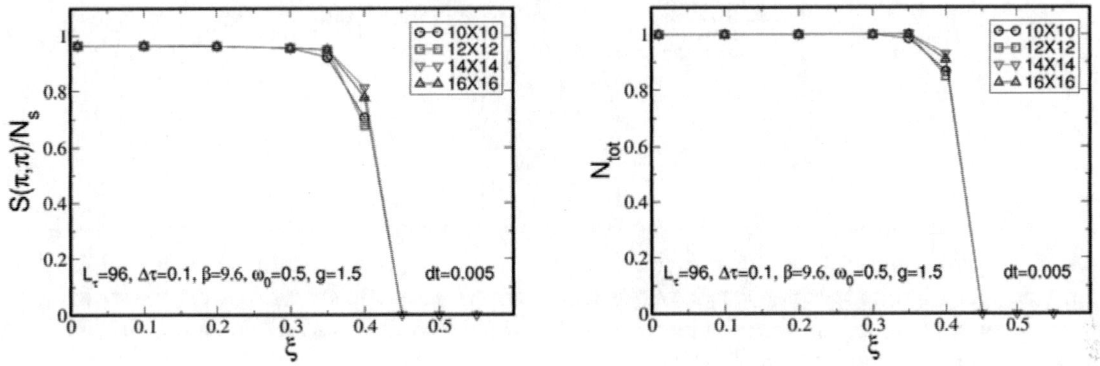

Figure 5. (Color online) Left: The structure factor, $S(\pi,\pi)$ as a function of the range of the interaction, ξ, keeping the chemical potential tuned to half filling. When the interaction is large enough, $S(\pi,\pi)$ drops to zero due to phase separation. Right: Same as left panel but showing the total number of particles. For $\xi \gtrsim 0.4$ the occupation vanishes, indicating phase separation. See also Fig. 6

4. Phase diagram: Long range interaction

Having validated the Langevin algorithm by demonstrating its convergence properties and agreeement with DQMC, we now use it to determine the ground state and finite temperature phase diagrams for the electron-phonon model with long range interaction. Such studies are basically unfeasible with DQMC due to the N_s^4 scaling of simulation time.

We fix $\omega_0 = 1/2$ and low enough temperature to obtain ground state values ($\beta = 9.6$), and study the properties of the system for different values of g and ξ to map the phase diagram. The left panel of Fig. 5 shows $S(\pi,\pi)$ as a function of the interaction range, ξ, for several system sizes. We see that for $\xi < 0.4$ the system is in a CDW phase which ends rapidly at $\xi \approx 0.4$. At the same the particle density suddenly drops from 1 (half filling) to 0, signaling phase separation. By making several such cuts in phase space, we map the ground state phase diagram, which is presented in Fig. 6.

In addition to ground state properties, the Langevin algorithm, like DQMC, provides information about finite temperature behavior. To determine the finite temperature phase transition from disordered at high temperature, to CDW as the temperature is lowered, we perform finite size scaling. Since the transition breaks a discrete (Z_2) symmetry, the phase transition is expected to be in the universality class of the two-dimensional Ising model. In the left panel of Fig. 7 we show the data collapse for $S(\pi,\pi)$ when Ising exponents are used. The collapse indicates that for the chosen parameters, the critical inverse temperature is $\beta_c = 6.01 \pm 0.05$. Repeating this kind of study for several values of ξ, we obtain $\beta_c(\xi)$ which is shown in the right panel. We see that as the range of the interaction increases, the tendency is to remain disordered down to lower temperatures before phase separation takes place at $\xi_c \approx 0.55$.

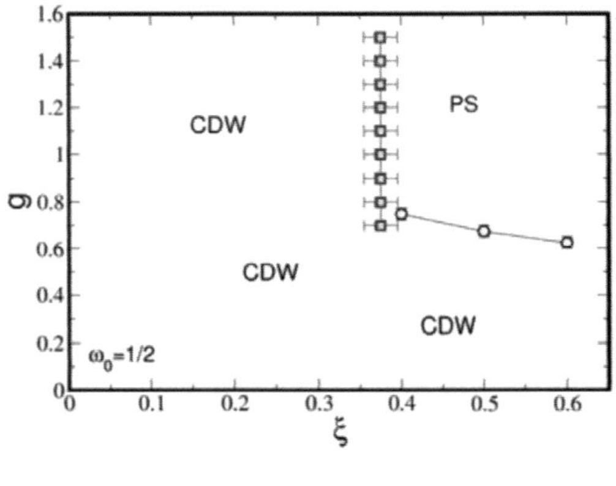

Figure 6. (Color online) The phase diagram in the g-ξ plane at half-filling and $\beta = 9.6$. The phonon frequency is fixed at $\omega_0 = 1/2$. At small ξ, the Holstein limit CDW correlations form. This occurs even at weak coupling owing to the Fermi surface nesting on the square lattice. As ξ increases, phase separation occurs.

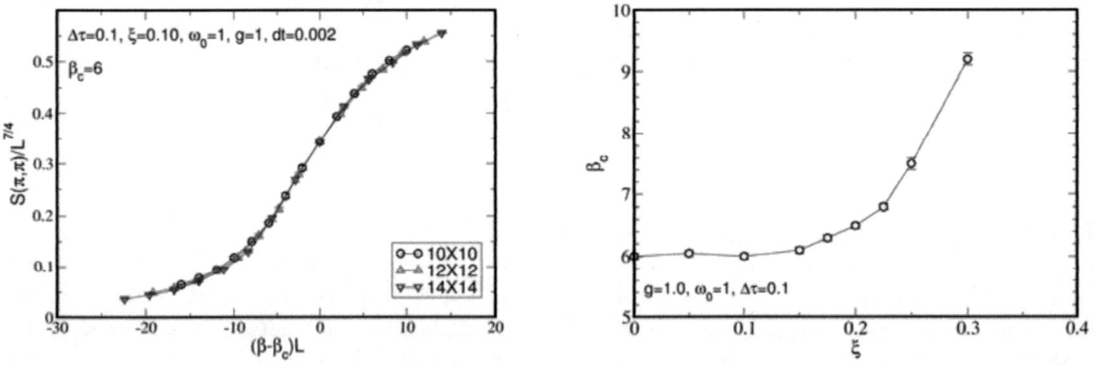

Figure 7. (Color online) Left: Scaling of the structure factor near the finite temperature transition from disordered to CDW phase. The error bars are smaller than the symbols. Right: The critical inverse temperature, β_c, as a function of the range of the electron-phonon interaction, ξ.

5. Summary

Langevin, and related, simulations of the Hubbard model have been attempted a number of times over the past several decades, with only limited success. In this paper we have shown that the smoothing effect of a phonon kinetic energy term makes these algorithms promising for electron-phonon problems. Specifically, we show that the iterative conjugate gradient solver converges in a reasonable number of steps, and that this number does not grow significantly with spatial size, inverse temperature, or coupling constant, as has bedeviled potential Hubbard model applications.

In addition to focusing on that fundamental bottleneck, we also explored various systematic effects, including those associated with discretization of the Langevin equation and of imaginary time. These results, with associated comparisons with DQMC, both verify the correctness of the algorithm and identify the choices of dt and $\Delta\tau$ which provide acceptable systematic effects (e.g. smaller than the statistical sampling error bars).

There are two immediately promising applications of our method. The first is to the Holstein model in three dimensions, where T_c is not known because sufficiently large lattices are not accessible. The second is to problems with a long range (momentum dependent) electron-phonon coupling. Such couplings are of interest in explaining the behavior of a number of systems exhibiting novel CDW and superconducting phases [29, 30, 31, 32], but, like 3D, are out of reach of current approaches except in low dimension [33]. With the latter possibility in mind, we have presented results for the Frohlich model, and generated its phase diagram as a proof-of-principle calculation.

Another promising line of investigation concerns the Hubbard model and Holstein-Hubbard type of models. In the presence of the Hubbard U, the interaction term needs to be decoupled via the Hubbard-Stratonovitch (HS) transformation which introduces an additional auxiliary field which should be updated in the Monte Carlo simulation. In some cases, due to certain symmetries in the problem, the product of the two determinants resulting from tracing out the up and down fermionic operators, is positive definite and the simulation can proceed using well known methods such as DQMC. However, in most cases for $U > 0$, the product is not positive leading to the well known "fermionic sign problem". A possible way to deal with this problem is use a complex HS transformation and Langevin dynamics governed by the resulting complex effective action. This leads to the complex Langevin (CL) equation which has been investigated in the context of LGT[34]. In particular, this method was recently applied to the one-dimensional Hubbard model[35] and ultra-cold fermionic atoms with unequal masses[36]. For the Holstein-Hubbard model, where we now have two bosonic fields (the phonon and HS fields) the CL equation was shown to be efficient[37] in the interaction range $U > g^2/\omega_0$.

6. Acknowledgements

The work of RTS was supported by DOE-SC0014671. GGB is supported by the French government, through the UCAJEDI Investments in the Future project managed by the National Research Agency (ANR) with the reference number ANR-15-IDEX-01. We thank Steven Johnston, WeiTing Chiu, and Thomas Devereaux for useful conversations.

References

[1] Swendsen R H and Wang J-S, 1987 *Phys. Rev. Lett.* **58**, 86.
[2] Wolff U, 1989 *Phys. Rev. Lett.* **62**, 361.
[3] Hasenbusch M and Meyer S, 1990 *Phys. Lett.* **B241**, 238.
[4] Evertz H G, 2003 *Adv. in Phys.* **52**, 1.
[5] Sandvik A W, 1999 *Phys. Rev.* **B59**, R14157.
[6] Capogrosso-Sansone B, Söyler S G, Prokof'ev N, and Svistunov B, 2008 *Phys. Rev.* **A77**, 015602.
[7] Weber M, Assaad F F, and Hohenadler M, 2017 *Phys. Rev. Lett.* **119**, 097401.
[8] Scalettar R T, Scalapino D J, and Sugar R L, 1986 *Phys. Rev.* **B34**, 7911.
[9] Scalettar R T, Scalapino D J, Sugar R L, and Toussaint D, 1987 *Phys. Rev.* **B36**, 8632.
[10] Yamazaki I, Bai Z, Chen W, and Scalettar R T, 2009 *Numerical Mathematics: Theory, Methods and Applications* **2**, 469.
[11] Bai Z, Chen W, Scalettar R T and Yamazaki I, in "Multi-Scale Phenomena in Complex Fluids" edited by Thomas Y. Hou, Chun Liu and Jian-Guo Liu, pp. 1-110, Higher Education Press (2009).
[12] Broecker P, Carrasquilla J, Melko R G, and Trebst S, 2017 *Nature Sci. Rep.* **7**, 8823.
[13] Ch'ng K, Carrasquilla J, Melko R G, and Khatami E, 2017 *Phys. Rev.* **X7**, 031038.
[14] Costa N C, Hu W, Bai Z J, Scalettar R T, and Singh R R P, 2017 *Phys. Rev.* **B96**, 195138.
[15] Nomura Y, Darmawan A S, Yamaji Y, and Imada M, 2017 *Phys. Rev.* **B96**, 205152.
[16] Liu J, Qi Y, Meng Z Y, and Fu L, 2017 *Phys. Rev.* **B95**, 041101(R).
[17] Nagai Y, Shen H, Qi Y, Liu J, and Fu L, 2017 *Phys. Rev.* **B96**, 161102.
[18] Shen H, Liu J, and Fu L, arXiv:1801.01127.

IOP Conf. Series: Journal of Physics: Conf. Series **1290** (2019) 012004 doi:10.1088/1742-6596/1290/1/012004

[19] Liu J, Shen H, Qi Y, Meng Z Y, and Fu L, 2017 *Phys. Rev.* **B95**, 241104.
[20] Chen C, Xu X Y, Liu, J, Batrouni G G, Scalettar R T, and Meng Z Y, 2018 *Phys. Rev.* **B98**, 041102.
[21] Batrouni G G, Katz G R, Kronfeld A S, Lepage G P, Svetitsky B, and Wilson K G, 1985 *Phys. Rev.* **D32**, 2736.
[22] Batrouni G G, 1987 *Nucl. Phys.* **A461**, 351c.
[23] Davies C T H, Batrouni G G, Katz G R, Kronfeld A S, Lepage G P, Wilson K G, Rossi P and Svetitsky B, 1988 *Phys. Rev.* **D37**, 1581.
[24] Katz G R, Batrouni G G, Davies C T H, Kronfeld A S, Lepage G P, Rossi P, Svetitsky B, and Wilson K G, 1988 *Phys. Rev.* **D37**, 1589.
[25] Fröhlich H, 1954 *Adv. Phys.* **3**, 325.
[26] Scalettar R T, Scalapino D J, and Bickers N E, 1989 *Phys. Rev.* **B40**, 197.
[27] Blankenbecler R, Scalapino D J, and Sugar R L, 1981 *Phys. Rev.* **D24**, 2278.
[28] Risken H, Springer Series in Synergetics, Springer-Verlag (New York), 1996.
[29] Wang Y, Nakatsukasa Y, Rademaker L, Berlijn T, and Johnston S, 2016, *Supercond. Sci. Technol.* **29**, 054009.
[30] Weber F, Rosenkranz S, Castellan J-P, Osborn R, Hott R, Heid R, Bohnen K-P, Egami T, Said A H, and Reznik D, 2011, *Phys. Rev. Lett.* **107**, 107403.
[31] Swartz A G, Inoue H, Merz T A, Hikita Y, Raghu S, Devereaux T P, Johnston S, and Hwang H Y, 2018, PNAS 201713916.
[32] Devereaux T P, Shvaika A M, Wu K, Wohlfeld K, Jia C J, Wang Y, Moritz B, Chaix L, Lee W-S, Shen Z-X, Ghiringhelli G, and Braicovich L, 2016, *Phys. Rev.* **X6**, 041019.
[33] Hohenadler M, Assaad F F, and Fehske H, 2012, *Phys. Rev. Lett.* **109**, 116407.
[34] Aarts G, Seiler E, Sexty D, Stamatescu I-O, 2017, *J. High Energ. Phys.* **44**.
[35] Loheac C, Drut J E, 2017, *Phys. Rev.* **D95**, 094502.
[36] Rammelmüller L, Drut J E, Braun J, 2018, *J. Phys.: Conf. Ser.* **1041** 012006.
[37] Karakuzu s, Seki K, and Sorella S, arXiv:1808.07759.

IOP Conf. Series: Journal of Physics: Conf. Series **1290** (2019) 012005 doi:10.1088/1742-6596/1290/1/012005

Ground state phase diagram of the one-dimensional Bose-Hubbard model from restricted Boltzmann machines

Kristopher McBrian
Department of Physics and Astronomy, San Jose State University, San Jose, CA 95192, USA

Giuseppe Carleo
Center for Computational Quantum Physics, Flatiron Institute, New York, NY 10010, USA

Ehsan Khatami
Department of Physics and Astronomy, San Jose State University, San Jose, CA 95192, USA

Abstract. Motivated by recent advances in the representation of ground state wavefunctions of quantum many-body systems using restricted Boltzmann machines as variational ansatz, we utilize an open-source platform for constructing such ansatz called NetKet to explore the extent of applicability of restricted Boltzmann machines to bosonic lattice models. Within NetKet, we design and train these machines for the one-dimensional Bose-Hubbard model through a Monte Carlo sampling of the Fock space. We vary parameters such as the strength of the onsite repulsion, the chemical potential, the system size and the maximum site occupancy and use converged equations of state to identify phase boundaries between the Mott insulating and superfluid phases. We compare the average density and the energy to results from exact diagonalization and map out the ground state phase diagram, which agrees qualitatively with previous finding obtained through conventional means.

1. Introduction

In recent years, artificial neural networks have been employed to study the Bose-Hubbard model [1, 2], which describes the many-body systems of interacting bosonic particles on lattices [3]. In Ref. [1], the author approaches the problem using feedforward artificial neural networks where the inputs are Fock states (see below) and the output layer, consisting of two nodes, provides the Re and Im parts of the log of the wavefunction. The networks are trained following conventional supervised learning algorithms for feedforward networks with the cost function taken to be the ground state energy. Ref. [2] expands on this idea and examines the role multiple hidden layers or convolutional layers play in the accuracy of the output and the efficiency of the algorithm.

Restricted Boltzmann machines (RBMs) on the other hand have emerged as useful tools among artificial neural networks for providing a superior ansatz for representing quantum wave functions [4, 5, 6, 7, 8], or mimicking thermodynamics of classical and quantum many-body systems [9, 10, 11]. For example, in Ref. [4], it was shown that RBMs lead to lower variational ground state energy for classical and quantum magnetic models than the state-of-the-art ansatz. This in turn motivated several other studies in which the RBMs were used to represent topological [5, 6] or chiral [8] states, draw equivalences between RBMs and tensor network states [7], and to accelerate Monte Carlo simulations [12], to name a few.

Here, we use RBMs to represent ground state wavefunctions of the one-dimensional Bose-Hubbard model and study the extent to which its equilibrium properties can be captured in this architecture for a range of model parameters. It has been shown that multi-valued RBMs, such as the one we have used here, can represent a wide class of many-body entangled states efficiently [13], and to the best of our knowledge, our results are the first to demonstrate that. We perform the training using Monte Carlo sampling with the same goal of minimizing the ground state energy as in Refs. [1, 2]. We employ NetKet [14], an open-source platform for solving quantum many-body problems using artificial intelligence. We show that RBMs can be trained to represent the ground state of the bosonic system with a high degree of accuracy. We find that the energy and density both in the Mott insulating and superfluid phases rapidly converge to final values during the training. By calculating equations of state for different values of the interaction strength in the grand canonical ensemble, we map out the phase diagram of the model and show that it agrees well with one obtained in the past directly from quantum Monte Carlo simulations.

2. The Bose-Hubbard Model

The Bose-Hubbard Hamiltonian is expressed as

$$H = -t\sum_{\langle ij\rangle} b_i^\dagger b_j + \frac{U}{2}\sum_i n_i(n_i-1) - \mu\sum_i n_i, \tag{1}$$

where $b_i^\dagger$ and b_j are the bosonic creation and annihilation operators, respectively, $\langle ij\rangle$ indicates that i and j are nearest neighbors, t is the corresponding hopping amplitude, which we set to unity to serve as the unit of energy throughout the paper, U is the strength of the onsite repulsion, $n_i = b_i^\dagger b_i$ is the site occupation operator, and μ is the chemical potential. In this study, we consider the periodic one-dimensional (1D) geometry.

For large enough interactions, the ground state of the system is the interaction-driven Mott insulating phase while it remains a superfluid for smaller interactions. This was shown in a seminal paper by Fisher et al. [3] where the universality class of the model was explored and a mean-field solution was presented. Later, the model received much attention in the 1990's, mostly confirming numerically the transition in different dimensions [15, 16, 17, 18, 19, 20, 21, 22, 23, 24, 25]. The Mott insulator (MI) to superfluid (SF) transition was also observed in experiments with cold atoms in optical lattices [26].

3. Restricted Boltzmann Machine

3.1. The Architecture

The RBM is a two-layer fully-connected artificial neural network. The architecture is shown in Fig. 1. It has one visible layer, and one hidden layer. Each layer consists of several nodes with an external bias parameter associated with them (b_i for the visible and c_i for the hidden layer). Each node in a layer is connected to all the nodes in the other layer and a weight parameter (W_{ij}) is associated to every connection.

The RBM lends itself well to representing ground state wavefunctions of the Bose-Hubbard model because a probability distribution of possible Fock states of the system can then be calculated directly from the RBM parameters after training. This is similar to representing thermodynamics of a many-body system using an RBM where a probability distribution of microstates can be extracted [10].

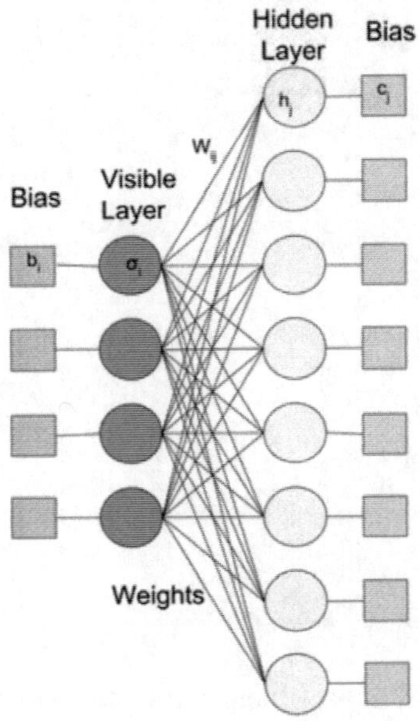

Figure 1. Sample architecture of the RBM used in this study. We choose visible layers of different sizes for $L = 4, 5, 6$, and 7, which also represent the system sizes and twice as many nodes in the hidden layer as in the input layer. Both the input nodes and hidden nodes take binary values; σ_i and h_i, respectively. Every site occupation $n_j = \{0, 1, ..., N_{max}\}$ is converted to $N_{max} + 1$ binary digits, all 0's except the n_jth digit, which is 1. Therefore, the number of input nodes is $(N_{max} + 1) \times L$. Each node has a bias value (shown as b_i or c_i) and nodes are fully connected between the two layers with weights w_{ij} associated with each connection. The weights and biases are determined through a stochastic optimization process that involves local Metropolis moves between Fock states where the expectation value of the Bose-Hubbard Hamiltonian is minimized.

Within NetKet, we define a wave function ansatz for the $q = 0$ momentum sector using a RBM with spin-1/2 hidden nodes and permutation symmetry. The input layer can be visualized as a two-dimensional $L \times (N_{max} + 1)$ array of binary numbers, where each row represents a site, and a 1 in column number n indicates that there are n bosons at that site. This so-called "one-hot" encoding of site occupations is not unique. For example, an alternative would be to use L non-binary input nodes whose values can vary from 0 to N_{max} instead. One advantage of the one-hot representation over the latter is that the number of network parameters is much larger and therefore, the training will presumably be more refined. We choose L and N_{max} values as large as 7 and 5, respectively, and a hidden layer with twice as many nodes as in the visible layer.

3.2. The Training

We write the ground state wave function as a linear combination of the Fock state basis

$$|\psi\rangle = \sum_n \psi_n |n\rangle , \tag{2}$$

where $|n\rangle = |n_1, n_2, \ldots, n_L\rangle$ are the particle number Fock states with site occupancies $n_i \in [0, N_{max}]$. The goal is to create a RBM that can stochastically represent $|\psi\rangle$ by producing the correct probability, $|\psi_n|^2$ for every input Fock state $|n\rangle$ through the input layer. For the Bose-Hubbard model, the coefficients ψ_n are known to be real and positive and are written for the RBM as

$$\psi_n = \sum_{\{h\}} e^{\sum_i b_i\sigma_i + \sum_j c_j h_j + \sum_{ij} W_{ij}\sigma_i h_j} . \tag{3}$$

where σ_i and h_i denote the input node and hidden node binary values, respectively.

The hidden degrees of freedom can be integrated out exactly for the simple architecture shown in Fig. 1,

$$\begin{aligned}\sum_{\{h\}} e^{\sum_i b_i\sigma_i+\sum_j c_jh_j+\sum_{ij} W_{ij}\sigma_ih_j} &= e^{\sum_i b_i\sigma_i}\sum_{\{h\}}\prod_j^{N_h} e^{c_jh_j+\sum_i W_{ij}\sigma_ih_j} \\ &= e^{\sum_i b_i\sigma_i}\prod_j^{N_h}\sum_{h_j=0,1} e^{c_jh_j+\sum_i W_{ij}\sigma_ih_j} \\ &= e^{\sum_i b_i\sigma_i}\prod_j^{N_h}(1+e^{c_j+\sum_i W_{ij}\sigma_i}) \\ &= e^{\sum_i b_i\sigma_i}\prod_j^{N_h} e^{\log\left(1+e^{c_j+\sum_i W_{ij}\sigma_i}\right)}, \end{aligned} \tag{4}$$

which results in [4, 9]

$$\psi_n = e^{\sum_i b_i\sigma_i+\sum_j \log\left(1+e^{c_j+\sum_i W_{ij}\sigma_i}\right)}. \tag{5}$$

The goal in the training of the RBM is to optimize its weights and biases in order to arrive at a probability distribution for ψ_n so that $|\psi\rangle$ represents the ground state. This is accomplished through the minimization of the expectation value of the Hamiltonian,

$$E = \frac{\langle\psi|H|\psi\rangle}{\langle\psi|\psi\rangle}. \tag{6}$$

NetKet adopts a Metropolis algorithm for importance sampling of $|n\rangle$'s during which E is minimized also using stochastic techniques. Since we are working in the grand canonical ensemble, we choose local moves in which a site is picked at random and its occupation is proposed to change to a value in the range $[0, N_{max}]$. The move is accepted with the probability $\min[1, |\psi_{n'}|^2/|\psi_n|^2]$, where n' and n are the new and old states, respectively. This guarantees that after the training is completed, the probability distribution is $P_n = |\psi_n|^2/\sum_m |\psi_m|^2 = |\psi_n|^2/\langle\psi|\psi\rangle$. The minimization of energy is done using the Stochastic Reconfiguration algorithm [27] with AdaMax optimizer in NetKet and the default parameters. Our sampler performs 1000 sweeps before sampling the probability distribution and updating the RBM parameters. More details about the method and the training parameters can be found in Ref. [14].

4. Results

In Figs. 2 and 3, we show the average energy, E, and the average density, ρ, as the training of our RBM progresses for two sets of values for the interaction strength and the chemical potential. The system with $\mu/U = 0.56$ (blue lines) is a Mott insulator with a density of one particle per site and the system with $\mu/U = 1.00$ (orange lines) is a superfluid for the system size we have considered ($L = 6$). Here, $N_{max} = 5$ is much larger than the average density. We observe that the energies and densities quickly converge to their final values regardless of whether the system is in the MI or SF phase. However, we find that the average density has larger fluctuations over the training steps inside the SF phase while it is pinned to unity when the system is in the first Mott lobe.

For comparison, we also show in Figs. 2 and 3 results from exact diagonalization as horizontal dashed and dotted lines. We find that the error in both the energy and the density is larger in the superfluid phase. The fluctuations in the density seem to extend to very long training times

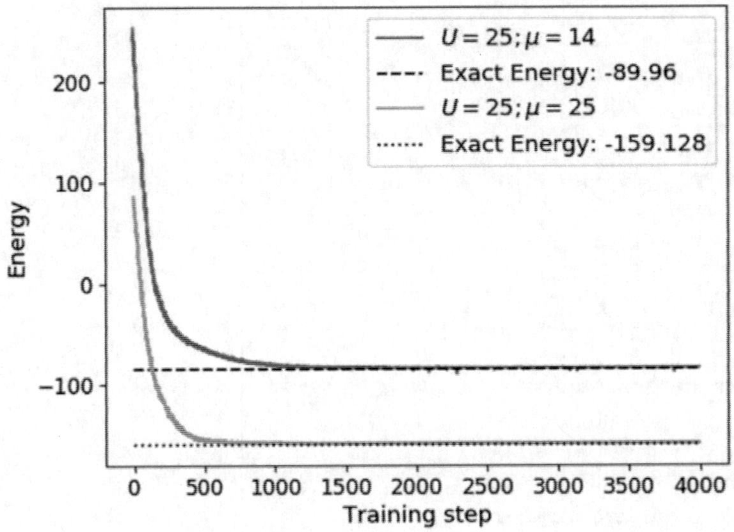

Figure 2. Training a RBM using stochastic reconfiguration aims to find the ground state of the system. This is done by minimizing the expectation value of the energy, which is plotted above as a function of training step. This convergence is an indication that the system has entered the ground state of the Hamiltonian. Here, $L = 6$ and $N_{max} = 5$.

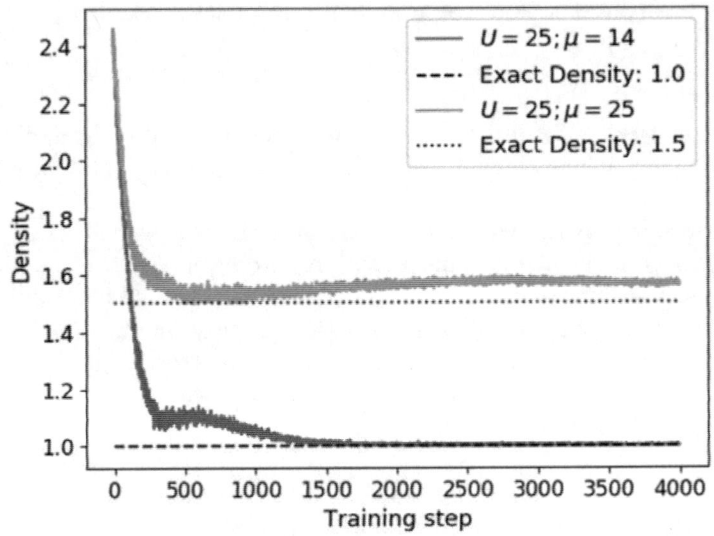

Figure 3. Same as in Fig. 2 but for the evolution of the average density over training step.

and point to the limitations of the method and the fact that more complex networks or better optimization may be needed to capture the physics in the superfluid phase.

Taking long-time values of the converged density as we vary the chemical potential at a fixed interaction strength, we obtain the equation of state. Figure 4 shows this property for $U/t = 15$ and several different combinations of L and N_{max}. We find the expected behavior [3, 15] where the density is pinned at integer numbers of bosons per site as we cross multiple SF and MI phase boundaries. Two density plateaus corresponding to the first and second Mott lobes are clearly visible in this plot. Increasing the system size from 4 to 7 seems to lead to smoother curves, signaling better convergence during the training, but no appreciable change in the onsets of the Mott regions. Moreover, the results do not change significantly by increasing N_{max} from 4 to

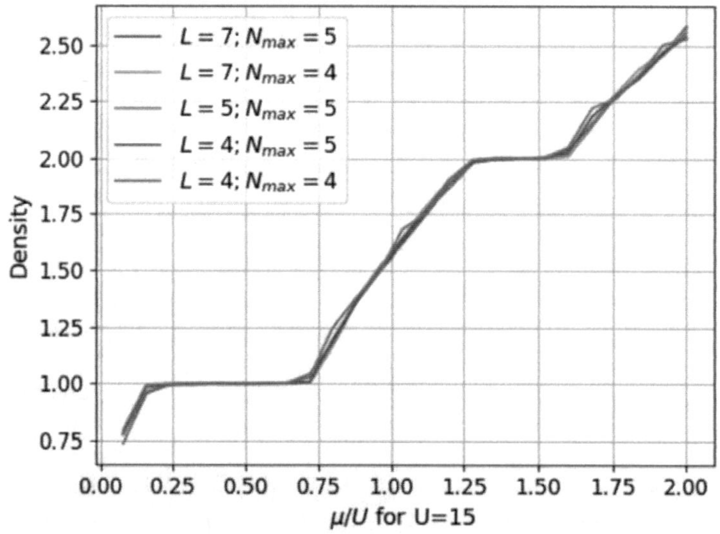

Figure 4. Equation of state for a fixed interaction strength of $U = 15$, different system sizes and different limits for the boson site occupation numbers. The density, ρ, does not show significant variations by changing the system size or the maximum occupation numbers as long as $N_{max} > \rho$.

5. These observations do not mean that our results are not suffering from finite size effects. Much larger clusters have been shown to give different results in conventional treatments of the model [15, 28].

By training different RBMs as we sweep the chemical potential at increasing values of the interaction strength, we are able to map out the ground state phase diagram of the model in the interaction-chemical potential space. The findings are summarized in Fig. 5. We extend the results up to the second Mott lobe. They qualitatively agree with those obtained using Monte Carlo algorithms and much larger system sizes (see e.g. Ref. [28]). As with other techniques, the accurate locating of the phase boundary near the tip of the Mott lobes are the most challenging due to the Kosterlitz-Thouless nature of the transition [29].

5. Summary

Using the open-source package NetKet, recently developed for the study of quantum many-body systems through machine learning algorithms, we design and train RBMs to represent the ground state wavefunction of the 1D Bose-Hubbard model on small chains. We demonstrate the degree of applicability of the RBM ansatz in various regions of the parameter space as MI or SF phases set in. We show that the optimization of free parameters of the RBMs during training in order to minimize the ground state energy leads to physical properties, including expectation values of the Hamiltonian and a phase diagram, that agree with known exact results from conventional treatments of the model.

6. Acknowledgements

KM acknowledges support from Undergraduate Research Grants at San Jose State University. EK acknowledges support from the National Science Foundation (NSF) under Grant No. DMR-1609560.

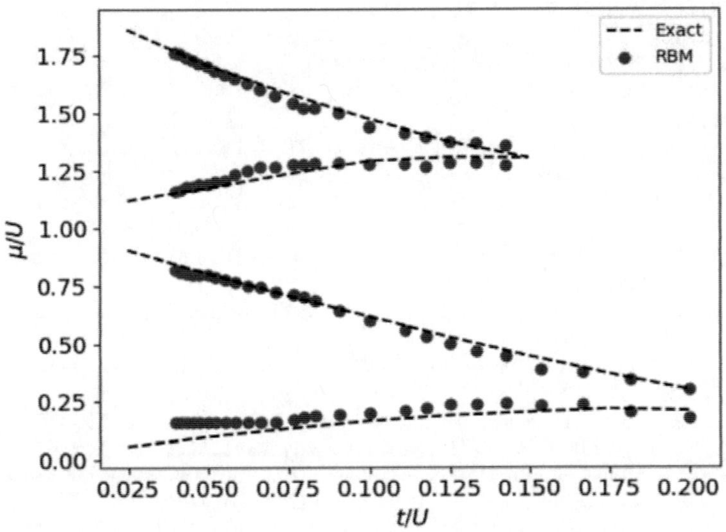

Figure 5. Ground state phase diagram of the Bose-Hubbard model for $L = 6$ and $N_{max} = 5$ in the t/U-μ/U space from the RBM. The first two Mott lobes for $\rho = 1$ and $\rho = 2$ are clearly captured in the strong-coupling region where $t/U \ll 1$. We use a tolerance of ± 0.05 for the density around integer values to determine the departure from the MI phases. Dashed lines are exact results for $L = 128$ from Ref. [28].

7. References

[1] Saito H 2017 *Journal of the Physical Society of Japan* **86** 093001 URL https://doi.org/10.7566/JPSJ.86.093001
[2] Saito H and Kato M 2018 *Journal of the Physical Society of Japan* **87** 014001 URL https://doi.org/10.7566/JPSJ.87.014001
[3] Fisher M P A, Weichman P B, Grinstein G and Fisher D S 1989 *Phys. Rev. B* **40**(1) 546–570 URL https://link.aps.org/doi/10.1103/PhysRevB.40.546
[4] Carleo G and Troyer M 2017 *Science* **355** 602–606 ISSN 0036-8075 URL http://science.sciencemag.org/content/355/6325/602
[5] Deng D L, Li X and Das Sarma S 2017 *Phys. Rev. B* **96**(19) 195145 URL https://link.aps.org/doi/10.1103/PhysRevB.96.195145
[6] Torlai G and Melko R G 2017 *Phys. Rev. Lett.* **119**(3) 030501 URL https://link.aps.org/doi/10.1103/PhysRevLett.119.030501
[7] Chen J, Cheng S, Xie H, Wang L and Xiang T 2018 *Phys. Rev. B* **97**(8) 085104 URL https://link.aps.org/doi/10.1103/PhysRevB.97.085104
[8] Huang Y and Moore J E 2017 (*Preprint* arXiv:1701.06246) URL https://arxiv.org/pdf/1701.06246.pdf
[9] Torlai G and Melko R G 2016 *Phys. Rev. B* **94**(16) 165134 URL https://link.aps.org/doi/10.1103/PhysRevB.94.165134
[10] Torlai G, Mazzola G, Carrasquilla J, Troyer M, Melko R and Carleo G 2018 *Nature Physics* **14** 447–450 ISSN 1745-2481 URL https://doi.org/10.1038/s41567-018-0048-5
[11] Nomura Y, Darmawan A S, Yamaji Y and Imada M 2017 *Phys. Rev. B* **96**(20) 205152 URL https://link.aps.org/doi/10.1103/PhysRevB.96.205152
[12] Huang L and Wang L 2017 *Phys. Rev. B* **95**(3) 035105 URL https://link.aps.org/doi/10.1103/PhysRevB.95.035105
[13] Lu S, Gao X and Duan L M 2018 (*Preprint* arXiv:1810.02352) URL https://arxiv.org/pdf/1810.02352v2.pdf
[14] Source code and tutorials by Giuseppe Carleo are available at https://www.netket.org/.
[15] Batrouni G G, Scalettar R T and Zimanyi G T 1990 *Phys. Rev. Lett.* **65**(14) 1765–1768 URL https:

//link.aps.org/doi/10.1103/PhysRevLett.65.1765

[16] Scalettar R T, Batrouni G G and Zimanyi G T 1991 *Phys. Rev. Lett.* **66**(24) 3144–3147 URL https://link.aps.org/doi/10.1103/PhysRevLett.66.3144

[17] Niyaz P, Scalettar R T, Fong C Y and Batrouni G G 1991 *Phys. Rev. B* **44**(13) 7143–7146 URL https://link.aps.org/doi/10.1103/PhysRevB.44.7143

[18] Rokhsar D S and Kotliar B G 1991 *Phys. Rev. B* **44**(18) 10328–10332 URL https://link.aps.org/doi/10.1103/PhysRevB.44.10328

[19] Krauth W, Trivedi N and Ceperley D 1991 *Phys. Rev. Lett.* **67**(17) 2307–2310 URL https://link.aps.org/doi/10.1103/PhysRevLett.67.2307

[20] Krauth W and Trivedi N 1991 *EPL (Europhysics Letters)* **14** 627 URL http://stacks.iop.org/0295-5075/14/i=7/a=003

[21] Krauth W, Caffarel M and Bouchaud J P 1992 *Phys. Rev. B* **45**(6) 3137–3140 URL https://link.aps.org/doi/10.1103/PhysRevB.45.3137

[22] Kampf A P and Zimanyi G T 1993 *Phys. Rev. B* **47**(1) 279–286 URL https://link.aps.org/doi/10.1103/PhysRevB.47.279

[23] Bruder C, Fazio R and Schön G 1993 *Phys. Rev. B* **47**(1) 342–347 URL https://link.aps.org/doi/10.1103/PhysRevB.47.342

[24] van Otterlo A, Wagenblast K H, Baltin R, Bruder C, Fazio R and Schön G 1995 *Phys. Rev. B* **52**(22) 16176–16186 URL https://link.aps.org/doi/10.1103/PhysRevB.52.16176

[25] Prokof'ev N, Svistunov B and Tupitsyn I 1998 *Physics Letters A* **238** 253 – 257 ISSN 0375-9601 URL http://www.sciencedirect.com/science/article/pii/S0375960197009572

[26] Jaksch D, Bruder C, Cirac J I, Gardiner C W and Zoller P 1998 *Phys. Rev. Lett.* **81**(15) 3108–3111

[27] Becca F and Sorella S 2017 *Quantum Monte Carlo Approaches for Correlated Systems* (Cambridge University Press)

[28] Ejima S, Fehske H, Gebhard F, zu Münster K, Knap M, Arrigoni E and von der Linden W 2012 *Phys. Rev. A* **85**(5) 053644 URL https://link.aps.org/doi/10.1103/PhysRevA.85.053644

[29] Carrasquilla J, Manmana S R and Rigol M 2013 *Phys. Rev. A* **87**(4) 043606 URL https://link.aps.org/doi/10.1103/PhysRevA.87.043606

IOP Conf. Series: Journal of Physics: Conf. Series **1290** (2019) 012006 doi:10.1088/1742-6596/1290/1/012006

Principal component analysis of the magnetic transition in the three-dimensional Fermi-Hubbard model

Ehsan Khatami

Department of Physics and Astronomy, San Jose State University, San Jose, CA 95192, USA

Abstract. Machine learning techniques have been widely used in the study of strongly correlated systems in recent years. Here, we review some applications to classical and quantum many-body systems and present results from an unsupervised machine learning technique, the principal component analysis, employed to identify the finite-temperature phase transition of the three-dimensional Fermi-Hubbard model to the antiferromagnetically ordered state. We find that this linear method can capture the phase transition as well as other more complicated and nonlinear counterparts.

1. Background and Introduction

We humans and other animals owe our often effortless recognition of objects, faces, voices and other cues to the highly complex neural network in our brain. While we cannot exactly model the tens of billions of neurons, each with a few thousand connections to other neurons, in our brains using the current computer technology, biological neural networks have certainly been the model and inspiration for the development of deep artificial neural networks (ANN), and more broadly, machine learning techniques. Their practical applications, mostly in the technology industry, such as Google's deepmind [1], Apple iPhone's voice and face recognition, Netflix's narrowing down of movie preferences for its customers, are often designed to perform specific tasks.

Machine learning paradigms in science have also been around for decades, for example, in the studies of connections between neural networks and spin glass models and how they can be used as memories [2, 3, 4, 5, 6, 7, 8]. However, it was the development of efficient algorithms for training deep networks that led to the rise of popularity of machine learning and its physical applications in recent years. For example, the seminal paper by Carrasquilla and Melko [9], which focused on learning algorithms for distinguishing phases of matter, or by Carleo and Troyer [10], which focused on using ANNs as superior variational ansatz for obtaining the ground state wavefunction of quantum many-body systems, both in 2016 paved the way for hundreds of other studies in the field of statistical mechanics and quantum many-body physics that followed.

In Ref. [9], the authors replace images of animals, objects, or handwritten digits, often used to showcase the power of novel machine learning algorithms in the industry, by Ising "images"; configurations of binary spins from Monte Carlo (MC) simulations of the Ising model on two-dimensional (2D) geometries. In one case, a simple fully-connected feed-forward ANN was trained, using these images at different temperatures, to identify the location of the critical temperature of the 2D Ising model. They not only showed that their network could locate the phase transition using images it had not seen during the training, but that the trained network could also be used to identify the same transition on other geometries (e.g., the triangular lattice) with over 99% accuracy. They showed that the network output in this case is essentially the magnetization.

In studies that followed immediately [11, 12], the question explored was whether similar techniques can be applied to quantum mechanical models, more specifically, to fermionic lattice models. For quantum lattice models, such as the Fermi-Hubbard model [13, 14], and their quantum Monte Carlo (QMC) treatments [15], the question is, what type of "images" can be used for training? Unlike for the Ising model, where MC simulations sample directly over configurations of the constituents (spin-1/2), the configurations sampled over in QMC simulations of fermionic models are auxiliary fields and involve an "imaginary time" dimension. In Ref. [11], it was shown that ANNs cannot be trained to identify the quantum phase transition between a semi-metal and an antiferromagnetic Mott insulator in the honeycomb lattice Hubbard model using these fields directly, at least not when geometrical information was not preserved. Instead, using calculated Green's functions as inputs for ANNs, they showed that the networks can learn to accurately predict the location of the transition even if training has been performed using data far from the transition point. In Ref. [12], it was shown that raw auxiliary field configurations can in fact be used by ANNs to learn to identify the finite-temperature transition of the three-dimensional (3D) Fermi-Hubbard model from a paramagnetic to an antiferromagnetic phase using ANN architectures that respect the geometry of the model and involve so-called convolutional layers for extracting local features. Moreover, it was shown that the knowledge learned during training can be transferred to locate the same phase transition in a model with a different strength for fermionic interactions or at incommensurate fillings.

The question of machine-assisted phase discovery was also explored early on using the so-called *unsupervised* machine learning algorithms for both classical [16, 9, 17, 18, 19] and quantum systems [20, 21, 22]. For this class of methods, labels for the data (e.g., whether or not they belong to the ordered phase) need not be provided by a "supervisor" during the training process. Instead, unsupervised algorithms are designed to learn about similarities and dissimilarities among the data on their own during the training and categorize data based on that information. The categorization is meant to reduce the huge dimensionality of the configuration space (e.g., 2^N for the Ising model, where N is the number of spins) to a few "latent" dimensions where the similarity of configurations manifests itself in the clustering of corresponding data points. Such visualizations can be clearly seen in Refs. [16, 9, 18] and other studies. One can then classify a new configuration that has not been involved in the dimension reduction routine by projecting it onto the latent space. A summary of these unsupervised techniques is provided in Ref. [21].

Here, the question is again, which techniques can see through quantum fluctuations and effectively classify auxiliary configurations obtained during QMC simulations? And, how does the classification depend on the nature of the phases under investigation? Features observed in the reduced dimensional space can provide insights into properties and symmetries of the system as a tuning parameter like the temperature is varied, and sudden changes can hint at phase transitions. Several nonlinear algorithms have already been shown to exhibit nontrivial features in dimensionally-reduced QMC data for the 2D and 3D Hubbard models [21]. However, a linear technique called the principal component analysis (PCA) [23] has not shown a consistent performance. For example, no discernible features could be found for the quantum many-body localization transition of the random-field Heisenberg model [20], whereas transitions to certain magnetic or charge density wave phases could be captured for the Fermi-Hubbard model [22].

Here, we consider the case of the half-filled 3D Fermi-Hubbard model and its finite-temperature transition to the long-range antiferromagnetic phase and examine to what extent this transition can be captured using the PCA. As a benchmark, we first reproduce results from

IOP Conf. Series: Journal of Physics: Conf. Series **1290** (2019) 012006 doi:10.1088/1742-6596/1290/1/012006

the PCA for the 2D Ising model.

2. Models

2.1. 2D Ising Model

The energy of the ferromagnetic Ising model on the square lattice is written as

$$E = -J \sum_{\langle ij \rangle} \sigma_i \sigma_j, \tag{1}$$

where $\sigma_i = \pm 1$, $\langle ij \rangle$ denotes that i and j are nearest neighbors, and J is the strength of the coupling, which we set to 1 as the unit of energy for this model. The critical temperature for this model is $T_c = \frac{2}{\ln(1+\sqrt{2})} = 2.27$.

We use a Markov chain MC algorithm with the Metropolis algorithm to obtain 10,000 configurations, 100 configurations per each of the 100 temperature values on a uniform grid between 1.5 and 4.0 for a 20×20 system. Starting with a random configuration, in this algorithm, a site i is picked at random and a move $\sigma_i \to -\sigma_i$ is proposed. The move is accepted with the probability $\min[1, e^{-\beta \Delta E}]$, where β is the inverse temperature, and ΔE is the total energy after the spin slip minus that before the flip. We employ a simulated annealing algorithm in which the temperature is reduced adiabatically using an exponential protocol during the simulations. This will allow the system to settle into one of the two degenerate ferromagnetic ground states that break the time reversal symmetry (all up spins or all down spins).

2.2. 3D Fermi-Hubbard Model

The Fermi-Hubbard Hamiltonians is expressed as

$$H = -t \sum_{\langle ij \rangle \sigma} c_{i\sigma}^{\dagger} c_{j\sigma} + U \sum_{i} n_{i\uparrow} n_{i\downarrow} \tag{2}$$

where $c_{i\sigma}$ ($c_{i\sigma}^{\dagger}$) annihilates (creates) a fermion with spin σ on site i, $n_{i\sigma} = c_{i\sigma}^{\dagger} c_{i\sigma}$ is the number operator, U is the onsite Coulomb interaction, and t is the hopping amplitude, which we take to be unity. We consider the model with an average of one fermion per lattice site (half-filled limit) on a cubic lattice where it has been shown that the critical temperature for the transition to the antiferromagnetic phase is maximal around $T_N = 0.35$ for $U \sim 9$ [24, 25].

We use the determinant QMC algorithm [15, 26, 27] on a $N = 4 \times 4 \times 4$ cluster and 32 temperatures ranging from 0.10 to 0.65. As mentioned above, the auxiliary spin configurations here live in a $D+1$ dimensional space, where D is the physical dimensionality (here 3). The Metropolis algorithm, which is performed separately for each temperature, proceeds with the same Markov chain as for the Ising model. However, the acceptance probability for a spin flip has to be computed using "fermion determinants" [15, 26]. In Fig. 1, we show a $D+1$ dimensional spin configuration for $D = 1$. The vertical axis β is discretized with a spacing $\Delta\tau$. In order to keep the number of auxiliary spin variables ("features" in our PCA analysis below) fixed, we fix the number of imaginary time slices to 200. As a result, $\Delta\tau$ varies between 0.050 and 0.008 in our temperature range, rendering systematic errors negligible. Details of the method and other parameters we have used can be found in Ref. [12].

3. Principal Component Analysis

PCA is essentially a transformation of the coordinate system in a high-dimensional space to a set of principal axes determined through diagonalizing the covariance matrix of the data. The eigenvectors corresponding to the largest eigenvalues of the covariance matrix indicate the principal directions along which the data have the largest variance. After the PCA is performed we often project the data onto the two principal directions with the largest eigenvalues.

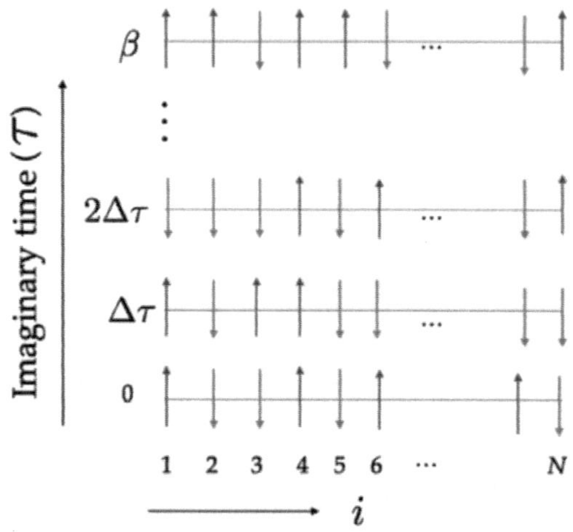

Figure 1. A random configuration of auxiliary spins during the QMC simulation of the one-dimensional Fermi-Hubbard model with a system of N sites at an inverse temperature of β. The horizontal axis represents the spatial dimension and the vertical axis represents the imaginary time dimension that extends from 0 to β. The latter is discretized with the step $\Delta\tau$.

3.1. The Algorithm
The PCA algorithm is as follows:

(i) Center the data. Each data point is a collection of "features", or components in the original space. For example, an Ising configuration can be thought of as a data point in a N dimensional space, where N is the system size and components are σ_i. In this step, the mean value of every feature among all the data is subtracted.

(ii) Create the matrix of data X. Each data point forms a row, columns represent features,

(iii) Compute the covariance matrix of data: $X^T X$.

(iv) Compute eigenvalues λ_n and normalized eigenvectors v_n of the covariance matrix. Eigenvalues represent the variance of each principal axis.

(v) Project data onto the eigenvectors: $P_n = X \cdot v_n$. These are the principal components.

4. Results
4.1. 2D Ising Model
In Fig. 2, we show results after applying the above PCA algorithm to the MC Ising configurations. The size of the covariance matrix, and therefore, the number of principal components is $N = 400$. The left panel in this figure shows the eigenvalues of the covariance matrix, where the largest eigenvalue is an order of magnitude larger than the rest. This implies that the data are mostly spread out along the first principal direction. We note that we perform the PCA on data from all the temperatures at once as opposed to a separate PCA calculation for every temperature.

Projecting the configurations onto the principal directions corresponding to the largest two eigenvalues of the covariance matrix of data, we can visualize any patterns developing as the temperature dips below T_c. In the right panel of Fig. 2, we plot the data from all temperatures in the space of the first two principal components. The color of symbols denotes the temperature indicated on the color bar. We observe a well-defined disk of data points corresponding to high temperatures above T_c. As the temperature decreases to near the critical value, the data spread out in both the P_1 and P_2 directions. At lower temperatures below T_c, we find that an overwhelming number of data points gather in a cluster at large P_1. We infer that these data

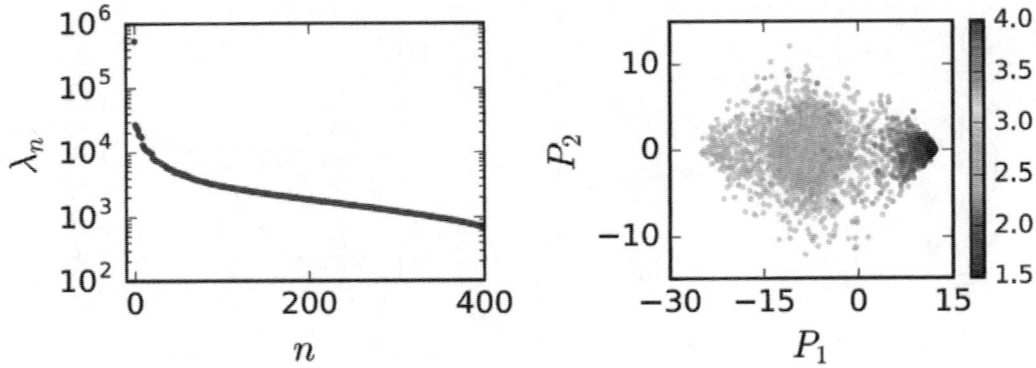

Figure 2. PCA applied to the 2D Ising model. Left: Eigenvalues P_i of the covariance matrix of spin configurations on a 20×20 square lattice. We find that the second eigenvalue is more than an order of magnitude smaller than the first eigenvalue. Right: The projection of data to the space of first two principal components. Colors denote the temperature. The high-temperature disk of data points suddenly changes to a cluster for low-temperature data points at larger P_1 as one crosses the critical temperature.

points correspond to the ordered phase. We point out that the reason for observing a single low-temperature cluster as opposed to two, e.g., as in Refs. [16, 9, 18], is that we have used simulated annealing which causes the system to gradually settle in one of the two degenerate ferromagnetic ground states of the model, as opposed to the possibility of systems at close temperatures falling into phases with magnetizations of opposite sign.

4.2. 3D Fermi-Hubbard Model

We now turn to the quantum model and apply the same technique to data obtained from QMC for the auxiliary spin configurations of the 3D Fermi-Hubbard model. Figure 3 summarizes our findings. In this case, the dimension of the covariance matrix (the number of features) is $N \times 200 = 12,800$. In the left panel of Fig. 3, we show the first 3,000 eigenvalues of the covariance matrix. The largest is more than an order of magnitude larger than the rest of the eigenvalues, again suggesting that there is mostly one special direction in the high-dimensional configuration space where the data have a relatively large variance. Finding the first two principal components allows us to project the data in a two-dimensional space and visualize them.

The results are shown in the right panel of Fig. 3. There, the evolution of clustering of data points as a function of temperature is less dramatic than in the Ising model across the critical temperature. As the temperature decreases the data simply spread out along P_1 and they do so in a symmetric way; the number of points with $P_1 > 0$ does not seem to be significantly different form those with $P_1 < 0$. The smoother evolution is partly a reflection of the fact that the determinant QMC used in this work is formulated using the fermion spins represented in the z basis, whereas the model has spin rotational symmetry and does not have a preferred direction; the antiferromagnetic ordering below T_N can happen in the xy plane for some instances, which will manifest itself essentially as randomness in the configuration of (auxiliary) spins represented in the z basis. We point out that although the auxiliary spins do not exactly represent particle spins, it can be shown that their correlation functions are proportional to the correlation functions of the actual fermion spins [28, 29].

We believe that another contributor to the lack of distinct low-temperature clusters in the Hubbard data shown in Fig. 3 is the existence of quantum fluctuations in the system, i.e., preventing a perfectly anti-aligned spin configuration in any spin direction from being the exact

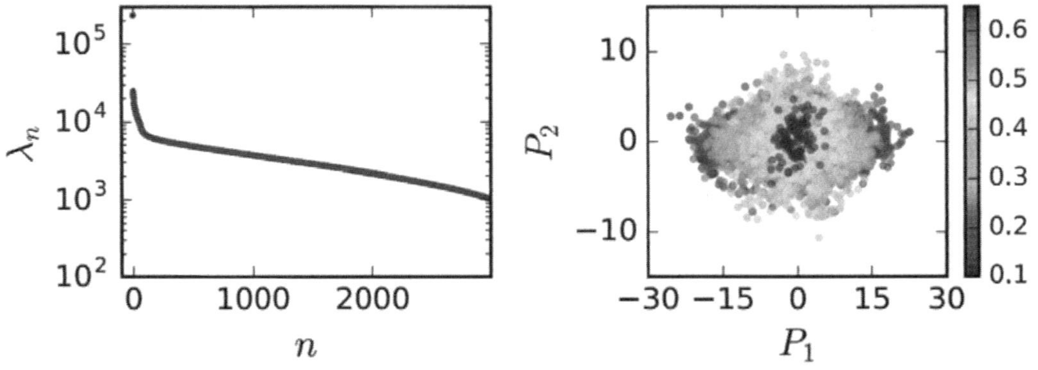

Figure 3. Same as Fig. 2 but for the half-filled 3D Hubbard model with $U = 9$. We show only the first 3,000 eigenvalues in the left panel. Similarly to the Ising case, the largest eigenvalue is at least an order of magnitude larger than the rest. Here, the evolution of clustering of data points as a function of temperature is less dramatic than in the Ising model across the critical temperature. As the temperature decreases the data points simply spread out along P_1.

ground state of the model, unlike for the classical model. Nevertheless, it has been shown that indicators that quantify the transformation of the distribution of data in the reduced-dimensional space as the temperature is varied can represent physically meaningful properties such as the magnetic structure factor [21]. The distribution in the right panel of Fig. 3 obtained through a linear machine learning algorithm seems to be as useful in gaining insight into the phase transition of the model as those obtained through more sophisticated and nonlinear methods applied to the same data [21].

5. Summary

Considering two classical and quantum many-body systems, both exhibiting a continuous finite-temperature phase transition to magnetically ordered phases, we study how a linear machine learning technique, the principal component analysis, can provide clues for the location and nature of the transitions using configurations generated during MC or QMC simulations of the model and without the calculation of any conventional properties. We find that, similar to what was found in past studies with the 2D Ising model, the PCA can signal the transition for this model with a high degree of accuracy. Around T_c, we observe a sudden shift in the distribution of dimensionally reduced Ising configurations in the space of the first two principal components from a central disk at high temperatures to a different cluster at $T \lesssim T_c$. This is consistent with the use of simulated annealing in our MC simulations. We find that the picture is different for the data corresponding to the 3D Fermi-Hubbard model. Due to the specific basis chosen for the QMC simulations and the existence of quantum fluctuations, in that case, the dimensionally-reduced data simply spread out from a central region as the temperature is decreased. However, despite that, quantifying the spread can likely results in reliable indicators for the phase transition.

6. Acknowledgments

I acknowledge support from the National Science Foundation under Grant No. DMR-1609560.

7. References

[1] `https://deepmind.com/`.

[2] J. J. Hopfield. Neural networks and physical systems with emergent collective computational abilities. *Proceedings of the National Academy of Sciences*, 79(8):2554–2558, 1982.
[3] Daniel J. Amit, Hanoch Gutfreund, and H. Sompolinsky. Spin-glass models of neural networks. *Phys. Rev. A*, 32:1007–1018, Aug 1985.
[4] G. Toulouse, S. Dehaene, and J. P. Changeux. Spin glass model of learning by selection. *Proceedings of the National Academy of Sciences*, 83(6):1695–1698, 1986.
[5] N. Parga and M. A. Virasoro. The ultrametric organization of memories in a neural network. *J. Phys. France*, 47(11):1857–1864, 1986.
[6] G. Parisi. Asymmetric neural networks and the process of learning. *Journal of Physics A: Mathematical and General*, 19(11):L675, 1986.
[7] M. Mézard, J. P. Nadal, and G. Toulouse. Solvable models of working memories. *J. Phys. France*, 47(9):1457–1462, 1986.
[8] R. Scalettar and A. Zee. Perception of left and right by a feed forward net. *Biological Cybernetics*, 58(3):193–201, Feb 1988.
[9] Juan Carrasquilla and Roger G. Melko. Machine learning phases of matter. *Nat Phys*, 13(5):431–434, May 2017. Letter.
[10] Giuseppe Carleo and Matthias Troyer. Solving the quantum many-body problem with artificial neural networks. *Science*, 355(6325):602–606, 2017.
[11] Peter Broecker, Juan Carrasquilla, Roger G. Melko, and Simon Trebst. Machine learning quantum phases of matter beyond the fermion sign problem. *Scientific Reports*, 7(1):8823, 2017.
[12] Kelvin Ch'ng, Juan Carrasquilla, Roger G. Melko, and Ehsan Khatami. Machine learning phases of strongly correlated fermions. *Phys. Rev. X*, 7:031038, Aug 2017.
[13] N. F. Mott. The basis of the electron theory of metals, with special reference to the transition metals. *Proceedings of the Physical Society. Section A*, 62(7):416, 1949.
[14] J. Hubbard. Electron correlations in narrow energy bands. *Proceedings of the Royal Society of London A: Mathematical, Physical and Engineering Sciences*, 276(1365):238–257, 1963.
[15] R. Blankenbecler, D. J. Scalapino, and R. L. Sugar. Monte carlo calculations of coupled boson-fermion systems. i. *Phys. Rev. D*, 24(8):2278–2286, Oct 1981.
[16] Lei Wang. Discovering phase transitions with unsupervised learning. *Phys. Rev. B*, 94:195105, Nov 2016.
[17] Sebastian J. Wetzel. Unsupervised learning of phase transitions: From principal component analysis to variational autoencoders. *Phys. Rev. E*, 96:022140, Aug 2017.
[18] Wenjian Hu, Rajiv R. P. Singh, and Richard T. Scalettar. Discovering phases, phase transitions, and crossovers through unsupervised machine learning: A critical examination. *Phys. Rev. E*, 95:062122, Jun 2017.
[19] Ce Wang and Hui Zhai. Machine learning of frustrated classical spin models. i. principal component analysis. *Phys. Rev. B*, 96:144432, Oct 2017.
[20] Evert P. L. van Nieuwenburg, Ye-Hua Liu, and Sebastian D. Huber. Learning phase transitions by confusion. *Nat Phys*, 13(5):435–439, May 2017. Letter.
[21] Kelvin Ch'ng, Nick Vazquez, and Ehsan Khatami. Unsupervised machine learning account of magnetic transitions in the Hubbard model. *Phys. Rev. E*, 97:013306, Jan 2018.
[22] Natanael C. Costa, Wenjian Hu, Z. J. Bai, Richard T. Scalettar, and Rajiv R. P. Singh. Principal component analysis for fermionic critical points. *Phys. Rev. B*, 96:195138, Nov 2017.
[23] I. Jolliffe. *Principal component analysis*. John Wiley and Sons, Ltd, 2002.
[24] R. Staudt, M. Dzierzawa, and A. Muramatsu. Phase diagram of the three-dimensional Hubbard model at half filling. *The European Physical Journal B - Condensed Matter and Complex Systems*, 17(3):411–415, 2000.
[25] Ehsan Khatami. Three-dimensional Hubbard model in the thermodynamic limit. *Phys. Rev. B*, 94:125114, Sep 2016.
[26] S. R. White, D. J. Scalapino, R. L. Sugar, E. Y. Loh, J. E. Gubernatis, and R. T. Scalettar. Numerical study of the two-dimensional Hubbard model. *Phys. Rev. B*, 40:506–516, Jul 1989.
[27] *Codes available at* `http://quest.ucdavis.edu/` *and* `https://code.google.com/p/quest-qmc`.
[28] J. E. Hirsch. Discrete Hubbard-Stratonovich transformation for fermion lattice models. *Phys. Rev. B*, 28:4059–4061, Oct 1983.
[29] J. E. Hirsch. Connection between world-line and determinantal functional-integral formulations of the Hubbard model. *Phys. Rev. B*, 34:3216–3220, Sep 1986.

IOP Conf. Series: Journal of Physics: Conf. Series **1290** (2019) 012007 doi:10.1088/1742-6596/1290/1/012007

The Angular Fractal Dimension in Galaxy's Distributions

Loay Khalifa[1], **Jesùs Pando**[1]

[1] Department of Physics, DePaul University, 2219 N Kenmore Ave, Chicago, IL, US

E-mail: lkhalifa@mail.depaul.edu

E-mail: jpando@depaul.edu

Abstract. The study of large scale structure (LSS) of the Universe has entered a precision era due to all-sky surveys and numerical simulations. The new data has provided a way to bring new methods to bear to analyze the cosmology as probed by large scale structure. We use wavelet packets to investigate fractal point-processes on galactic scales. In particular, we develop a method to calculate the angular fractal dimension of galaxy distributions as a function of cosmological comoving distance. Taking advantage of the self-similarity and localization properties of discrete wavelets, we compute the angular fractal dimension of galaxies in narrow redshift bins. The narrow bins assure that dynamical evolution in the range being studied has not occurred to a significant extent. We use both real and simulated data from the Baryon Oscillation Spectroscopic Survey (BOSS) and the Mock Galaxy Catalogs produced by the Sloan Digital Sky Survey (SDSS). Using the wavelet packet power spectrum, we find areas in the galaxy distribution which have power law like behavior indicating fractal processes are present. The exponent of the power law is the Hurst exponent H, which is directly related to the fractal dimension of spatial point processes. We find the fractal dimension ranges from $D = 1.1$ to $D = 1.4$ for BOSS Galaxies while it ranges from $D = 1.4$ to $D = 1.8$ for Mock Galaxy Catalogs. The results are mildly dependent on the number of galaxies present in each redshift bin and less so on the resolution at which the data is binned. We conclude that this method can be used to characterize large scale structure and its evolution as a function of redshift. There are hints that the galaxy distribution may be fractal at higher redshifts than previously reported, however more data is necessary before a firmer conclusion is reached.

1. Introduction

Fractals were introduced to the modern reader by the ground breaking work of Mandelbrot[1]. This work spurred some early work to determine if fractals might be present in spatial distribution of galaxies [2, 3]. These early studies strongly suggested that at small distances, $r < 10h^{-1}$ Mpc, the galaxy distribution was fractal. These early studies were limited in both depth and coverage of the sky, but did establish that at small scales the cosmological principle, that the universe is both isotropic and homogeneous, did not hold. With the advent of all sky surveys that probed the matter field deeper in redshift as well as across the entire celestial sphere, fractal studies could be extended to probe the scales at which these results hold.

In this work, we calculate the angular fractal dimension of the angular galaxy distributions using wavelet packet transform methods at redshifts between $0.5 < z < 1$. The novel feature of this study is the use of the fractal dimension as a function of redshfit (or comoving distance) to probe the evolution of large scale structure. To do so, we calculate the angular fractal dimension in narrow redshift bins. To find the fractal dimension we make use of the both the self-similar and compact support properties of wavelet packets. Unlike the original studies, we have access to a rich data set that covers most of the northern sky and at much deeper redshifts. Specifically, we use the Sloan Digital Sky Survey's Baryonic Oscillation Spectroscopic Survey (SDSS BOSS) [4]and SDSS Mock Galaxy Catalogs. This survey covered 10,000 square degrees of the sky and found the redshifts of about 1.5 million luminous galaxies and gives one of the most complete galaxy surveys in existence.

2. Fractals

The exact definition of a fractal varies from study to study but all contain the following traits. First, a fractal is self-similar. That is, the distribution is copy of itself when seen at different scales and/or when translated. When seen at a single scale, the geometry appears irregular and not smooth. The associated idea of fractal dimension is a dimension d, which is a description of the capacity of the data set to fill the embedding space.

Secondly, the measures used to determine the scaling behavior, themselves scale. This is the origin of the oft-used box counting dimension in which a distribution is covered by boxes of one scale, then covered again with boxes of a different scale. The behavior of number of boxes versus scale is a measure of the fractalness of the distribution. It is this property that leads us to use wavelets to determine the existence of fractal behavior in the galaxy distribution.

Generally, fractal behavior is suggested if the power spectrum follows a power law over some range of frequencies. That is, there is a strong indicator of fractal behavior if the spectrum of system has a form

$$S(f) \approx f^{-\alpha} \tag{1}$$

where $S(f)$ is the power at frequency f and α is called the power law index. It is this fact that we will use in subsequent sections to determine any fractal behavior in the galaxy distributions. In essence, determining the power spectrum of our process will let us know if there are fractals, and the index α will be related to the fractal dimension. One of the important features of this work is exactly how we find the power spectrum.

The use of the fractal dimension in the study of large scale structure is not the most common measure. The power spectrum and the two point correlation function are established statistical measures to classify large scale structure since the earliest days of galaxy surveys. Following the development of fractal geometry, fractal analysis of galaxies has been added to the statistical measures of cosmology to better understand the large scale structure. In principle, all statistical measures contain the same information so long as one has complete access to all the data and multiple data sets are available. Neither of these is the case in cosmology where there is only one universe and all observational data is limited due to the constraints of the measuring instruments. This means that in practice different statistical measures extract different kinds of information. To form a complete picture of the universe therefore, multiple lines of inquiry are needed. From a cosmological perspective as applied to galaxy distribution, the fractal dimension is a measure of the degree of inhomogeneity in the distribution, i.e., how irregular the distribution is. It provides information about the dominance of voids in structure, hence, it is an important tool in studying matter clustering. In addition, the transition from the homogeneous universe to the irregular universe can be used as guide as to when nonlinear effects must be accounted for.

3. Wavelet Transform

The wavelet transform is an integral transform that has gained increasing use in all types of signal analysis. In the study of large scale structure, wavelets have been used to study the interstellar medium [5], the distribution of quasars [6, 7], galactic clustering and sub-structure [8, 9], and even to analyze the cosmic microwave background [10, 11]. The main reason for this is that

IOP Conf. Series: Journal of Physics: Conf. Series **1290** (2019) 012007 doi:10.1088/1742-6596/1290/1/012007

wavelets can analyze a signal and get both frequency and spatial information simultaneously. Further, the basis functions that are used to perform the transform can built to fit a particular set of boundary conditions. This has led to a myriad of wavelet-like tools such as needlets, spherical wavelets, maximal overlap wavelet, etc. For good reviews on wavelets, see for example [12, 13], and for reviews on the statistical applications of wavelets [14, 15]

While wavelets come in various forms, they can be categorized into two broad classes, discrete and continuous. The difference between the two is in how the basis functions treat scaling. Continuous wavelets use exponential-like scaling with powers less than 2. The discrete wavelets analyzes the scale information of a distribution in powers of 2. This difference determines the choice of general wavelet to used. The discrete wavelets form an orthogonal basis which make perfect reconstruction possible and statistical analysis much more convenient. Continuous wavelets allow for finer localization and frequency resolution, but are over redundant and there is significant overlap between wavelets at each scale and between scales. Because our interest are in analyzing the galaxy as a function scale in order to detect fractal behavior, the fact that the discrete wavelet keeps the information between wavelets independent determines our choice to use the discrete wavelet transform.

Here we will use the language of signal analysis to describe discrete wavelets in more detail. Typically discrete wavelets are constructed using two functions, the wavelet functions which act like high pass filters and the scaling functions that work like low pass filters. These two filters work on a signal producing two new sub-signals, one which contains the low frequency information and one that contains the high frequency information. Another way to look at this is that the low pass filter generates a signal that contains the local density information, while the high pass filters generates a signal that contains the local fluctuations.

In the discrete wavelet transform, the high and low pass filters are translated in discrete steps, l across the signal at a particular scale j. The resulting coarse grained local density that results from the action of the low pass filter is then subjected to the filtering process again. Two new sub-signals emerge. This process is repeated and forms the basis for multi-resolution analysis and can be used to define a wavelet spectrum.

One can consider signals to be points in a vector space. Here we are concerned with the wavelet basis, $\phi_(t)$s for this vector space so that any signal, $s(t)$ can be written as

$$s(t) = \sum_{-\infty}^{\infty} \langle \phi(k)s(k) \rangle \, \phi(t) \tag{2}$$

where $\langle \phi(k)s(k) \rangle$ is the inner product. These coefficients are called the wavelet coefficients and the ϕ are made up of two separate filters. The filters can vary and depend on the length extent that the filter is designed to span. The simplest filters are the *Haar* filters and in the following we focus on these.

The expansion in Eq.(2) can be recast in terms of the low and high pass filters as

$$s(t) = \sum_{-\infty}^{\infty} \langle \phi(k)s(k) \rangle \, \phi(t) = \sum a_j \phi_j(t) + \sum d_j \psi_j(t) \tag{3}$$

where $\phi_j(t)$ is now the low pass filter and $\psi_j(t)$ the high pass filters and the wavelet coefficients of expansion are now a_j and d_j.

As with the Fourier transform, the coefficients contain the crucial information we seek and in the case of the Haar wavelet are particularly simple to construct. The coefficients are given

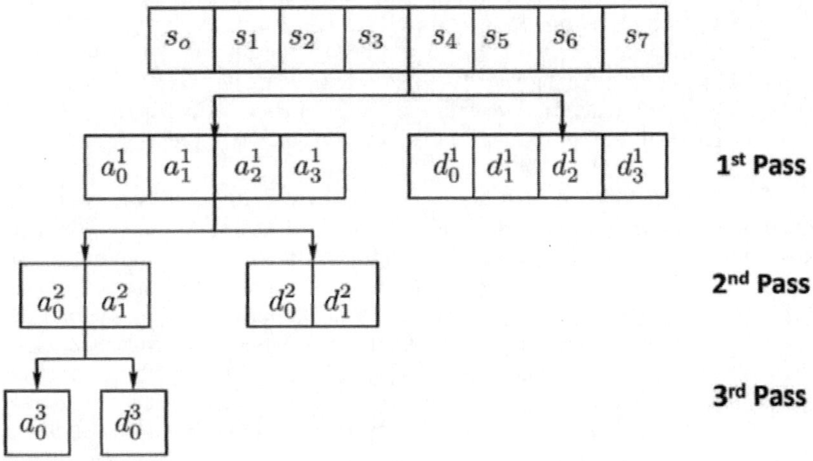

Figure 1. The Discrete Wavelet Tree. The original signal in the top row is acted on by the low pass filter $\mathcal{G}$ and the high pass filter $\mathcal{H}$. The result is two sub-signals as represented by the wavelet coefficients. The a- coefficients capture the local density, while the d capture the local fluctuations.

by

$$a_j = \frac{s_j + s_{(j+1)}}{2} \tag{4}$$
$$d_j = \frac{s_j - s_{(j-1)}}{2} \tag{5}$$

where s_j is the $j$the data point. We see immediately that the approximation coefficients, a_j are the average of the signal at two successive points j and j_1, and the detail coefficients, d_j are the difference of the two points. In general, the approximation and detail coefficients are obtained via a convolution between the filter and the signal. That is,

$$a_j = f(x) \circledast \mathcal{G} \tag{6}$$
$$d_j = f(x) \circledast \mathcal{H} \tag{7}$$

This gives two sub-signals characterized by the coefficients, a_l^1, d_l^1, where l gives the location of sub-signal, and 1 is the pass that the signal has been acted on by the wavelet. The process now continues, with wavelet operating on the sub-signal composed of the a_l^1 coefficients. The process is illustrated in figure 1. At each pass, the wavelet generates a coarse grained signal that reveals features, both local densities and local fluctuations, at that scale. The resolution of the signal is reduced each pass by a half.

3.1. The Wavelet Packet Transform

While the wavelet transform is useful as is, it does suffer from rather low frequency resolution as each pass reduces the resolution at which the signal is analyzed by one-half. This coarse graining if often too large to capture important features in a signal.

The wavelet packet transform is one way in which we can keep the useful features of the wavelet transform, while increasing the frequency resolution at which one can analyze a signal. In the wavelet packet transform, the filters are applied to *both* approximation and detail coefficients after the first level of decomposition. In other words, another branch appears underneath the initial d_l^j coefficients shown in figure (1) as shown now in figure (2). Each box in figure 2 represents a set of coefficients that capture the information in a frequency band. The total frequency band covered by the boxes (called *leafs* or *nodes*) ranges from 0 to the Nyquist frequency. The bands are evenly divided in this range.

In figure 2, the coefficients in each node are labeled as being acted on by the low pass filter, a's or high pass filters, d's. In this example, the original signal contains 8 data points labeled s_0-s_7. After the first pass, we have four coefficients, a_0-a_3 that contain information on the lower frequencies of the signal, and four coefficients $d_0 - d_3$ that contain information on the higher frequencies. After the second pass, things get more interesting. The original approximation coefficients are subjected to both filters as was the case with the wavelet transform. However, the detail coefficients are also passed through the high and low pass filters. At the second pass, $j = 2$ the first leaf shows two coefficients labeled, $a(a_o^1, a_1^1)$ and $a(a_2^1, a_3^1)$. This means that the the low pass filter a has acted on each pair of coefficients and produced 2 new coefficients. These new coefficients cover the lowest frequency range of the signal. The second node is similar but instead the high pass filter is applied to the coefficients. This is similar to the wavelet case. The third node has two coefficients labeled $a(d_o^1, d_1^1)$ and $a(d_2^1, a_3^1)$. This means that the low pass filter has been acted on each pair of *difference* coefficients. Here we see the real difference between wavelets and wavelet packets. The process continues in a similar manner as shown in the rest of the figure.

The entire frequency range analyzed by the wavelet packet transform is 0 to the Nyquist frequency. In figure 2, each node at $j = 2$ contains information on 1/4 this range, while at $j = 3$, each node contains information at 1/8 the range. Thus the more passes (higher j) the finer the frequency resolution. However at some j, the number of coefficients in each node become so few that no statistically valid information can be obtained.

3.2. Wavelet Packet Spectrum

It is well known in Fourier analysis that using Parseval's theorem one can relate the power (or variance) of a signal as a function of frequency to the Fourier coefficients. That is, the power, $P(k)$ as function of frequency, k, is

$$P(k) = \sum \left|\hat{\delta}(k)\right|^2 \tag{8}$$

where $\hat{\delta}(k)$ are the Fourier coefficients in a frequency band centered at k.

Because the wavelet transform and wavelet packet transform form an orthogonal basis, Parseval's theorem hold and it can be shown that wavelet packet power spectrum [16]

$$P_j = \sum \left|\tilde{\delta}_j\right|^2 \tag{9}$$

where the $\tilde{\delta}_j$ are the wavelet coefficients at node j

A signal that is self-similar will have a power spectrum that scales over some frequency range. That is, power law behavior in the power spectrum is a strong indicator of fractal behavior. In terms of the wavelet packet transform, using Eq. (9) fractals are indicated when,

$$P_j = \sum \left|\tilde{\delta}_j\right|^2 \approx j^{\alpha}. \tag{10}$$

It is this probe that in subsequent sections, we will use to look for fractal behavior.

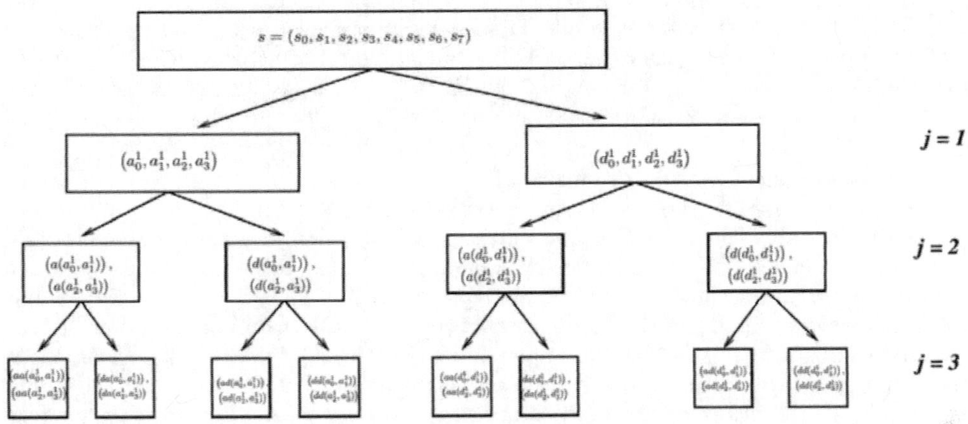

Figure 2. The WPT tree. The vector S is the original signal. After the first pass or sweep $j = 1$, the original signal vector is transformed into approximation coefficients a_i^1 and detail coefficients d_i^1. The approximation coefficients are the result of the action of the low pass filter on the signal. The detail coefficients are the result of the action of the high pass filter on the signal. At the next pass, *both* sets of coefficients are passed through the filtering process

4. Data

We used the Sloan Digital Sky Survey Data Release DR 13'th Baryon Oscillation Spectroscopic Survey (BOSS) [4]. The redshift covered in these surveys is $0 < z < 0.8$. The surveys covers two continuous regions of sky covering approximately 10400 deg^2, with a total of 2.5 million spectra. Once the data were obtained from the target objects, the spectra were processed by the BOSS data pipeline to classify the objects types (star, galaxy, quasar, etc.).

4.1. Mock Galaxy Catalogs

In addition the BOSS survey, we used also the Mock Galaxy Catalogs produced by the SDSS [17]. Mock Catalogs are used to optimize the survey, to create the covariance matrix for the clustering measurements, and to test the accuracy of the analysis. The technique to create the mock catalogs is the N-body simulation. The catalogs are created using the quick particle mesh method with the benchmark cosmological model as outlined in [17]. The model is flat Λ CDM model with $\Omega_m = 0.274$, $\Omega_b = 0.046$, $\Omega_\Lambda = 0.726$, $H = 70$ km s $^{-1}$ Mpc, $n = 0.95$, and $\sigma_8 = 0.8$. The assumptions made in creating the mock catalogs are: (1) Galaxies form and remain in the potential wells of the dark matter halos, (2) Halos represent over dense regions of the mass field that is (100-300) times greater than the mean density that arise form the non-linear gravitational collapse. The steps of creating the mock catalogs are:

(i) Predicting the evolution of the mass field.
(ii) Locating the dark matter halos and characterize its properties.
(iii) Populating the halos with mock galaxies.
(iv) Apply survey characteristics to the galaxies box.

An updated version of the TreePM^2 code was used to evolve $(3000)^3$ particles with mass a of $(5.9 * 10^{10}h^{-1}M$ in a box of side $(2750)h^{-1}$ Mpc. The number of data points we used was

IOP Conf. Series: Journal of Physics: Conf. Series **1290** (2019) 012007 doi:10.1088/1742-6596/1290/1/012007

approximately 32 million. The initial conditions of this simulation were: (1) Displacing the particles from a regular grid using 2nd order Lagrangian Perturbation theory, (2) $z = 25$, (3) The RMS displacement is 1.4 Mpc, (4) Time steps are sampled using the 2nd order leap frog method (friends of friends algorithm), (5) In populating galaxies by evolving the density field, the particles are selected such that they match the one and two point statistics of the dark matter halos.

5. Estimating the Fractal Dimension D

We used the raw SDSS galaxies data described earlier as identified using the SDSS data processing pipeline. The data files contain the right ascension, declination and red shift. The distribution of galaxies lies on the celestial sphere. We pixelize this data into matrices that span right ascension and declination in a specific redshift range. We use these matrices as inputs to the signal matrix to perform the wavelet packet transform. We estimate the power spectrum by radially averaging the two dimensional power spectrum using Eq. (9). If fractals are present, the power spectrum should have a power law behavior like that shown by Eq. (10). Fitting a log-log plot of the power spectrum gives the power law index, α. This exponent is related to quantity known as the Hurst exponent, H. Finally, we calculate the fractal dimension D from the Hurst exponent.

So the procedure for calculating the fractal dimension is:

(i) Using the HealPix and SDSSPix schemes, create the 2-d matrices that represent the galaxies distribution on the celestial sphere.

(ii) Apply the WPT on $2^n * 2^n$ squared sampled matrices, and estimate the radially averaged power spectrum

$$P_i = var[C_i] \Rightarrow var(C_i, k) = \frac{1}{2^j} \sum_{k=0}^{2^j-1} |C_i, k|^2 \tag{11}$$

where (C_i, k) are the wavelet packet coefficients at level i and frequency k.

(iii) Plot the log-log plot of the power spectrum vs. the frequency, and fit a line to the curve over the appropriate frequencies,

(iv) Calculate the Hurst exponent H from the slope of the line α. If $\alpha = (-1, -3)$, so H is given by

$$H = \frac{\alpha + 3}{2} \tag{12}$$

If $\alpha = (-1, 1)$, so H is given by

$$H = \frac{\alpha + 1}{2} \tag{13}$$

The Hurst exponent is a dimensionless parameter that indicates self-similarity present in the auto-correlation of a data set. Thus, it can also be defined as a measure of the self-similarity of a time series.

(v) From the Hurst exponent H, calculate the fractal dimension by

$$D = n + 1 - H \tag{14}$$

where n is known as the embedding dimension, i.e., (the minimum dimension where the object can lives in). Since we are treating galaxies as points (dimension 0) our embedding dimension is $n = 1$. Therefore, equation 14 takes the form, $D = 2 - H$.

In our analysis, we slice the galaxy distribution into redshift (or comoving) spherical shells and compute the *angular* fractal dimension as a function of cosmological comoving space.

The term angular fractal dimension is used because the analysis is being done on spherical shells with no depth. Thus, the roughness of the distribution as measured by the Hurst exponent and fractal dimension represent roughness across the spherical shell.

6. Results

We used five different resolutions in creating the two dimensional matrices and three particular types of wavelets, Daubechies "db" , Symlets "sym" and Coiflets "coif", with different filters. The three wavelets are orthogonal and compactly supported but they have different vanishing moments. The Symlets are remarkably similar to the Daubechies wavelets. They are nearly symmetrical wavelets proposed by Daubechies as modifications to the "db" family, and they are also the least asymmetric Daubechies wavelets. The Coiflets were introduced by Coifman. The wavelet function has 2N moments equal to 0 and the scaling function has $2N-1$ moments equal to 0. The two functions have a support of length $6N-1$. Finally, they are the compactly supported wavelets with highest number of vanishing moments for the basis functions for a given support width. We expect our results to be similar regardless the wavelet type used as long as the wavelet is orthogonal and compactly supported.

Each resolution generated pixels spanning different solid angles. The smaller the resolution, the larger the solid angle and the more galaxies that were binned into a pixel. Clearly a small resolution would cause too many galaxies to loose their spatial resolution. On the other hand, a large resolution meant a very sparse matrix with many pixels having no galaxies. This generated a large amount of shot noise that overwhelmed the signal. We chose a range of resolutions to show that, within a reasonable range, our results were not resolution dependent.

In dealing with the issue of edge distortions, we used different ways to extend the signal at the boundaries. There was very little effect on our results as a function of the boundaries. After trying out the different options, we settled down to the simplest choice: symmetric extension. In this scheme, the signal at these boundaries is replicated by symmetric values from the original signal. The powerful localization property of wavelets kept the error in the replicated signal, if any, extended to a few coefficients.

In our analysis, we used the proper distance instead of the redshift to handle the data properly. We computed the proper distance at corresponding redshifts. In reproducing the distributions of galaxies, we used a redshift bin width of $z = 0.04$ which corresponds to (167.9885) Mpc at redshift 0.5. We tested different bin widths such as $z = 0.01$ and $z = 0.1$. This bin width was chosen because it was a good compromise between the 2-D projection onto a plane not being too deep while still having a good number of galaxies to process.

In fitting the line, we used the correlation coefficient R^2 to determine the goodness of the fit. We used only data that had at least an $R^2 > 0.8$. After calculating the different values of H at different resolutions, we calculated the fractal dimension using the relation $D = 2 - H$. We present our results of the fractal dimension plotted against the proper distance computed using different wavelets types and filters at three different resolutions, 16, 32 and 64.

In figures 3 and 4 we show a typical power spectrum result. Figure 4 shows the log-log plot of the power spectrum in which the power nature of the power spectrum is evident over several epochs. At large frequencies, the plot flattens out as the distribution becomes more smooth. This region is not fractal. However at the lower frequencies, the power law behaviour is evident and indicates the presence of fractals at these scales. Unfortunately, the resolutions with which we could examine the data does not allows us to go below about 8 Mpc. Nevertheless, this particular spectrum shows clear power law behavior on scales of 26 Mpc to 8 Mpc, strongly hinting at a fractal structure at these scales.

At resolution 8, the values of the Hurst exponent for the BOSS Galaxies is $H = 0.6 \pm 0.2$. For the Mock Catalogs Galaxies, the Hurst exponent is almost constant around $H = 0.4 \pm 0.1$. The fractal dimension for BOSS galaxies is $D = 1.4 \pm 0.2$ and it has the lowest value towards the

IOP Conf. Series: Journal of Physics: Conf. Series **1290** (2019) 012007 doi:10.1088/1742-6596/1290/1/012007

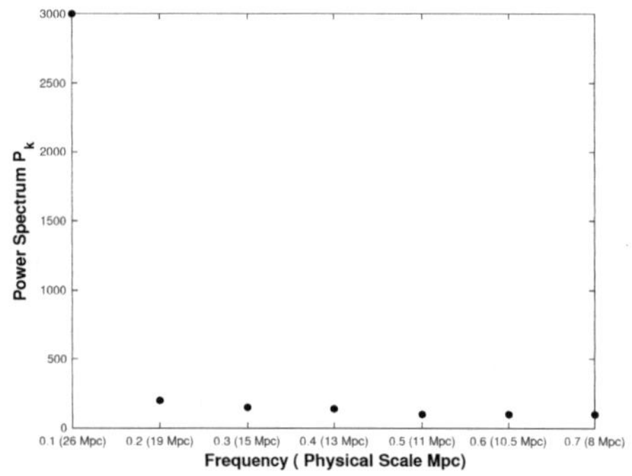

Figure 3. The power spectrum plotted against the physical scale obeys the power law distribution at redshift $z = 0.55$..

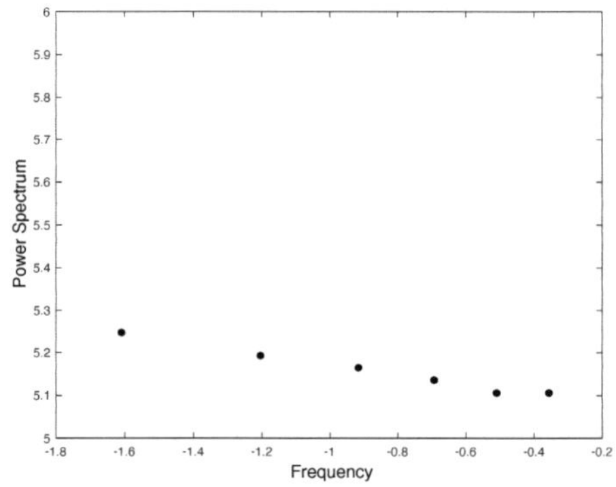

Figure 4. log- log plot of the power spectrum vs. frequency.

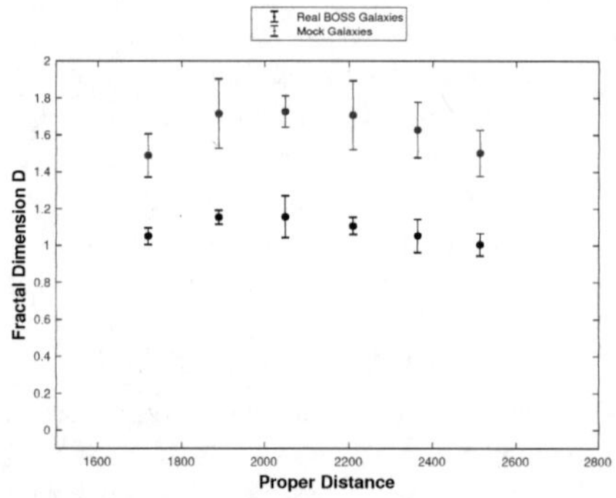

Figure 5. D at resolution 32 using "db1".

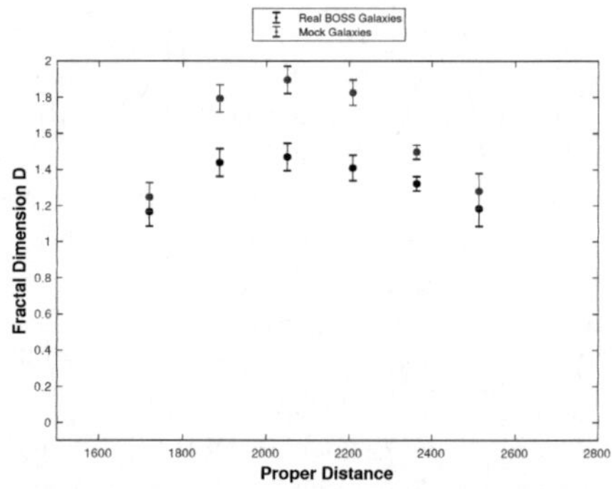

Figure 6. D at resolution 64 using "db1".

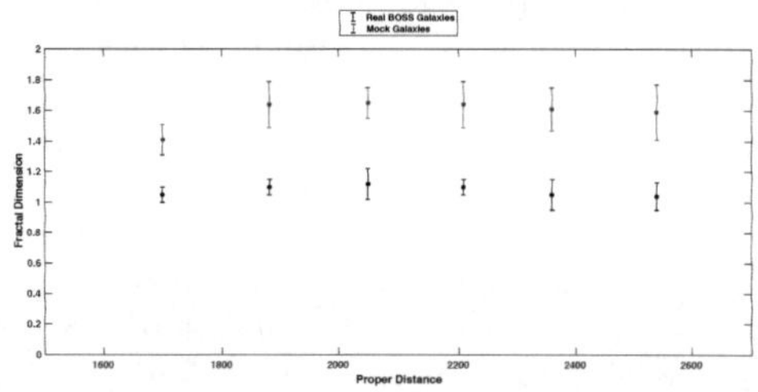

Figure 7. D at resolution 32 using "sym1".

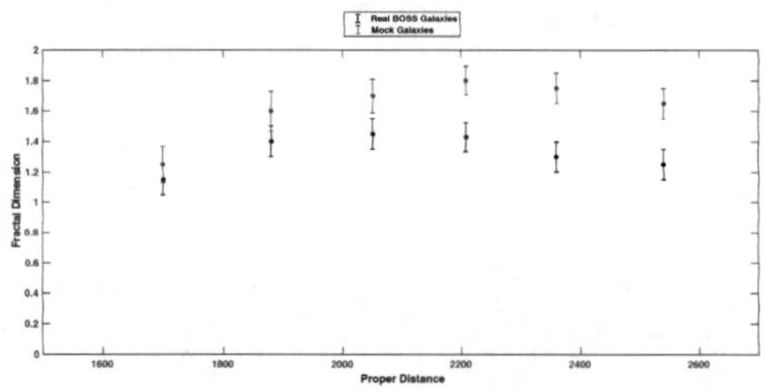

Figure 8. D at resolution 64 using "sym1".

highest proper distance. For Mock Catalogs Galaxies, the fractal dimension is almost constant

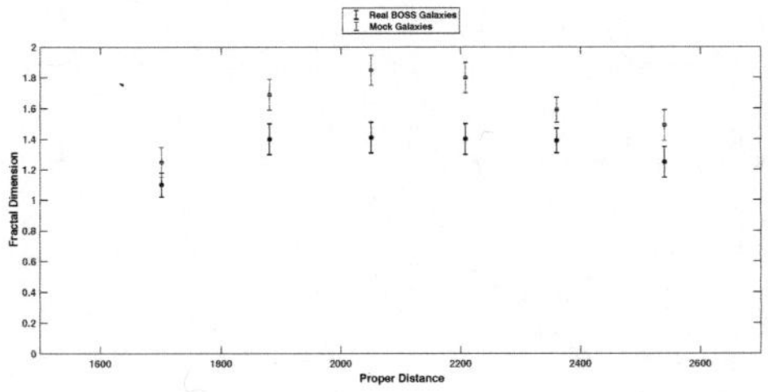

Figure 9. D at resolution 64 using "coif1".

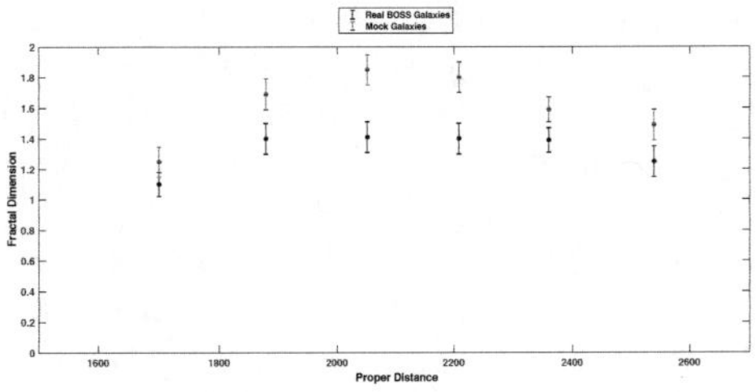

Figure 10. D at resolution 64 using "coif1".

around $D = 1.6$. This is the lowest resolution we present. Again the spatial extant of each pixel is quite large at this resolution and this result may suffer from too large a coarse graining due to pixelization size used. We address this issue by increasing the resolution and note differences.

At resolution 16, the Hurst exponent for BOSS Galaxies is $H = 0.7 \pm 0.1$. For the Mock Catalogs Galaxies, it is $H = 0.4 \pm 0.1$. The fractal dimension of the Mock Catalogs Galaxies is almost constant around $D = 1.6 \pm 0.1$. For BOSS Galaxies, the fractal dimension is almost constant about $D = 1.3 \pm 0.1$. Despite doubling the resolution, or decreasing the pixel size, the results are remarkably close to those generated the previous resolution.

At resolution 32, the Hurst exponent of BOSS Galaxies is $H = 0.8 \pm 0.1$. For Mock Catalogs Galaxies, the Hurst exponent is $H = 0.3 \pm 0.2$. The fractal dimension for BOSS Galaxies is relatively constant around $D = 1.2 \pm 0.1$. For Mock Catalogs Galaxies, it has a higher fractal dimension around $D = 1.7 \pm 0.2$. At this resolution, there are differences in the fractal dimension than those generated at the previous resolutions, although the results are consistent within error bars. At this resolution, there are many more empty pixels than at the previous resolutions and this is probably affecting our results. The results produced by the Symlets and Coiflets wavelets are similar to the results of Daubechies wavelets as expected. Using the Coiflets wavelets, the fractal dimension for BOSS Galaxies is around $D = 1.1 \pm 0.15$. For Mock Catalogs Galaxies, it has a higher fractal dimension around $D = 1.55 \pm 0.2$. For the Symlets wavelets, the fractal dimension for BOSS Galaxies is around $D = 1.1 \pm 0.1$. For Mock Catalogs Galaxies, it has a higher fractal dimension around $D = 1.6 \pm 0.2$. The results using all three wavelets are in excellent agreement with each other. Moreover, the fractal dimension computed using Symlets and Coiflets wavelets lies with in the error bars of results produced by Daubechies wavelets. Changing the filter from 1 to 3 to 6 has produced a minor difference in the value of D. The results of the three wavelets are consistent with each other regardless the filter number used. This coherence in results regardless the wavelet type and filter used did establish that, wavelets can detect the fractal structure as long as they are orthogonal and compactly supported. The fractal dimension does not evolve as one would expect when it reaches the maximum near $D = 2$. At higher redshifts, we expect the distribution to be smoother and the fractality of the structure to be suppressed. For the mock data, the fractal dimension is near 2 at a proper distance of 2200 Mpc and we cannot conclude that galaxies are fractals. Before then, D is not 2 but because there are fewer galaxies so it is difficult to disentangle the effects of small number of galaxies.

At low redshifts, one would expect the distribution to become rougher and irregularities are more pronounced. Our results indicate that at low redshifts, the galaxies do appear more fractal. While this fractality might be a real effect, it is difficult again to confirm what causing this because the number of galaxies decreases. We did not generate resolutions that probed galaxies at scales down to 1 Mpc. To generate such a data at co-moving distances, we would have to reproduce the galaxy distribution using very high resolutions. As mentioned earlier, the

IOP Conf. Series: Journal of Physics: Conf. Series **1290** (2019) 012007 doi:10.1088/1742-6596/1290/1/012007

generated matrix would be extremely sparse and the shot noise generated swamps the signal.

Our results strongly hint that at least at some redshifts, galaxies are fractals probably out to distances about $10 - 20$ Mpc as figure 3 indicates. The lack of data at higher redshifts present challenges to our analysis. However, our method does show that it is possible to study the redshift evolution of a fractal dimension in the galaxy distribution if enough data is available.

One important observation from the results is that the chosen resolution has only a small effect - the results are consistent within error bars regardless the resolution used. If our results were dependent on the chosen resolution, it would seriously raise questions on our method.

Another consistency in all the cases is that the Hurst exponent and fractal dimension values lie within the same range. The Hurst exponents are in the range of $H = (0.5 \pm 0.3)$ for BOSS Galaxies and $H = (0.4 \pm 0.2)$ for Mock Catalogs Galaxies. For the fractal dimension, the values of D are always in the range of $(1 - 2)$ for both the BOSS and Mock Catalogs Galaxies. This consistency in ranges at all resolutions is another indication of the robustness of the statistics we used.

7. Conclusion

We developed a novel way of using wavelet transform to calculate the angular fractal dimension of galaxy distributions as a function of cosmological redshift. The fractal dimension as applied to galaxies quantifies the degree of homogeneity or regularity in the structure. We used data from the BOSS survey and Mock Galaxies Catalogs of the SDSS. We estimate the power spectrum of the galaxy distribution using wavelet packets and found that the power spectrum, generally, followed a power law. We extracted the power law index, also called the Hurst exponent, and from that the fractal dimension. The Hurst exponent is $H = (0.5 \pm 0.3)$ for BOSS Galaxies and $H = (0.4 \pm 0.2)$ for Mock Galaxies Catalogs. The fractal dimension reported for BOSS Galaxies is $D = (1.3 \pm 0.2)$ and $D = (1.6 \pm 0.2)$ for Mock Catalogs. The consistency of the results at all resolutions indicates the robustness of the statistics we used. What we conclude is the distribution may be fractal but more data is needed before we can make a more firm conclusion.

References

[1] Mandelbrot B 1982 *The Fractal Geometry of Nature* (San Francisco: W.H. Freeman)
[2] Peebles P *Astron.Astrophys* **68** 345
[3] Efstathiou G, Fall S and Hogan C *Mon. Not. R. Astr. Soc* **189** 203
[4] Dawson *et al.* 2013 *Astonomical Journal* **145** 10
[5] Cambrésy L 1999 *Astronomy and Astrophysics* **345** 965
[6] Schild R E 1996 *Astrophysical Journal* **464** 125
[7] Pando J and Fang L 1998 *Phys Rev E* **57** 3593
[8] Pando J, Fang L, Lipa P and Greiner M 1998 *Astrophysical Journal* **496** 9
[9] Kalinkov M, Valtchanov I and Kuneva I 1998 *Astrophysical Journal* **506** 509
[10] Cayon L e a 2001 *MNRAS* **326**(4) 1243
[11] Pando J, Valls-Gaboud D and Fang L 1998 *Phys. Rev.Lett.* **81** 4568
[12] Kasier G 1994 *A Friendly Guide to Wavelets* (Berlin: Mirkhauser)
[13] Daubechies I 1992 *Ten Lectures on Wavelet* (Philadelphia: SIAM)
[14] Vidakovic B 1999 *Statistical Modeling by Wavelet* (New York: Wiley-Interscience)
[15] Ogden R 1997 *Essential wavelets for statistical applications and data analysis* (Berlin: Birkhauser)
[16] Pando J and Fang L 1997 *Current Topics in Astrofundamental Physics* (Singapore: World Scientific) p 616
[17] Alam *et al.* *Mon. Not. R. Astr. Soc.* **470** 2617

IOP Conf. Series: Journal of Physics: Conf. Series **1290** (2019) 012008 doi:10.1088/1742-6596/1290/1/012008

Data Reduction Using Lossy Compression for Cosmology and Astrophysics Workflows

Jesus Pulido[1,2], Zarija Lukic[3], Paul Thorman[4], Caixia Zheng[5], James Ahrens[2], Bernd Hamann[1]

[1]Department of Computer Science, University of California, Davis, CA 95616, USA
[2]Los Alamos National Laboratory, P.O. Box 1663, Los Alamos, NM 87545, USA
[3]Lawrence Berkeley National Laboratory, 1 Cyclotron Rd, Berkeley, CA 94720, USA
[4]Haverford College, 370 Lancaster Avenue, Haverford, PA 19041, USA
[5]Northeast Normal University, 2555 Jingyue Street,Changchun 130117, China

E-mail: [1]jpulido@ucdavis.edu

Abstract. This paper concerns the use of compression methods applied to large scientific data. Specifically the paper addresses the effect of lossy compression on approximation error. Computer simulations, experiments and imaging technologies generate terabyte-scale datasets making necessary new approaches for compression coupled with data analysis. Lossless compression techniques compress data with no loss of information, but they generally do not produce a large-enough reduction when compared to lossy compression methods. Lossy multi-resolution compression techniques make it possible to compress large datasets significantly with small numerical error, preserving coherent features and statistical properties needed for analysis. Lossy data compression reduces I/O data transfer cost and makes it possible to store more data at higher temporal resolution. We present results obtained with lossy multi-resolution compression, with a focus on astrophysics datasets. Our results confirm that lossy data compression is capable of preserving data characteristics very well, even at extremely high degrees of compression.

1. Introduction

The scientific datasets generated today via computer simulations are too large for direct processing and analysis. Effective and efficient methods are becoming increasingly important to compress datasets, keeping in mind the need to preserve relevant scientific behavior in compressed representations as much as possible. Processing extraordinarily large datasets exposes limitations in current hardware, making data analysis challenging. As more powerful computer and imaging systems are being developed, these challenges grow exponentially. Currently, processing such large datasets introduces massive strain on data transfer and storage systems. On the observational side, one massive-dataset-generating system is the Large Synoptic Survey Telescope (LSST), a wide-field survey telescope currently under construction [1]. It is estimated to produce hundreds of gigabytes of raw data per night. Regarding computer simulations, the AMReX/Nyx computational cosmology code produces massive N-body and gas dynamics simulation outputs consisting of hundreds of terabytes of data for a typical large-scale run [2].

For these two specific applications, transferring and processing raw data creates many challenges as network and I/O (input/output) systems are strained. Previous efforts have

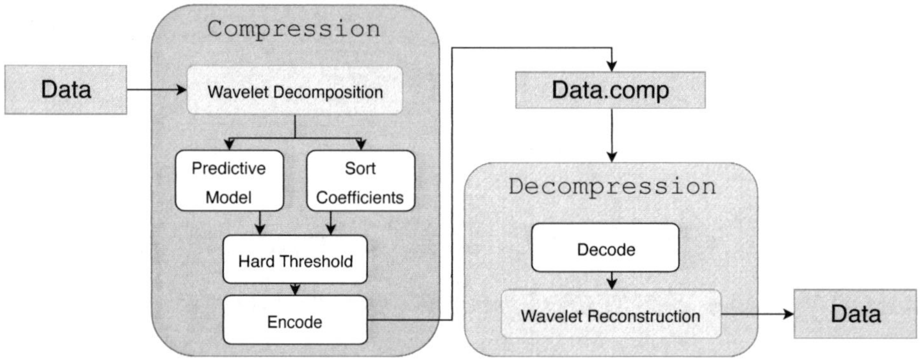

Figure 1. Compression and decompression pipeline.
Simplified pipeline shows data products (green), compression (red), and decompression (blue) operations.

utilized mainly lossless data compressors to reduce data size in memory before conducting I/O operations but they have generally not achieved significant data reduction [3]. Both transferring data between on-site systems and sharing data for collaborative research imposes significant bottlenecks, making adaptive lossy data reduction methods more viable. These benefits have been demonstrated in domains with similar data properties, such as simulations of turbulence, enabling collaborative remote visualization [4]. The consensus is that lossless compression methods are unable to reduce massive datasets sufficiently to offset I/O and memory bottlenecks.

As lossy methods have become necessary with the presented data challenge, it is important to understand the errors introduced by lossy data reduction for varying degrees of compression. We introduce a lossy multi-resolution compression framework and provide quantitative and qualitative results concerning error for regular-grid datasets, specifically for cosmology and astronomical image datasets.

2. Method

Lossy data compression can be achieved by selecting a wavelet basis, performing a data decomposition, then removing the lowest-magnitude coefficients via hard thresholding. This selection process can be either done with a predictive model or by sorting the coefficients. A hard threshold operation is performed by setting-to-zero those values that are below a certain coefficient magnitude, effectively throwing them away. The resulting coefficients are followed up by an encoding process to achieve additional amounts of data reduction by establishing several levels of numerical precision and applying dictionary-based methods for packaging. Decompression is a much simpler operation that simply decodes the data stream and reconstructs the remaining set of wavelet coefficients. Fig. 1 describes the complete compression and decompression pipeline.

2.1. Algorithms

It is possible to achieve high degrees of data reduction through discrete wavelet sampling. Wavelets are a generalization of the Fourier transform by using a basis that represents both location and spatial frequency [5]. A typical wavelet representation contains several vanishing moments that makes possible a sparse and highly accurate representation via dataset-sampling based on a small number of coefficients. The underlying basis functions support a discrete representation of a continuous signal, designed to produce responses for different frequencies. When applying a wavelet basis to a dataset, the result is a set of coefficients for that basis

IOP Conf. Series: Journal of Physics: Conf. Series **1290** (2019) 012008 doi:10.1088/1742-6596/1290/1/012008

function, capturing behavior at various resolutions. Wavelets that are commonly used include Haar wavelets, Daubechies wavelets, bi-orthogonal spline wavelets, and coiflet wavelets.

Pulido et al. [6] analyzed and compared in great detail many multi-resolution representation methods in the context of feature extraction and data analysis. Among those tested, higher-order B-spline wavelets were the most effective in terms of the ability to capture a broad range of quantities relevant for the representation of turbulence structures with a reduced set of coefficients. When viewing these results in the context of regular-grid and continuous datasets, the higher-order B-spline wavelet families consistently scored near the top of the compression schemes considered; therefore, biorthogonal cubic B-spline wavelets [7] were used for generating the results presented in this paper.

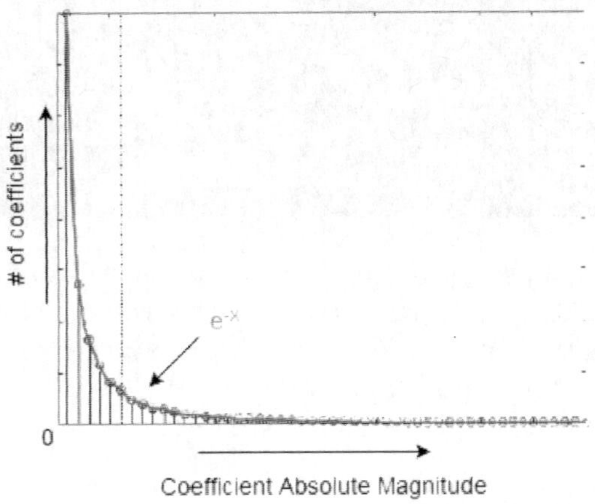

Figure 2. Absolute magnitude coefficient distribution.
An exponential decay behavior describes the relationship between coefficient magnitudes and number of coefficients. A small number of high-magnitude coefficients holding the most entropy can be kept while many small-valued coefficients can be removed to achieve effective data reduction.

Once a cubic B-spline wavelet signal is sampled via a regular-grid dataset, wavelet coefficients are produced using a dimension-by-dimension approach, i.e., in one step only one dimension is considered for the computation of coefficients. Although this method takes advantage of and is designed for regular structured grids, it can handle non-regular grids and simulated particle data by treating the data as a single, one-dimensional data stream. To preserve the most relevant features during lossy compression, one must keep coefficients with highest magnitudes and discard those with relatively very low magnitude. Datasets in many scientific domains produce a distribution of coefficients, i.e., a distribution of coefficients by magnitude, where the number of coefficients decreases as coefficient magnitude increases. This distribution can be modeled with an exponentially decreasing function, see Figure 2. Considering this behavior, a value for hard-thresholding must be selected. This can either be done by using a parallel quick-sort algorithm applied to the absolute magnitudes of the coefficients to select an exact value, or a predictive model for approximation for an understood specific scientific application.

When using a predictive model for approximation, significant time can be saved during compression by avoiding an expensive sorting step. By providing minimum, maximum and target compression values (% of coefficients or compression ratio), this model can be adjusted on-the-fly per application domain.

As final encoding step we perform data re-quantization and use an off-the-shelf dictionary compressor for packaging, e.g., LZ4 [8]. Wavelet coefficients are derived as floating-point,

IOP Conf. Series: Journal of Physics: Conf. Series **1290** (2019) 012008 doi:10.1088/1742-6596/1290/1/012008

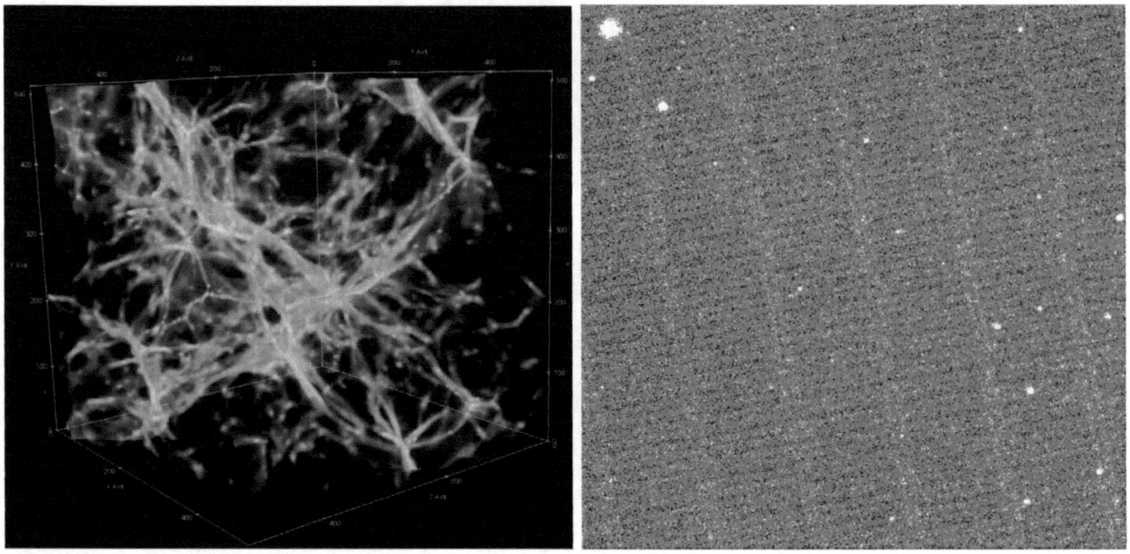

Figure 3. Cosmology and Astronomy features of interest.
3D-grid cosmology simulation showing baryon density volume (left) demonstrates clearly defined density clusters and interconnects to preserve during compression. A 2D-image sky-survey simulation (right) presents elliptical bright objects as stars and galaxies, combined with simulated noise.

numerical values – presenting an additional challenge when data is converted from an integer (astronomical images) to floating-point format. Traditionally, integer values lead to higher degrees of compression than floating-point values when using dictionary-based lossless compressors. By using floating-point-to-integer re-quantization, it is possible to preserve the majority of a wavelet coefficient's entropy at the cost of a small reduction in numerical precision. This is achieved by selecting a fixed number of significant digits and preserving them during the conversion to integer numerical space.

Decompression requires significantly less effort compared to compression. To retrieve compressed data, a file is decoded, re-quantized back to floating-point precision and reconstructed using a wavelet signal. A lossy representation of the original data is then retrieved.

2.2. Features and Implementation

Using wavelet-based methods provides several benefits such as de-noising, region-specific reconstruction (decompression), data streaming, and coefficient-space data analysis capability, without the need of full decompression. The primary component of this compressor uses a wavelet CPU implementation based on a heavily modified version of the GNU Scientific Library (GSL) [9]. Modifications were applied to the open-source library, including: support for 3D/N-D data, support for non-square and non-powers-of-two resolution data, higher-order B-spline basis function support, handling of boundary conditions, and parallel computation. An early open-source implementation of this compression code in both C++ and Matlab is available, see [10].

3. Results

3.1. Multi-dimensional Cosmology

The lead Nyx Lyman-alpha simulation significantly furthers the state-of-the-art in the field, using as many as 8192^3 grid cells. The simulation discussed in this paper for compression is a smaller, test simulation with 512^3 cells with six scalar components. The only difference between the lead and the test simulation is the physical size, which is 250 megaparsecs (Mpc) for the 8192^3 run and 15.625 Mpc for the smaller test case. The smaller simulation has the same physical resolution as the original, performed with the same physics and using the same choice of cosmological parameters: $h = 0.675$, $\Omega_m = 0.31$, $\Omega_\Lambda = 0.69$, $\Omega_b = 0.0487$, $\sigma_8 = 0.82$, $n_s = 0.965$, and $w_{\text{de}} = -1.0$.

The parallel, shared memory version of this compressor was tested on Intel's Haswell and Xeon Phi (KNL) architectures on Cori (NERSC, Lawrence Berkeley National Laboratory) to determine scaling performance. The two datasets have the same physical simulation box, but they vary in cell resolutions. For a physical grid size of 512^3, we found that a single Haswell node with 64 threads completes the main wavelet decomposition in 1.594 seconds on average, for a single scalar field of 512MB size. When using a resolution of 2048^3, the same routine requires 59.653 seconds on average for completetion, for a single scalar field of 32GB size. A single KNL node with 256 threads processes the 512^3 physical box in 2.334 seconds and a 2048^3 resolution in 88.491 seconds on average. These performance results are preliminary results, as it is still possible to further optimize our implementation.

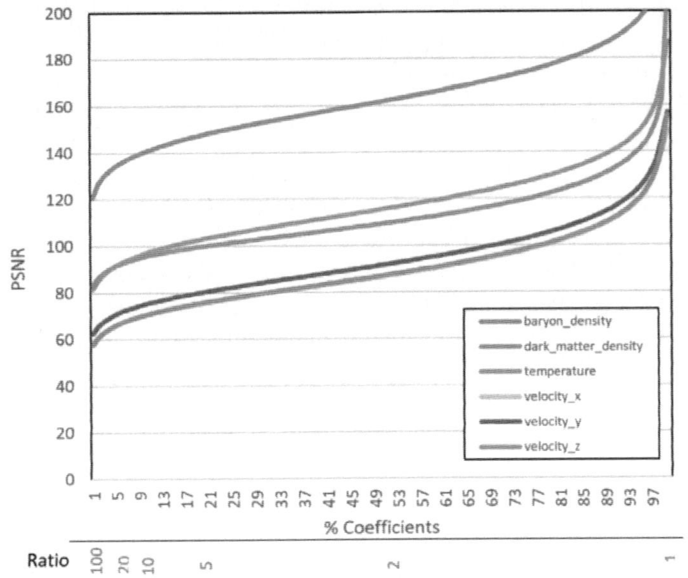

Figure 4. PSNR behavior on lossy compression.
PNSR comparisons show compression behavior across the six scalar fields; densities compressing the best compared to velocities. An inflection point is reached between 5%-7% with diminishing returns in quality. Velocity components compress similarly with a small positive bias in the y-axis.

Figure 4 shows the compression capabilities of the method, demonstrating a high degree of control over the quality of a lossy compression. Using the 512^3 dataset to represent the initial physical simulation box size, the peak-signal-to-noise (PSNR) plot shows the relative compression quality for each scalar component. Baryon density can be compressed the most, while velocity components can be compressed the least. The y-direction velocity component can be compressed slightly more than the x- and z-directions. This fact could be the result of the specific initial conditions used in the simulation.

The same PSNR plot shows that generally all velocity components contain the same amount of error when compressed. The most important factor to consider is the observed inflection point that occurs for thresholding around 5-7 % coefficients (20:1 - 15:1 compression ratios). The inflection point generally signifies a "point of compromise"; using more coefficients for compression would only yield diminishing returns in quality. Therefore, 20:1 to 15:1 compression ratios are recommended for this specific cosmological dataset.

Figure 5 shows the progression from 1%, to 5%, and eventually 10% coefficients by rendering contours for dark matter density (logarithmic scale). For reference, a 10% visualization is "visually identical" to the original data. In our numerical results, compression degrees between 5% and 10% represent the inflection point. Visually, there is very little change from the latter two. By using 1% coefficients for compression (100:1), we see a deterioration of the contours in high-density regions.

Figure 6 shows the effect of different compression levels for the power spectra of dark matter density (left panel) and gas density (right panel). We see that the power spectra are in excellent agreement, especially considering that there is little cosmological interest in the $k > 10\,\mathrm{Mpc}^{-1}h$ regime. In particular, we observe that compression works extraordinarily well on baryon (gas) density. This is a physical explanation: Dark matter is pressure-less, thus forming a structure even on the smallest scales resolvable in a simulation. Gas, on the other hand, is described by the Euler equations, and is pressure-smoothed, with the smoothing scale depending on gas temperature. For this reason, density fluctuations are filtered on scales of a few cells, suppressing small structures and enabling a high degree of compression.

3.2. Astronomical Images

The LSST dataset [11] is a simulated image dataset that attempts to emulate observational features of the actual telescope, the LSST. Using GalSim, the lens properties and other characteristics of the telescope can be simulated to produce expected outputs for the future observational datasets. The sample set of FITS files are of size 4000 x 4000 containing several frequency bands. For sky survey simulations, noise is typically added to replicate characteristics of the actual telescope. Each pixel in an image represents about 0.2" of physical space, i.e., a single image represents up to 800" x 800" or 0.05 square degree depth of coverage.

As a pre-processing step, a grid-based de-noising scheme is used to remove the obvious observable artifacts, where otherwise a stacking the 2D image set would already remove them. Data compression and object detection can then be performed for this data.

Section 2.1 discusses the possibility of using alternative lossless compression schemes to encode coefficients of basis functions. To explore this issue, Figure 7 compares the use of stand-alone lossless compression and combinations of lossy compression with the methods we have introduced in this paper. We examine the use of traditional used Gzip [12] in standard FITS file compression, LZ4 [8], and BZip2 [13] as stand-alone compressors. Typically, computation times for a FITS file of this size vary from a few seconds to tens of seconds. When sorting methods from fastest to slowest, one obtains the following order: LZ4, GZIP, Wavelets with LZ4, Wavelets with BZip2, and BZip2.

In summary, we can employ lossy compression by preserving up to 50% wavelet coefficients and combine it with Bzip2 encoding to achieve file sizes as low as 9.8% of the original size. This result compares favorably relative to using stand-alone methods such as LZ4 (28%), Gzip (23.5%), and BZip2 (14.9%). More extreme levels of compression can be achieved by reducing the percentage of coefficients preserved, at the cost of numerical precision. Ideally, we would like to take advantage of fast decompression by pairing wavelets and LZ4, still producing compression

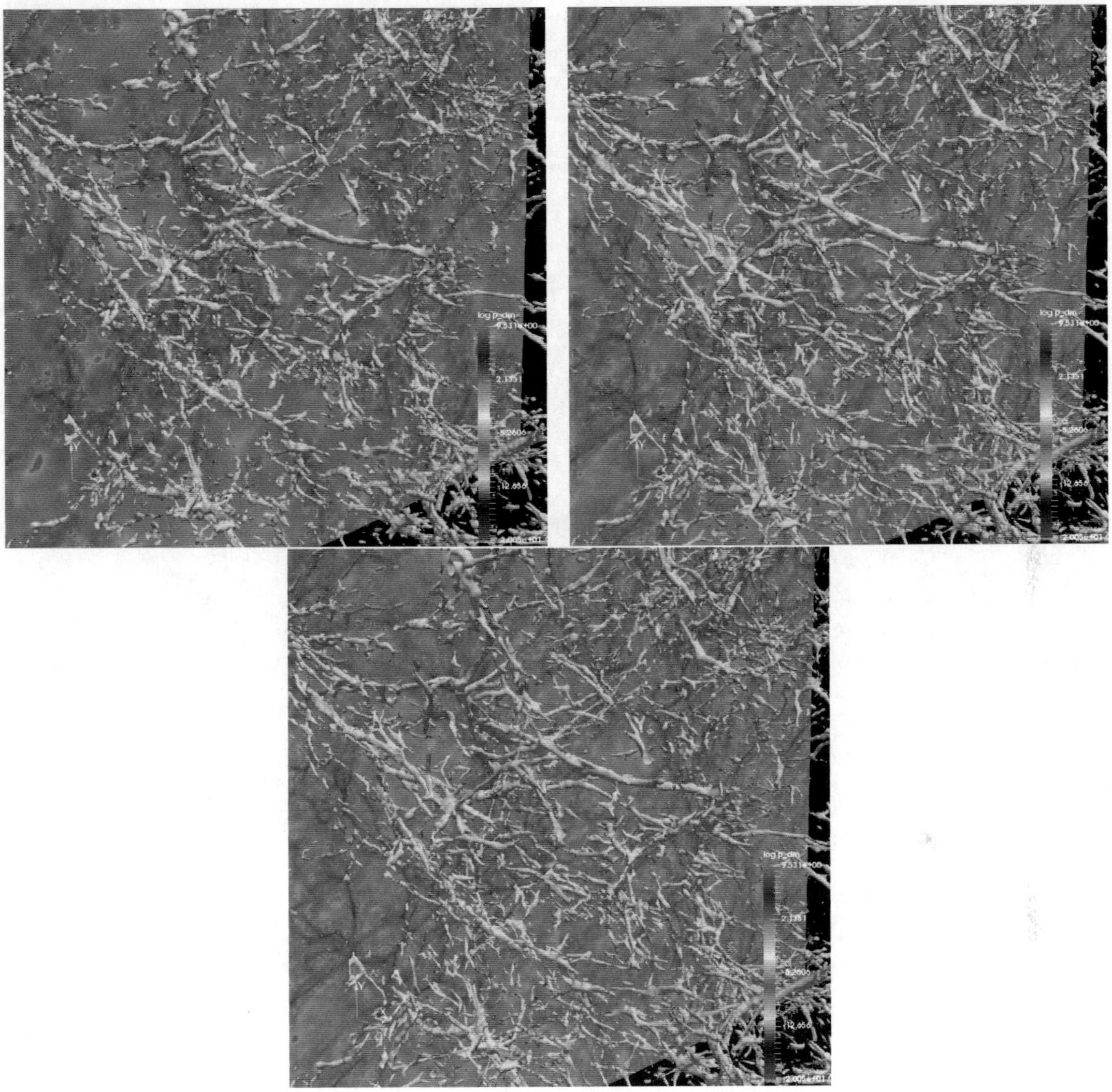

Figure 5. Dark matter density contour visualization comparisons.
Visual iso-contour structures can be compared at various compression levels. Left to Right: 1%, 5%, 10% coefficients. Areas of high-density clustering and general web-like structures are well preserved at high levels of compression. At extreme compression (1%), the web-like structure begins to deteriorate slightly.

sizes smaller than that obtained with traditional standalone Gzip. The choice of percentage of coefficients has the most impact on lossy compression quality.

A highly effective method for object detection by Zheng et al. [14] is used to evaluate the quality of lossy compression and object detection at various levels. In the original data, a total of 401 objects can be extracted with this method on a 64MB FITS container. By only using less than half of the wavelet coefficients (40%), it is possible to preserve at best 98% of the originally detected objects, with only 9.5% of the original file size. At more extreme levels, 90% object detection is still achievable when using only 15% of coefficients, effectively reducing the file size

IOP Conf. Series: Journal of Physics: Conf. Series **1290** (2019) 012008 doi:10.1088/1742-6596/1290/1/012008

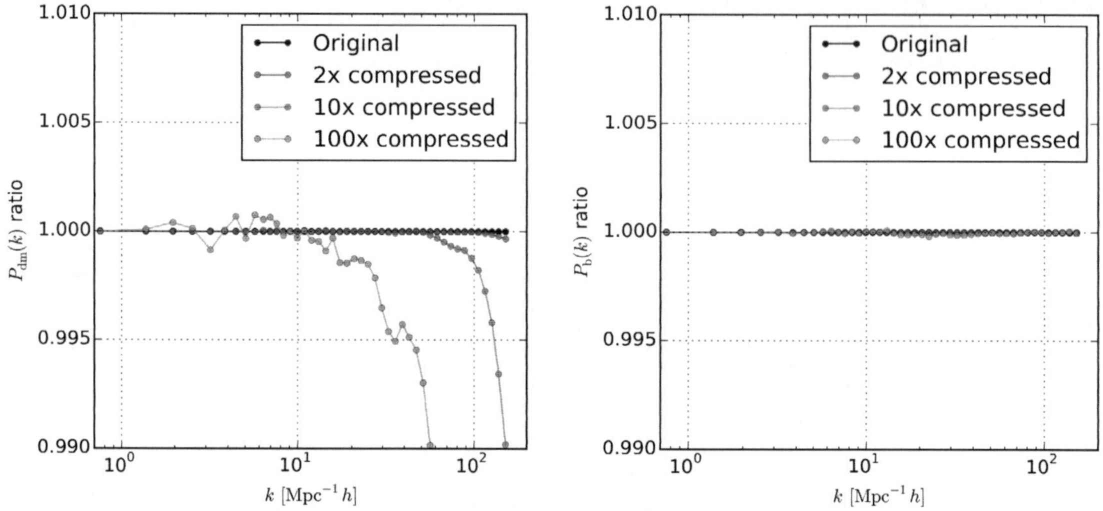

Figure 6. Nyx relative power spectra for different compression levels. Dark matter density (left) shows well-preserved energy ranges across all levels up to 10x compression, and low-power behavior up to 100x. Baryon density (right) shows excellent preservation even at extreme 100x compression.

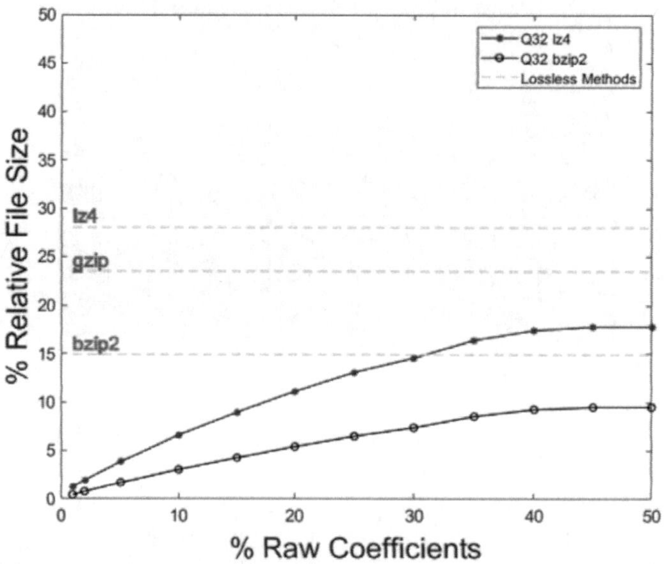

Figure 7. Comparison between compression methods. The dotted green lines show file sizes using three standard lossless compression methods while the others are lossy with wavelets (Q32). When lossy compression is used, data sizes can be significantly reduced by several magnitudes more.

to 5% of the original, see Figure 8 and Figure 9.

Figure 9 shows a comparison of the magnitudes of detected objects ("R_mag" in LSST catalog) for all ranges. These results show that compression at extreme levels affects the

IOP Conf. Series: Journal of Physics: Conf. Series **1290** (2019) 012008 doi:10.1088/1742-6596/1290/1/012008

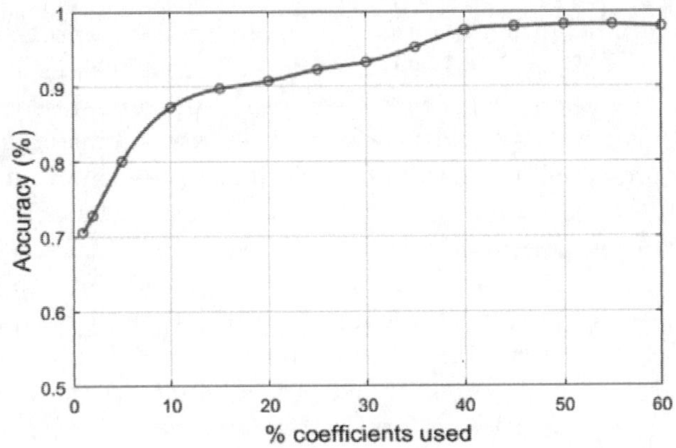

Figure 8. Object detection.
Trends in object detection show that using as little as 40% coefficients preserves over 98% of detected objects. Using an extreme data reduction that targets a 5% relative file size to the original, choosing to keep 15% coefficients can preserve 90% of detected objects.

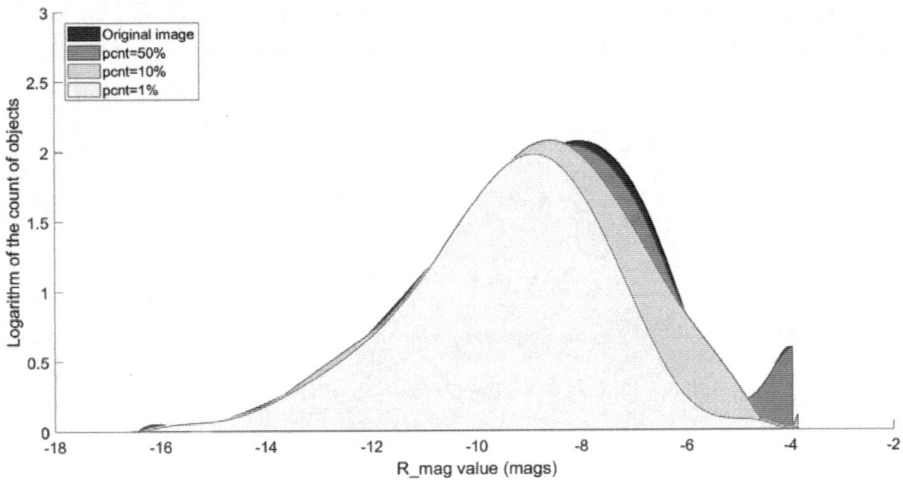

Figure 9. R_mag accuracy.
Coefficients by 1% (yellow), 10% (cyan), 50% (orange), and original (blue) show binned distributions of detected objects. A polynomial fit shows that bright and large objects are preserved well even at extreme compression percentages while smaller, usually fainter objects are lost.

detection of R_mag values representative of typically small objects. This result is consistent with reduced numerical precision, but it achieves better performance through the use of wavelets. Considering the representation with 50% coefficients, the R_mag ranges are nearly consistent with the originally detected objects, and even when using a representation with only 10% coefficients most properties are preserved.

IOP Conf. Series: Journal of Physics: Conf. Series **1290** (2019) 012008 doi:10.1088/1742-6596/1290/1/012008

4. Conclusions

We have demonstrated the capabilities of a lossy data compression approach and implementation for workflows arising in cosmology and astrophysics. Large degrees of data reduction are possible with the approach, without significantly compromising domain-specific features relevant for specific applications. For regular-grid astrophysics datasets, clustered density features are visually and numerically preserved, even at high levels of compression. Considering astronomical images, a majority of objects can still be detected and properly characterized even when high degrees of data reduction have been used.

We plan to conduct memory usage optimization tests for in-situ applications, where compression methods must be performed within a severely limited computation budget. The presented compression scheme is capable of performing temporal compression for 3D stacked image data or 4D simulated data. In order to better evaluate the computational capabilities of the compression approach, we need to perform strong and weak scaling tests, since computer systems are becoming increasingly parallel, and homogeneous architectures allow us to use both CPUs and GPUs for greatly accelerated computations.

Acknowledgments

We thank the DOE Exascale computing project for providing partial funding of this work at Los Alamos National Laboratory, under contract no. DE-AC52-06NA25396. We also thank the Argonne National Laboratory ALCF and Lawrence Berkeley National Laboratory NERSC groups for making their computer resources available for this project.

References

[1] Ivezić Ž and the LSST Science Collaboration 2013 Lsst science requirements document URL `http://ls.st/LPM-17`

[2] Almgren A S, Bell J B, Lijewski M J, Lukić Z and Van Andel E 2013 **765** 39 (*Preprint* `1301.4498`)

[3] Zeyen M, Ahrens J, Hagen H, Heitmann K and Habib S 2017 12–16

[4] Pulido J, Livescu D, Kanov K, Burns R C, Canada C, Ahrens J P and Hamann B 2018 *J. Parallel Distrib. Comput.* **120** 115–126

[5] Daubechies I 1992 *PA: SIAM*

[6] Pulido J, Livescu D, Woodring J, Ahrens J and Hamann B 2016 *Computers and Fluids* **125** 39 – 58 ISSN 0045-7930

[7] Cohen A, Daubechies I and Feauveau J C 1992 *Communications on Pure and Applied Mathematics* **45** 485–500

[8] Collet Y 2011 Lz4 - extremely fast compression URL `http://lz4.github.io/lz4/`

[9] Gough B 2009 *GNU Scientific Library Reference Manual - Third Edition* 3rd ed (Network Theory Ltd.) ISBN 0954612078, 9780954612078

[10] Pulido J 2018 Lossywave: A lossy, wavelet-based compressor for scientific simulation data. URL `https://github.com/lanl/VizAly-LossyWave/`

[11] 2014 Photon simulator (phosim) URL `https://www.lsst.org/scientists/simulations/phosim`

[12] GNU 1997 Gzip URL `https://www.gnu.org/software/gzip/`

[13] Steward J 1996 Bzip2 file compression URL `http://www.bzip.org`

[14] Zheng C, Pulido J, Thorman P and Hamann B 2015 *Monthly Notices of the Royal Astronomical Society* **451** 4445–4459 ISSN 0035-8711

IOP Conf. Series: Journal of Physics: Conf. Series **1290** (2019) 012009 doi:10.1088/1742-6596/1290/1/012009

A regression-based model evaluation of the Curie temperature of transition-metal rare-earth compounds

Duong-Nguyen Nguyen[1], Tien-Lam Pham[1,2,3], Viet-Cuong Nguyen[4], Anh-Tuan Nguyen[5], Hiori Kino[2,3], Takashi Miyake[2,3,6], and Hieu-Chi Dam[1,2,7]

[1]Japan Advanced Institute of Science and Technology, 1-1 Asahidai, Nomi,Ishikawa 923-1292, Japan
[2]CMI[2], NIMS, Tsukuba 305-0047, Japan
[3]ESICMM, NIMS, Tsukuba 305-0047, Japan
[4]HPC Systems Inc., Japan
[5]Hanoi Metropolitan University, 98 Duong Quang Ham, Cau Giay, Hanoi, Vietnam
[6]CD-FMat, AIST, Tsukuba 305-8568, Japan
[7]JST, PRESTO, 4-1-8 Honcho, Kawaguchi, Saitama, 332-0012, Japan

E-mail: dam@jaist.ac.jp

Abstract. The Curie temperature (T_C) of RT binary compounds consisting of $3d$ transition-metal (T) and $4f$ rare-earth elements (R) is analyzed systematically by a developed machine learning technique called kernel regression-based model evaluation. Twenty-one descriptive variables were designed assuming completely obtained information of the T_C. Multiple kernel regression analyses with different kernel types: cosine, linear, Gaussian, polynomial, and Laplacian kernels were implemented and examined. All possible descriptive variable combinations were generated to construct the corresponding prediction models. As a result, by appropriate combinations between descriptive variable sets and kernel formulations, we demonstrate that a number of kernel regression models can accurately reproduce the T_C of the RT compounds. The relevance of descriptive variables for predicting T_C are systematically investigated. The results indicate that the rare-earth concentration is the most relevant variable in the T_C phenomenon. We demonstrate that the regression-based model selection technique can be applied to learn the relationship between the descriptive variables and the actuation mechanism of the corresponding physical phenomenon, i.e., T_C in the present case.

1. Introduction

The development of strong permanent magnets is an urgent technological issue as well as a fundamental challenge in materials science. Most strong permanent magnets are rare-earth magnets that mainly consist of transition-metal and rare-earth elements. To date, various rare-earth transition-metal compounds have been synthesized. The strongest magnetic compound, $Nd_2Fe_{14}B$, which is the main phase of neodymium magnets, was developed by Sagawa et al. [1]. Because of its relatively low Curie temperature (T_C) of 586 K, its coercivity at high temperatures is not sufficient for technological usage; therefore, dysprosium (Dy) is added to ensure sufficient coercivity. In fact, Sagawa et al. intended to increase the T_C of R_2Fe_{17} (where R is a rare-earth element). However, the T_C was much lower than that of Sm_2Co_{17}, which was the strongest magnet before the invention of neodymium magnets. By adding boron to Nd_2Fe_{17} ($T_C = 335$ K), they obtained a new compound with a higher T_C, $Nd_2Fe_{14}B$. However, the T_C values are still much lower than those of Sm_2Co_{17} (1,193 K) and α-Fe (1,043 K). Owing to this complexity, it is highly desirable to establish a technique to predict accurate T_C values and to clarify the controlling parameters (variables) that influence this property for the development of a new strong magnet.

From a theoretical point of view, an accurate description of T_C is a demanding task. The spin magnetic moment at each atomic site and the magnetic exchange coupling between them can be evaluated within the framework of density functional theory [2]. T_C can be obtained by solving a derived classical spin model [3, 4]. This approach works reasonably well for α-Fe, whereas it gives a value that deviates significantly from the experimental observation for Ni, because the longitudinal spin fluctuation is not captured in the classical spin model [5]. In the case of rare-earth compounds, the $4f$ electrons add further complications. Both the electron correlation and spin–orbit interaction are strong in these compounds; hence, an advanced theoretical treatment is needed for a reliable description of the magnetic properties.

The recent rapid development of data-driven approaches in materials research offers another possibility [6, 7, 8, 9, 10]. At present, T_C data can be obtained for many compounds from databases or literature. Machine learning may be utilized to predict the T_C of a new compound from existing data. In the present work, we focus on the T_C of transition-metal rare-earth bimetals. We explain the Kernel ridge regression (KRR) technique with various forms of the kernel functions and how to select variables to represent the collected data. We demonstrate how the T_C value of a compound can be reproduced from the existing data by the KRR. We also analyze the role of variables for obtaining a high prediction accuracy. Next, a comparison between the T_C value estimated by KRR and the experimentally observed T_C value, for the new compounds is studied. Through a systematic analysis, we show that insights into the actuating mechanisms of physical phenomena can be obtained through the determination of a set of physically meaningful parameters by employing machine learning algorithms.

2. Methodology

2.1. Data collection and representation

We collected experimental data of 101 binary compounds consisting of transition metals and rare-earth metals from the Atomwork database of NIMS [11, 12], including the crystal structure of the compounds and their observed T_C values. Our task was to develop a model for estimating the T_C of a new compound from the training data of known compounds. For this purpose, one of the most important steps is the choice of an appropriate data representation that reflects the knowledge of the application domain, i.e., a model of the underlying physics. On the other hand, a data representation that derives a good estimation model may imply the discovery of the underlying physics. To represent the structural and physical properties of each binary compound, we use a combination of 21 variables. We divide all the 21 variables into three groups. The first and second categories contain descriptive variables that describe the atomic properties of the transition-metal (T) and rare-earth (R) constituents. The properties were as follows: (1) atomic number (Z_R, Z_T); (2) covalent radius (r_{covR}, r_{covT}); (3) first ionization (IP_R, IP_T); and (4) electronegativity (χ_R, χ_T). In addition, descriptive variables related to the magnetic properties were included: the (5) total spin quantum number (S_{3d}, S_{4f}); (6) total orbital angular momentum quantum number (L_{3d}, L_{4f}); and (7) total angular momentum (J_{3d}, J_{4f}).

The selection of these variables originates from the physical consideration that the intrinsic electronic and magnetic properties will determine the $3d$ orbital splitting at transition-metal sites. To capture the effect of the $4f$ electrons better, especially the strong spin-orbit coupling

IOP Conf. Series: Journal of Physics: Conf. Series **1290** (2019) 012009 doi:10.1088/1742-6596/1290/1/012009

effect, we add three additional variables to describe the properties of the constituent rare-earth metal ions, (15) the projection of the total magnetic moment onto the total angular moment ($J_{4f}g_j$), and (16) the projection of the spin magnetic moment onto the total angular moment ($J_{4f}(1-g_j)$) of the $4f$ electrons. The selection of these features is based on the physical consideration that the magnitude of the magnetic moment will determine the $T_{\rm C}$. It has been well established that information related to the crystal structure is very valuable for understanding the physics of binary compounds with transition metals and rare-earth metals. Therefore, we design a third group with structural variables that roughly represent the structural information at the transition metal and rare-earth metal sites, which are (17) the concentration of the transition metal (C_T), and (18) the concentration of the rare-earth metal (C_R). Note that if we use the atomic percentage for the concentration, the two quantities are not independent. Therefore, in this work, we measured the concentrations in units of atoms/Å^3. This unit is more informative than the atomic percentage, as it contains information on the constituent atomic size. As a consequence, (C_T) and (C_R) were not completely dependent on each other. Other structure variables were also added: the mean radius of the unit cell (19) between two rare-earth elements r_{RR}, (20) between two transition metal elements r_{TT}, and (21) between transition and rare-earth elements r_{TR}. We set the experimentally observed $T_{\rm C}$ as the target variable.

2.2. Kernel regression analysis

To learn a function for predicting the values of $T_{\rm C}$ of compounds from data represented by using vectors of the variables, we apply the KRR technique [13], which has recently been applied successfully in many materials science studies [14, 15, 16]. In our study, we apply the KRR with five kernel functions, the forms of which are defined as follows:

Cosine kernel:

$$k(\boldsymbol{x}_i, \boldsymbol{x}_j) = \frac{\boldsymbol{x}_i \cdot \boldsymbol{x}_j}{||\boldsymbol{x}_i|| \, ||\boldsymbol{x}_j||} \tag{1}$$

Linear kernel:

$$k(\boldsymbol{x}_i, \boldsymbol{x}_j) = \boldsymbol{x}_i \cdot \boldsymbol{x}_j + c \tag{2}$$

Polynomial kernel:

$$k(\boldsymbol{x}_i, \boldsymbol{x}_j) = (\alpha \boldsymbol{x}_i \cdot \boldsymbol{x}_j + c)^d \tag{3}$$

Gaussian kernel:

$$k(\boldsymbol{x}_i, \boldsymbol{x}_j) = \exp(\frac{-||\boldsymbol{x}_i - \boldsymbol{x}_j||^2}{2\sigma^2}) \tag{4}$$

Laplacian kernel:

$$k(\boldsymbol{x}_i, \boldsymbol{x}_j) = \exp(\frac{-||\boldsymbol{x}_i - \boldsymbol{x}_j||}{\sigma}) \tag{5}$$

where $||\boldsymbol{x}_i - \boldsymbol{x}_j|| = \sum_{a=1}^{m} |x_i^a - x_j^a|$; $\boldsymbol{x}_i \cdot \boldsymbol{x}_j = \sum_{a=1}^{m} x_i^a x_j^a$, and m is the number of dimensions; σ, α are the variance of the Gaussian/Laplacian and the hyper parameter of polynomial kernel functions, respectively; d is the polynomial order of the polynomial kernel—we set ($d = 3$) in all our experiments; c is a constant.

For a given new data instance $\boldsymbol{x}_*$, the predicted property $\hat{f}(\boldsymbol{x}_*)$ is expressed as the weighted sum of the kernel functions:

$$\hat{f}(\boldsymbol{x}_*) = \sum_{i=1}^{N} c_i k(\boldsymbol{x}_*, \boldsymbol{x}_i) \tag{6}$$

where N is the number of training data points, and the weighting coefficients c_i for the corresponding data point $\boldsymbol{x}_i$ are determined by minimizing

$$\sum_{i=1}^{N} [\hat{f}(\boldsymbol{x}_i) - y_i]^2 + \lambda \sum_{i=1}^{N} ||c_i||_2^2. \tag{7}$$

IOP Conf. Series: Journal of Physics: Conf. Series **1290** (2019) 012009 doi:10.1088/1742-6596/1290/1/012009

The regularization parameters λ and hyper parameters (σ, α) are chosen with the help of cross-validation[17, 18], i.e., by excluding some of the data instances during the training process and maximizing the coefficient of determination R^2:

$$R^2(y^{vld}_{\text{predict}}, y^{vld}_{\text{observed}}) = 1 - \frac{\sum_{j=1}^{p_{vld}} \left[\hat{f}(\boldsymbol{x}_j) - y_j\right]^2}{\sum_{j=1}^{p_{vld}} \left[\bar{y} - y_j\right]^2} \tag{8}$$

where p_{vld} is the number of validation data points used to evaluate how the predicted values agree with the actual observed values, and $\bar{y}$ is the average of the validation set used to compare the values predicted for the excluded compounds with the known observed values. By employing the cross-validation framework, the evaluation of the prediction accuracy is carried out by using data that are not involved in the training process to avoid overfitting the training data. In this study, we use the maximized R^2 as a measure of prediction accuracy. To obtain a good estimation of the prediction accuracy, we carried out KRR with leave-one-out cross-validation using the collected data, of which T_C is considered as a response variable, and the other variables are considered as descriptive variables. It should be noted that in KRR, the kernel metric represents a formulation to measure the similarity between data instances in an abstract description space [19]. Therefore, KRR analyses with different kernel metrics will provide a better understanding of the data set.

2.3. Kernel regression-based variable evaluation

To find the most appropriate set of variables for the prediction of T_C and the hidden knowledge regarding the determination of the values of T_C of the compounds, we utilize an exhaustive search to evaluate all possible variable combinations [20, 21, 22] to identify and remove irrelevant and redundant variables [23, 24, 25]. KRR models with various kernel types described previously are applied as the predictors. For a given variable combination, we determine the regularization parameters λ and other hyperparameters corresponding to the specific kernel setting in each experiment (detail in Section 2.2) to maximize the prediction accuracy for T_C of the KRR model. By training the KRR models for all combinations of the variables, we can obtain new data for all possible sets of variables. By using the variables in that set, the KRR model can achieve the best prediction accuracy. To analyze this kind of data, we focus on two research objectives: (1) to find the set of variables that yield the best prediction accuracy for the prediction of T_C, (2) to find the variables that are relevant for the prediction of T_C.

One can easily presume that incorporating irrelevant variables into the KRR model may impair its prediction accuracy. Therefore, we define the prediction ability $PA(S)$ of a set S by the maximum prediction accuracy that the KRR model can achieve by using the variables in a subset s of S as follows:

$$PA(S) = \max_{\forall s \subset S} R^2_s; \quad s_{PA} = \arg\max_{\forall s \subset S} R^2_s, \tag{9}$$

where R^2_s is the value of the coefficient of determination R^2 [26] achieved by the KRR using a variable set s as the independent variables. s_{PA} is the variable subset of S that yields the prediction model having the maximum prediction accuracy.

2.4. Variable relevance analysis

On the basis of Eq.(9), we can evaluate the relevance [27, 28] of a variable for the prediction of T_C using the expected reduction in the prediction ability resulting from the removal of this variable from the full set of variables. Let D be a full set of variables, d_i a variable, and $D_i = D - \{d_i\}$ the full set of variables after removing the variable d_i. The degree of the relevance of variables can be formalized as follows:

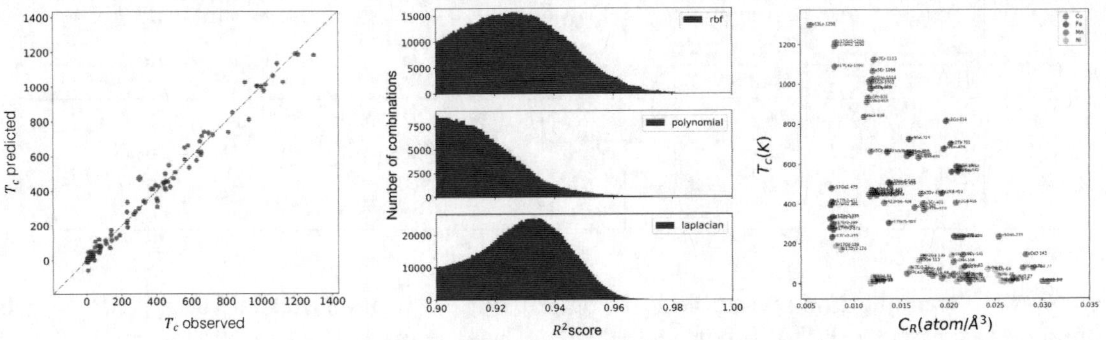

(a) (b) (c)

Figure 1: (a) Observed and predicted T_C for 101 transition–rare-earth bimetal alloy compounds by an ensemble Gaussian kernel regression. The prediction model is constructed by taking ensemble averaging of top 5 models that yield highest R^2 score after kernel regression-based model evaluation (Sect.2.3). The prediction accuracy of this model, R^2 score of 0.984, MAE: 31.21 (K), achieves a higher prediction accuracy than PA of all our designed variable sets. (b)The distribution of the KRR models that perform prediction accuracy with R^2 scores larger than 0.90 in kernel regression-based model evaluation with the target variable T_C. Different kernel types, including Gaussian, polynomial ($d = 3$), and Laplacian kernels, are shown. The PA values measured by the R^2 score are 0.982, 0.982, and 0.973, respectively, with the three kernels. The total number of models in the R^2 score range is 892, 612 models; 284, 649 models; and 1, 317, 193 modes, respectively, with the three kernels. (c) The dependence of T_C on the concentration of the rare-earth metal (C_R) in binary alloy compounds.

1. Strong relevance: a variable is strongly relevant if and only if

$$PA(D) - PA(D_i) = \max_{\forall s \subset D} R_s^2 - \max_{\forall s \subset D_i} R_s^2 > 0. \tag{10}$$

Among the strongly relevant variables, a variable that causes a larger reduction in the prediction ability when it is removed can be considered to be a strong variable. The degree of relevance of a strongly relevant variable can be computationally estimated by using the leave-one-out approach, i.e., by leaving out a variable in the currently considered variable set for the KRR analysis and evaluating the extent to which the prediction accuracy is impaired.

2. Weak relevance: a variable is weakly relevant if and only if

$$PA(D) - PA(D_i) = \max_{\forall s \subset D} R_s^2 - \max_{\forall s \subset D_i} R_s^2 = 0$$

and

$$\exists D_i' \subset D_i \text{ s.t } PA(\{d_i, D_i'\}) - PA(D_i') > 0. \tag{11}$$

It is clearly seen from Eq. (11) that estimation of the degree of relevance for the weakly relevant variables cannot be carried out in a straightforward manner, as with the strongly relevant variables. Weakly relevant variables are variables that are relevant for prediction, but they can be replaced by other variables.

2.5. Ensemble learning for prediction

In this step, we select a set of variable combinations, that yields top highest R^2 score KRR models, and consider as valuable variable combinations for capturing the T_C phenomenon. We

Table 1: Prediction accuracy of ensemble learning model with different kernel matrices

	Cosine kernel	Linear kernel	Polynomial kernel	Gaussian kernel	Laplacian kernel
PA	0.572	0.569	0.982	0.982	0.973
s_{PA}	Z_R, J_{4f}, Z_T, r_{covT}, S_{3d}, r_{TT}, C_R	r_{covR}, J_{4f}, r_{covT}, L_{3d}, r_{RR}, r_{TT}	Z_R S_{4f}, L_{3d}, J_{3d}, r_{RR}, C_T, C_R	Z_R, χ_T, J_{3d}, r_{TR}, C_T, C_R	χ_R, χ_T, $J_{4f}\,(1-g_j)$, Z_T, r_{covT}, IP_T, S_{3d}, L_{3d}, J_{3d}, C_R

then apply a technique from ensemble learning [29, 30, 31] to integrate all these KRR models to build a prediction model for predicting the T_C of new compounds.

In statistical theory, the prediction error of a model in the prediction of the target variable y for a new data instance $\boldsymbol{x}$ can be decomposed into three factors: bias, variance, and irreducible factors [29], as follows:

$$\mathbb{E}[(y - \hat{f}(\boldsymbol{x}))^2] = \left(Bias\left[\hat{f}(\boldsymbol{x})\right]\right)^2 + Var\left[\hat{f}(\boldsymbol{x})\right] + \sigma_{irr}^2 \tag{12}$$

with $Bias$ and Var are :

$$Bias\left[\hat{f}(\boldsymbol{x})\right] = \mathbb{E}\left[f(\boldsymbol{x}) - \hat{f}(\boldsymbol{x})\right] \tag{13}$$

$$Var\left[\hat{f}(\boldsymbol{x})\right] = \mathbb{E}\left[\hat{f}(\boldsymbol{x})^2\right] - \left(\mathbb{E}\left[\hat{f}(\boldsymbol{x})\right]\right)^2 \tag{14}$$

where $f(\boldsymbol{x})$ is an actual function associated with the new data points $y = f(\boldsymbol{x}) + \epsilon$, the noise ϵ has zero mean and variance σ_{irr}; $\hat{f}(\boldsymbol{x})$ is the estimated function (or predicted values of discrete points $\boldsymbol{x}$).

The irreducible factor σ_{irr} depends on the information of the target variable stored in the prediction variables, which were used for describing the training data instances. It should be noted that increasing the number of training data points does not guarantee that the prediction ability of the learning model will improve. This error can only be reduced by improving the similarity with the training data or by improving the designed descriptive variable. On the other hand, the reducible factors, bias and variance, depend on the nature of the modeling method, i.e., the KRR in these experiments. In particular, the variance factor only measured the movement of the KRR around its mean. In this research, we empirically investigate the probability distribution of the predicted values of the target variable for new data instances by utilizing an ensemble learning and bootstrap aggregating–bagging [32] algorithm. In more detail, we vary both the combination of descriptive variables and the training set by sampling with replacement from the original data uniformly. By doing so, we can indirectly observe the behavior of both the bias and the irreducible factors. An prediction model for predicting the T_C of new compounds can be constructed from the obtained distribution of prediction values.

3. Results and discussion

3.1. Kernel regression-based variable evaluation

To investigate the scientific connection [20] between the variables and the actuation mechanisms of the physical phenomenon of T_C in these bimetal systems, we examine whether T_C can be predicted by using the designed variables of the compounds. A screening was conducted for all possible variable combinations, $2^{21} - 1 = 2,097,151$ that respectively derive the same number of prediction models. As described in section 2.2, a given kernel metric formula associates with a method to measure the similarity between compounds. Therefore, five kernel metrics- cosine,

IOP Conf. Series: Journal of Physics: Conf. Series **1290** (2019) 012009 doi:10.1088/1742-6596/1290/1/012009

linear, Gaussian, polynomial, and Laplacian are all implemented and analyzed in this section. Leave-one-out cross-validations [17, 18] is performed to evaluate the prediction accuracy of these models.

The prediction abilities PA (equation 9) of the designed descriptive variables set for different kernel-type predictors are summarized in the table 1. The highest PA - R^2 score 0.982 is achieved by two models deriving from variable combinations $s_{PA} = \{Z_R, \chi_T, J_{3d}, r_{TR}, C_T, C_R\}$ with the Gaussian kernel and $s_{PA} = \{Z_R, S_{4f}, L_{3d}, J_{3d}, r_{RR}, C_T, C_R\}$ with the polynomial kernel. The PA value with the Laplacian kernel experiment achieves an R^2 score of 0.973, with its $s_{PA} = \{\chi_R, \chi_T, J_{4f}\,(1-g_j)\,, Z_T, r_{covT}, IP_T, S_{3d}, L_{3d}, J_{3d}, C_R\}$. In this screening result, there are 892,612 variable combinations associated with the Gaussian kernel; 284,649 variable combinations associated with the polynomial kernel and 1,317,193 variable combinations associated with the Laplacian kernel that yield regression models with R^2 scores exceeding 0.90. It is noted that, even with the same kernel metric, several regression models achieved a similar excellent prediction accuracy because the designed variables are not independent variables.

On the other hand, the PA values with linear and cosine kernels were lower than 0.8, i.e., the T_C variable cannot be predicted by our designed descriptive variable set with these kernel regression models.

Finally, the PA analysis indicates that with all the designed variables, it is possible to accurately predict the values of T_C of the rare-earth transition bimetal alloy compounds by using Gaussian, polynomial, and Laplacian KRR models with the designed variables. In the subsequent sections, we will discuss the method to improve the maximum prediction accuracy of the models as well as the physical meaning of strongly relevant variables.

3.2. Strongly relevant and weakly relevant features

Figure 2 shows the dependence of the best prediction accuracy PA—red lines on the number of variables recruited in the Gaussian-polynomial-Laplacian kernel regression models. In general, the prediction accuracy in all the experiments reaches the highest value with the number of descriptive variables from 6 to 8, and then gradually decreases when the number of recruited variables increases. The small subset of the designed descriptive variables and the large number of high-accuracy models described above originate from the fact that the overuse of many weakly relevant variables [23, 24, 25] weakens the correlation between the similarity of the compounds, which is measured using the kernel of the variables, and the differences in their T_C values.

Next, we evaluate the relevance of each variable for the prediction of T_C. We compare $PA(D)$ of the full set of variables D and $PA(D - \{d_i\})$ for all the variables d_i. We found that most of the variables are weakly relevant, and the prediction accuracy does not significantly change, except in one case—when removing the variables of the concentration of the rare-earth metal C_R. It is clearly seen in Figure 2 that the absence of C_R in the Gaussian and polynomial kernel model results in a dramatic decrease in the accuracy: $PA(D) < PA(D - \{C_R\})$; therefore, C_R is surely assigned as a strongly relevant variable in terms of the prediction of T_C (see Section 2.4).

This result is consistent with the understanding so far that the values of T_C of binary alloys consisting of $3d$ transition-metal and $4f$ rare-earth metals are mainly determined by the magnetic interaction in the transition-metal sublattice. In terms of molecular field theory [33], we have

$$T_C \sim n_{TT} M_T^2 / 3k_B, \tag{15}$$

where M_T and n_{TT} are the magnetization and molecular field coefficients of the transition-metal sublattice, respectively. Both M_T and n_{TT} strongly depend on C_R. The dependencies of M_T and n_{TT} on C_R are different in compounds with different combinations of rare-earth metal (R) and transition metal (T). In Co-based and Ni-based compounds, M_T and n_{TT} tend to decrease when C_R increases. This leads to a rapid decrease in T_C with an increase in C_R. For example, in

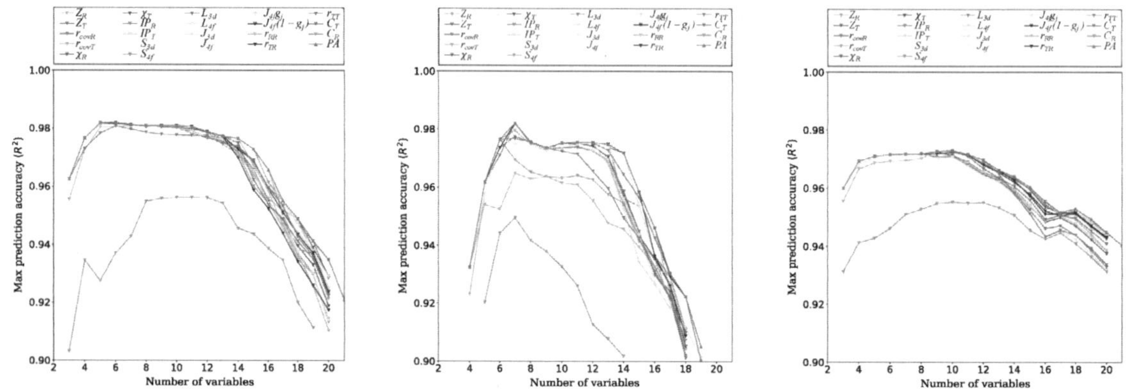

(a) (b) (c)

Figure 2: Dependence of the best prediction accuracy on the number of variables in the Gaussian(a) - polynomial(b) - Laplacian(c) kernel regression model. The red line represents the maximum prediction ability $PA(\mathrm{D})$ of the full descriptive variable set D with respect to the different numbers of variables. The other lines represent the trend of $PA(\mathrm{D}-\{d_i\})$ by removing variable $\{d_i\}$ from D in the same manner. The significant decrease of $PA(\mathrm{D}-\{C_R\})$ shows that the concentration of the rare-earth element C_R is strongly relevant to T_C.

Gd–Co compounds, T_C decreases from 1,404 K to 143 K with the increase in the concentration of Gd from 0% to 75%. In contrast, in R–Fe compounds, M_T and n_{TT} tend to increase with increasing C_R. This leads to an increase in T_C with increasing C_R. Indeed, in Gd–Fe compounds, T_C rapidly increases from 0 K to 827 K with the increase in the concentration of Gd from 0% to 33%. Detailed correlation between C_R and T_C in different transition metal-based compound is shown in Figure 1c.

3.3. Prediction of T_C for new compounds

In this section, we discuss the ability of ensemble learning using the obtained KRR models in the prediction of T_C for new compounds. The test set of new compounds includes five Fe–based compounds. The T_C of two of these compounds has been determined experimentally: $SmFe_{12}$ – 555(K)[34], YFe_{12} – 483(K) [35]. No T_C information is available for $NdFe_{12}$, $GdFe_{12}$, and $DyFe_{12}$.

From the discussion in Section 3.1, we know that C_R is the most strongly relevant to T_C. The C_R values of the five compounds in the test set are approximately 0.006 atoms/Å^3, which is much lower than the smallest C_R value for all the other Fe–based compounds (which is 0.0075 atoms/Å^3 of Fe_{17} compounds–based) (see Figure 1c). The only compound having a similar C_R value is $LaCo_{13}$ with C_R of 0.0055 atoms/Å^3. Further, the compounds in the test set are recently synthesized and have a crystal structure significantly different from that of all the other compounds in the data set. Therefore, the prediction for the T_C of these compounds could be seem closer to an *extrapolative* than an interpolative prediction problem.

Figure 3 shows the predicted value distributions for all test compounds (black line histogram) and the observed values (red dash line). It is easy to recognize that the distributions of the predicted T_C for these test compounds can be approximated by a mixture of three separate Gaussian distributions. We represent these distributions using red, blue and green, in increasing order of their mean values. From this Gaussian distribution decomposition, we can suggest that at least three distinguish functions $\hat{f}(\boldsymbol{x})$ can be regressed from the observed data to model the T_C phenomenon. Further, since the functions are learned from sub-groups of all the data set

IOP Conf. Series: Journal of Physics: Conf. Series **1290** (2019) 012009 doi:10.1088/1742-6596/1290/1/012009

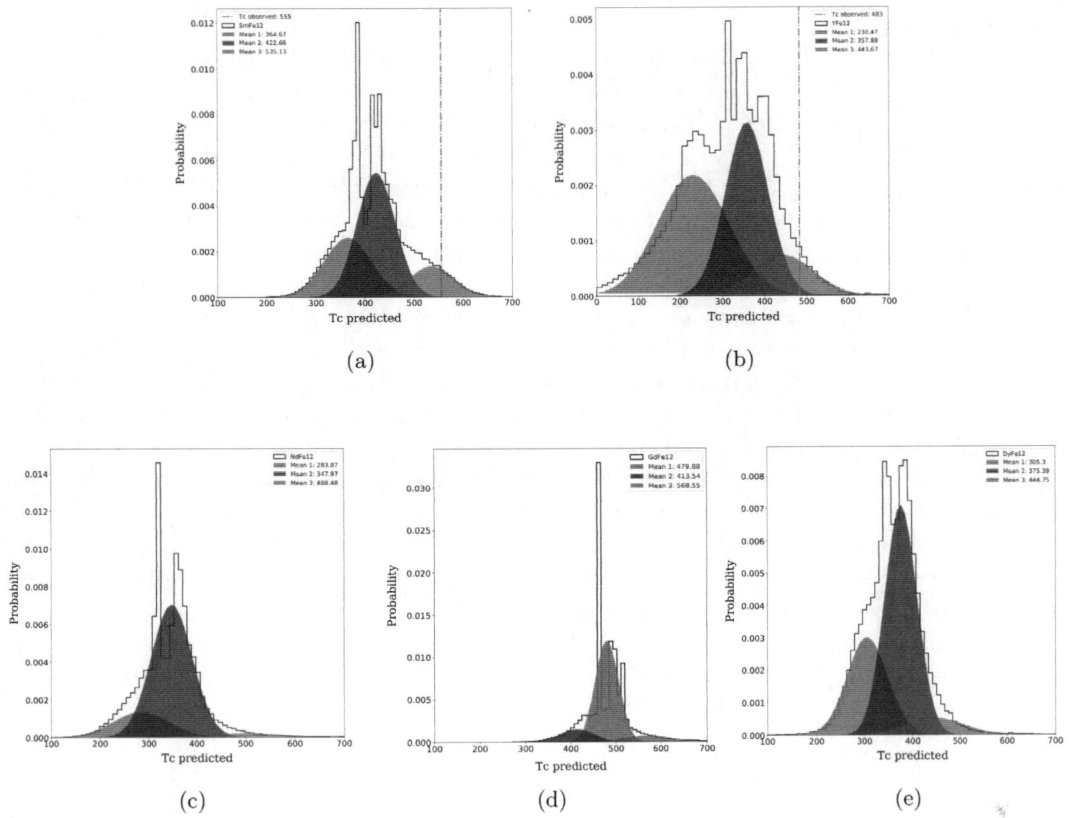

Figure 3: Predicted value distribution of test compounds by using a bagging model constructed from top–5 highest prediction accuracy kernel regression model with Gaussian and Laplacian kernels. The distributions appear as a mixture of several Gaussian distribution components. This serves as evidence to show that there are a number of functions that generate such components. The black dashed lines show the predicted value obtained by taking the average of all possible values. The red dashed lines show the observed values.

by the bagging method, we can suggest that there are at least three sub-groups of compounds corresponding to these three models.

Next, we analyze the relation of the experimentally observed T_C of the two test compounds $SmFe_{12}$ and YFe_{12} to their predicted T_C distribution. It is obvious that both the experimentally observed values correspond to the green Gaussian distribution, i.e., the largest T_C group (see Figure 3a, 3b). The mean value of these green Gaussian distributions is close to the observed value of 535.1 K, as compared with 555.0 K of $SmFe_{12}$, and 443.7 K as compared with 483.0 K of YFe_{12}. From the fact that all the five test compounds are Fe_{12}–based compounds with the same crystal structure, we can infer that the experimental values of the remaining three compounds also correspond to the green distribution. Therefore, we predict that the T_C of $NdFe_{12}$, $GdFe_{12}$, and $DyFe_{12}$ are 488.5 K, 568.5 K, and 444.7 K, respectively. There is also the potential for further experimental study on the T_C of these compounds.

Lastly, all the above investigations show that further research is required to extract hidden functions in data as well as an alternative method of integrating predicted results from ensemble learning algorithms.

4. Conclusion

In this study, we analyzed the T_C target variable of binary alloys consisting of $3d$ transition-metal and $4f$ rare-earth elements by a machine learning technique called kernel regression-based variable evaluation. Twenty-one descriptive variables were designed assuming completely obtained information of the target variable. Multiple kernel regression with different kernel types: cosine, linear, Gaussian, polynomial, and Laplacian kernels were implemented as predictor models. All possible descriptive variable combinations were generated to construct the corresponding prediction models. The prediction accuracy of T_C was maximized with different combinations of the designed descriptive variables with Gaussian, polynomial, and Laplacian kernel types. Further investigation the result of the screening prediction ability process shows that the rare-earth concentration is the most strongly relevant descriptive variable. Finally, we discuss the utilization of an ensemble model constructed by top highest prediction accuracy and a bagging algorithm to predict a test set of compounds. The final predicted T_C values for $NdFe_{12}$, $GdFe_{12}$, and $DyFe_{12}$ are 488.5 K, 568.5 K, and 444.7 K, respectively. This research also shows that there are at least three mechanisms that drive the T_C phenomenon. It also highlights the need to extract these mechanisms in future work. To conclude, in this research, we demonstrated how the kernel regression-based model evaluation can be utilized for ascertaining the scientific connection between the variable and the actuation mechanisms of a physical phenomenon by using both descriptive and predictive analyses.

Acknowledgments

This work was partly supported by PRESTO and by the "Materials Research by Information Integration" Initiative (MI^2I) project of the Support Program for Starting Up Innovation Hub, both from the Japan Science and Technology Agency (JST), Japan; by the Elements Strategy Initiative Project under the auspices of MEXT; and also by MEXT as a social and scientific priority issue (Creation of New Functional Devices and High-Performance Materials to Support Next-Generation Industries; CDMSI) to be tackled by using a post-K computer.

References

[1] Sagawa M, Fujimura S, Togawa N, Yamamoto H and Matsuura Y 1984 *J. Appl. Phys.* **55** 2083–2087
[2] Xu B Liu Z A 2011 *Computational Materials Science* **50** 841 – 846
[3] Wipf A 2012 *Statistical Approach to Quantum Field Theory* (Springer)
[4] PMyers H 1997 *Introductory Solid State Physics (Second Edition)* (CRC Press)
[5] Ruban A V, Khmelevskyi S, Mohn P and Johansson B 2007 *Physical Review B* **75**
[6] Yousef S, Da G, Thanh N, Scotty B, R C J and Wanda A 2012 *Phys. Rev. B* **85** 104104
[7] Yang S, Lach-hab M, Vaisman I I and Blaisten-Barojas E 2009 *Phys. Chem. C* **113** 21721
[8] Hautier G, Fischer C C, Jain A, Mueller T and Ceder G 2010 *Chem. Mater* **22** 3762
[9] C S J, Matthias R, Katja H, Müller K R and Kieron B 2012 *Phys. Rev. Lett* **108** 253002
[10] Isayev O, Fourches D, Muratov E N, Oses C, Rasch K, Tropsha A and Curtarolo S 2015 *Chem. Mater* **27** 735
[11] Villars P, Berndt M, Brandenburg K, Cenzual K, Daams J, Hulliger F, Massalski T, Okamoto H, Osaki K, Prince A, Putz H and Iwata S 2004 *Journal of Alloys and Compounds* **367**
[12] Xu Y, Yamazaki M and Villars P 2011 *Jpn. J. Appl. Phys.* **50**
[13] Murphy K P (ed) 2012 *Machine learning: a probabilistic perpective* (MIT Press)
[14] Matthias R 2015 *International Journal of Quantum Chemistry* **115** 1058–1073
[15] Botu V and Ramprasad R 2014 *Int. J. Quant. Chem.* **115** 1074–1083
[16] Pilania G, Wang C, Jiang X, Rajasekaran S and Ramprasad R 2013 *Scientific Reports* **3** 2810 EP –
[17] Stone M 1974 *Journal of the Royal Statistical Society. Series B (Methodological)* **36**
[18] Picard R R and Cook R D 1984 *Journal of the American Statistical Association* **79** URL `http://www.jstor.org/stable/2288403`
[19] Murphy K P (ed) 2012 *Chapter 14, Machine learning: a probabilistic perpective* (MIT Press)

IOP Conf. Series: Journal of Physics: Conf. Series **1290** (2019) 012009 doi:10.1088/1742-6596/1290/1/012009

[20] Dam H C, Nguyen V C, Pham T L, Nguyen A T, Terakura K, Miyake T and Kino H 2018 *Journal of the Physical Society of Japan* **87** 113801 (*Preprint* https://doi.org/10.7566/JPSJ.87.113801) URL https://doi.org/10.7566/JPSJ.87.113801
[21] Kohavi R and John G H 1997 *Artificial Intelligence* **97** 273 – 324 ISSN 0004-3702 relevance URL http://www.sciencedirect.com/science/article/pii/S000437029700043X
[22] Liu H and Yu L 2005 *IEEE Transactions on Knowledge and Data Engineering* **17** ISSN 1041-4347
[23] Duangsoithong R and Windeatt T 2009 Relevance and redundancy analysis for ensemble classifiers *Machine learning and data mining in pattern recognition* ed Perner P (Springer Berlin Heidelberg) pp 206–220
[24] Almuallim H and Dietterich T 1991 Learning with many irrelevant features *The Ninth National Conference on Artificial Intelligence* (AAAI Press, Menlo Park) p 547552
[25] Biesiada J and Duch W 2007 *Computer Recognition Systems 2, Advances in Soft Computing, Springer, Berlin, Heidelberg* **45**
[26] Kvalseth T O 1985 *The American Statistician* **39**
[27] Yu L and Liu H 2004 *J. Mach. Learn. Res.* **5** 1205–1224
[28] Visalakshi S and Radha V 2014 A literature review of feature selection techniques and applications: Review of feature selection in data mining *2014 IEEE International Conference on Computational Intelligence and Computing Research* pp 1–6
[29] Hastie Trevor Tibshirani Robert F J 2009 *The Elements of Statistical Learning* (Springer)
[30] TG D 2000 *Multiple Classifier Systems, Lecture Notes in Computer Science, Springer, Berlin, Heidelberg* **1857**
[31] Zhou Z H (ed) 2012 *Ensemble Methods: Foundations and Algorithms* (Chapman & Hall CRC)
[32] Breiman L 1996 *Machine Learning* **24** 123–140 ISSN 1573-0565 URL https://doi.org/10.1007/BF00058655
[33] PMyers H (ed) 1997 *Introductory Solid State Physics (Second Edition)* (CRC Press)
[34] Hirayama Y, Takahashi Y, Hirosawa S and Hono K 2017 *Scripta Materialia* **138** 62 – 65 ISSN 1359-6462 URL http://www.sciencedirect.com/science/article/pii/S1359646217302737
[35] Suzuki H 2017 *AIP Advances* **7** 056208 (*Preprint* https://doi.org/10.1063/1.4973799)

IOP Conf. Series: Journal of Physics: Conf. Series **1290** (2019) 012010 doi:10.1088/1742-6596/1290/1/012010

I-*V* Characteristics of In-Plate Graphene Nanoribbon/h-BN Heterojunctions and Resonant Tunneling

Mitsuyoshi Tomiya, Shoichi Sakamoto and Taiga Wakai

Faculty of Science and Technology, Seikei University, 3-3-1 Kichijyoji-Kitamachi, Musashino-shi, Tokyo, 180-8633, JAPAN

E-mail: `tomiya@st.seikei.ac.jp`

Abstract. We present here the first principle calculations of the electrical properties of in-plate heterojunctions of armchair graphene nanoribbon/h-BN(AGNR/h-BN)s. The calculations are carried out using SIESTA package, which is comprised of numerical codes of the density functional theory(DFT) and the non-equilibrium Green's function(NEGF). Especially, adopting the conductive (3n-1)-family of AGNR((3n-1)-AGNR) makes the lead parts on both side of the model metallic. Two transverse arrays of h-BN, which is a wide-gap semiconductor, are embedded in the middle of (3n-1)-AGNR and act as a double barrier system.

The quantum double barrier tunneling is found in the transmission functions(TF) and *I*-*V* characteristics of 8, 11, 14-AGNR/h-BN. The TF shows significantly spiky peaks in the neighborhood of the Fermi energy, and consequently, it results in step-sise *I*-*V* characteristics. Simple one-dimensional Dirac equation model for the double barrier system is also proposed to analyze numerical results. Our model reproduces most of the peaks of the transmission functions nearby the Fermi energy, as a result of quantum tunneling.

1. Introduction

Recently, single layer materials, i.e., graphene, hexagonal boron-nitride(h-BN), phosphorene, germanene, and III-nitrides[1-13] are considered as promising canditates to overtake the semiconductors in electric devices. Among these, graphene sheet(GS) is the representative material and shows metallic conduction properties[14]. Unfortunately, it does not have a band gap, it is always conductive. A graphene nanoribbon(GNR)[15, 16] is a narrow strip of graphene with a nanometer-scale width. GNR is expected to have metallic or semiconductor-like properties, depending on its chirality and width[17, 18].

Another promising material for manufacturing electronic devices is single-layered hexagonal crystal of boron nitride(h-BN), which is a wide gap semiconductor[19-22]. The same number of boron and nitrogen atoms form the boron-nitrogen nanostructure: h-BN. Its structure is almost the same as that of GS. Simply speaking, the carbon atoms at A-sites of graphene are replaced by boron atoms and those at the B-sites are replaced by nitrogen atoms. The heterojunction of GS and h-BN has already been realized[21]. It also has a high affinity to exchange parts of the GNR as thin transverse arrays to make a nanodevice(Figure 1).

We focus on the quantum tunneling through double barriers. The resonant tunneling diode(RTD) was developed more than four decades ago[23]. It utilizes the quantum tunneling

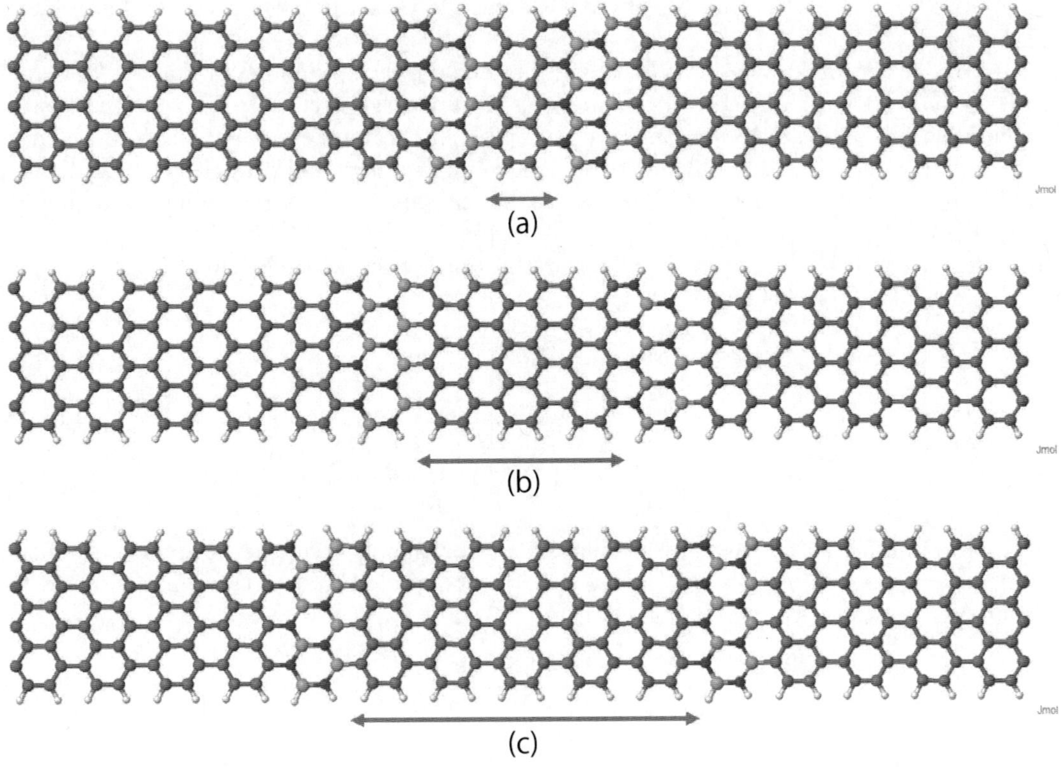

Figure 1. Schematic illustration of our in-plate 8-AGNR/h-BN heterojunction system in the cases of (a) $\Delta u = 1\text{U}$, (b) $\Delta u = 3\text{U}$, and (c) $\Delta u = 5\text{U}$. The arrows indicate the intervals Δu which are measured in the carbon honeycomb lattice as a unit: U.

resonance through double barriers made of semiconductors. The nano-structure of double barrier system is conceivable with a GNR and two embedded h-BN arrays, because the (3n-1)-family of armchair GNR(AGNR) is good conductor without any edge states around the Fermi energy and h-BN has wide band gap[17, 18]. Settling the two arrays in a AGNR(Figure 1) make them act like double barriers in graphene.

The one-dimensional(1D) Dirac fermion model that we introduced[24] is to analyze our first principle calculations. In the model, GNR is modelled by massless electron regions and barrier parts: h-BN arrays by massive electron region with the mass that is a half of the band gap of h-BN.

2. One-dimensional Dirac Equation Model

To examine the electrical behavior of our simulation, we borrow the idea of the tight-binding model, which is established as the reliable model for graphene[25], and approximate our model as the 1D Dirac equation. Then an electron should also be treated by the Dirac equation, despite it being massive and not relativistic inside the h-BN arrays, because the wave function of an electron has to be continuous in the whole material. The band structure of h-BN shows a quadratic dispersion relation with respect to the wavenumber and it contains a mass term. It has a large band gap of 4.6 eV[24]. Therefore, its mass can be estimated as half of the band gap $m_{\text{h-BN}} = 2.3$ eV $= 4.6/2$ eV. Our double barrier system engrafts two massive electron regions of h-BN in the massless region of (3n-1)-AGNR. The massive electron region strikingly prevents electrical conductivity, however, this system forms a pseudo-potential well in which the resonant tunneling takes place[26].

The finite mass then makes the h-BN regions into the effective barriers(cf. Figure 9 of [24]). Even when the barriers stop an electron considerably, if its energy is close to a pseudo-eigenstate and its wavefunction can last the exponential dumping in the barriers, then tunneling can be realized. The transmission probability also becomes significantly spiky and the peaks line up evenly(cf. Figure 10 of [24]). This feature is essentially the same as our first principle calculation of AGNR/h-BN.

Ignoring the transverse degree of freedom in the quite thin width of AGNR, it could be modeled as the 1D Dirac equation

$$(v_F \sigma_x p_x + m {v_F}^2 \beta + V(x) - E)\Phi(x) = 0, \tag{1}$$

which is set along the x axis (longitudinal current direction of AGNR). Here, $p_x = \frac{\hbar}{i}\frac{d}{dx}$ and two matices are set as

$$\sigma_x = \begin{pmatrix} 0 & 1 \\ 1 & 0 \end{pmatrix}, \qquad \beta = \begin{pmatrix} 1 & 0 \\ 0 & -1 \end{pmatrix}. \tag{2}$$

The wavefunction of the Dirac equation becomes a vector that is composed of two components $\Phi(x) = (\phi(x), \psi(x))$. Assuming the potential V to be constant, the first component ϕ and the second componet ψ exhibit the following relation

$$\phi(x) = \frac{i\hbar v_F}{m{v_F}^2 + V - E}\frac{d\psi}{dx}. \tag{3}$$

The mass is set as $m = 0$ for AGNR parts, $m = m_{\mathrm{h-BN}} = 2.3\mathrm{eV}$ for h-BN parts and in our model, and the Fermi velocity is set slightly lower ($v_F = c/350$) than the usual $c/300$ in this work.

Then, the continuity conditions for the wave functions come out for the wave functions at the boundaries between region I - region V (cf. Figure 9 of [24]);

$$\begin{array}{rll}
\mathrm{I}: & \phi_{\mathrm{I}}(x) = e^{ikx} + re^{-ikx} & x < -\frac{d}{2} - w \\
\mathrm{II}: & \phi_{\mathrm{II}}(x) = ae^{i\kappa x} + be^{-i\kappa x} & -\frac{d}{2} - w < x < -\frac{d}{2} \\
\mathrm{III}: & \phi_{\mathrm{III}}(x) = ce^{ikx} + de^{-ikx} & -\frac{d}{2} < x < \frac{d}{2} \\
\mathrm{IV}: & \phi_{\mathrm{IV}}(x) = fe^{i\kappa x} + ge^{-i\kappa x} & \frac{d}{2} < x < \frac{d}{2} + w \\
\mathrm{V}: & \phi_{\mathrm{V}}(x) = te^{ikx} & x > \frac{d}{2} + w,
\end{array} \tag{4}$$

and their counter parts $\psi_{\mathrm{I}}(x), \cdots, \psi_{\mathrm{V}}(x)$ are also required to satisfy the same kind of conditions(please refer to Eq.(4)). Here $k = \frac{E}{\hbar c}$, $\kappa = \frac{\sqrt{2m_{\mathrm{h-BN}}E}}{\hbar}$, w is the width of the wall of h-BN, d is the distance of the walls, r is the reflection coefficient, and t is the transmission coefficient. Therefore, there exist eight undetermined coefficients: r, a, b, c, d, f, g, and t, and eight independent equations at the boundaries also. Thus, we can have the unique solution of the coefficients in general.

3. Numerical Method

To execute the first principle calculation, we choose to use the general-purpose package SIESTA[27, 28], which is based on the density functional theory(DFT) method, and also employ the nonequilibrium Green function(NEGF) package TranSIESTA[29]. The cut-off energy is set at 350Ry. The samples of the wave-number of the first Brillouin zone are selected on 128 (= 4×4×8) points. Both the parameters control the precision of the calculation. The Kleinman-Bylander (KB) potential[30] is chosen for the pseudo-potential and the exchange correlation term for the PBE functional[31] of the generalized gradient approximation(GGA).

IOP Conf. Series: Journal of Physics: Conf. Series **1290** (2019) 012010 doi:10.1088/1742-6596/1290/1/012010

The numerical package TranSIESTA is developed for the simulation of non-equiliblium systems, e.g., an open system namely, the non-periodic boundary condition along the current (length) direction. In this work, both the ends of (3n-1)-AGNR are connected to the same kind of AGNRs at half infinity, which are also adopted for the electrodes and also known to be metallic. It makes the calculation of electric current flowing through the AGNR possible. The widths of the electrodes are set the same as that of the central part of the GNR to get rid of the dispersion of electric current at the junctions of the central domain and electrodes. Dispersion would significantly complicate the analysis of the simulation. Note that the dangling bonds of carbon atoms on the edges are terminated by hydrogen atoms to keep the nanoribbons flat. Otherwise, the bonds on the edges interact with each other and the ribbons often roll up or wave irregularly.

The reliable experimental carbon atomic distance in graphene is 1.42 Å; however, we use 1.43 Å as the basic value that is obtained by minimizing the total energy to stabilize the structure in the SIESTA simulation. It is known that GGA, which is introduced to evaluate an exchange term in DFT, usually tends to increase the lattice constant a little than that obtained from experimental data. Also by GGA in the DFT, the boron-nitrogen atomic distance in the h-BN sheet is set to 1.44 Å.

In this paper, we apply a unit U, which equals $\sqrt{3}$ times the in-plate lattice constant of graphene(Figure 1). Precisely speaking, the length of h-BN is slightly greater than that of graphene as mentioned above, and in nanoribbons the distances of atoms are all more or less different, unlike graphene sheet which has an infinitely wide extent. In this work, the structures of our materials obtained using SIESTA are always optimized before evaluating any electrical properties. Actually, all the atomic distances are dependent on their specific places where they are settled inside the ribbon. However, basic averaged distances are as follows. In nanoribbons C-C bond length is settled at 1.43 Å as above, and B-N bond length of the h-BN is 1.7 % longer than the C-C bond length. At the heterojunctions of AGNRs, the bond lengths of C-B, C-N, and B-N are evaluated as 1.50 Å, 1.40 Å, and 1.44 Å, respectively.

4. Electrical Properties of AGRN/h-BN

The electrical properties of in-plate AGNR/h-BN heterojunctions can be calculated using TranSIESTA. Since the number of carbon atoms in the transverse direction belong to the (3n-1)-family, we choose 8-, 11-, and 14-AGNR(n=3, 4, and 5 cases) which are known to have metallic properties[17, 18]. Then, h-BN arrays are embedded at two places in the AGNR as shown in Figure 1. The spacing between two h-BNs, Δu is changed from 1 to 7U. Our purpose is to find out the possibility of constructing nanodevices from the conductive (3n-1)-AGNR family. On the contrary, h-BN sheet is known to be a wide-gap semiconductor[20]. Their in-plate heterojunctions must have unique electrical characteristics. This kind of heterojunctions keep the materials two-dimensional (2D), unlike the multi-layered heterojunctions widely adopted recently. Ideally, the in-plate heterojunction is integrated in a single layer material.

Graphene is theoretically and experimentally known to have a Dirac cone shaped linear dispersion relation and exhibits metallic properties[14]. Its zero band gap is also numerically confirmed [24]. On the contrary, the h-BN sheet has a quadratic dispersion relation and a wide gap in its band structure. In addition, we have confirmed that the band gap is about 4.6eV in our simulation of DFT[24].

The I-V characteristics can be derived by the Landauer formula[32] and the NEGF method[33, 34]. Current I can then be calculated as

$$\begin{aligned} I = & \tfrac{2e}{h} \int T(E)[f_L(E) - f_R(E)]dE \\ = & \tfrac{2e}{h} \int T(E)[f(E-(E_F-\tfrac{eV}{2})) - f(E-(E_F+\tfrac{eV}{2}))]dE, \end{aligned} \quad (5)$$

where $f(E)$ represents the Fermi distribution function in the AGNR, E_F is the Fermi energy,

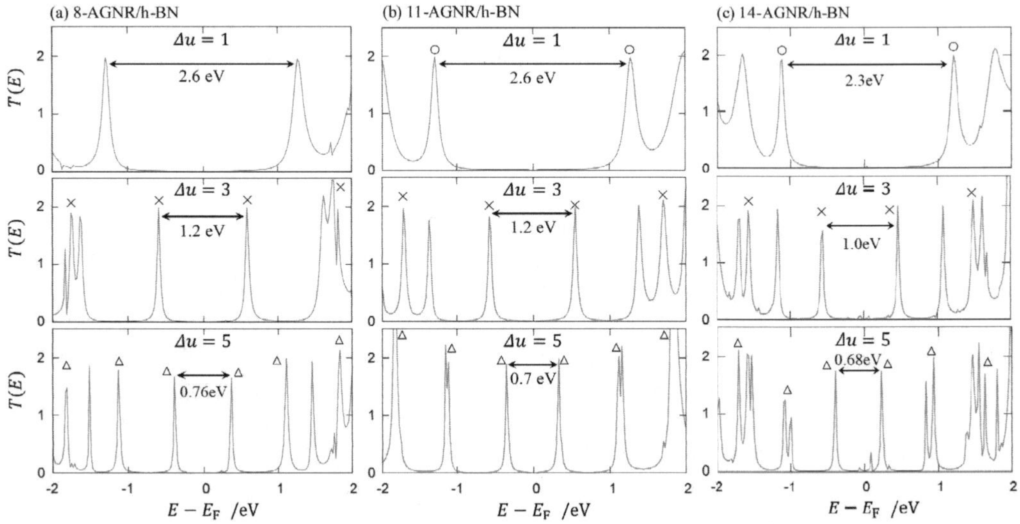

Figure 2. TFs of double barrier systems of in-plate heterojunctions of (a) 8-AGNR/h-BN, (b) 11-AGNR/h-BN, and (c) 14-AGNR/h-BN. Their shapes are actually simpler than that of GS/h-BN[24]. The peaks with symbols $\bigcirc$, $\times$, and $\triangle$ correspond to the peaks obtained from the 1D Dirac equation model shown in Figure 10 of [24].

$f_L(E)$ and $f_R(E)$ are the Fermi distributions of the left and the right leads, V is the external bias voltage, and e is the elementary charge.

The in-plate heterojunction of the two h-BNs in (3n-1)-AGNR makes TF discrete and spiky(Figure 2). Increase in the space between the two h-BN arrays in AGNR leads decrease in the energy gap of the TF: ΔE is thus obtained. The reciprocals of ΔE are plotted against Δu in Figure 3(a) for 11-AGNR/h-BN, and in Figure 4(a) for 14-AGNR/h-BN. They show that ΔE is almost inversely proportional to the distance Δu. Their I-V characteristics have step-wise shapes(Figures 3(b), and 4(b)). The I-V characteristics of the (3n-1)-AGNR/h-BN double barrier system shows that the current I begins to flow after the voltage V exceeds the threshold value, which is basically the same as ΔE. The threshold voltage can be read from its TF and I-V property. It suggests the possibility to adjust the threshold voltage to a certain degree by selecting Δu. Furthermore, if the voltage increases and approaches the other peaks in the TF to be utilized as additional conducting channels, the current also increases in steps. Next, we should clarify the mechanism that makes the simple structure of TF of the AGNR/h-BN a double barrier system around the Fermi energy.

5. Discussions

From the 1D Dirac equation model, the transmission probability(TP) $\tau = |t|^2$ can be obtained using the continuity conditions given in Eq.(4). The numerical result of τ also exhibits a clear spiky property as our first principle simulations[24]. Two transverse h-BN arrays embedded in (3n-1)-AGNR with exactly the same shape form a pseudo-well, which can have pseudo-eigenstates inside. This leads the TF to have sharp and discrete peaks. In AGNR, an electron is expected to behave as the 1D massless Dirac fermion, but as a massive fermion with a mass of 2.3 eV in h-BN. The potential itself is $V(x) = 0$ everywhere in our model. Then the shape of TP: τ is quite similar to our TF of the first principle simulations. In particular, around the Fermi

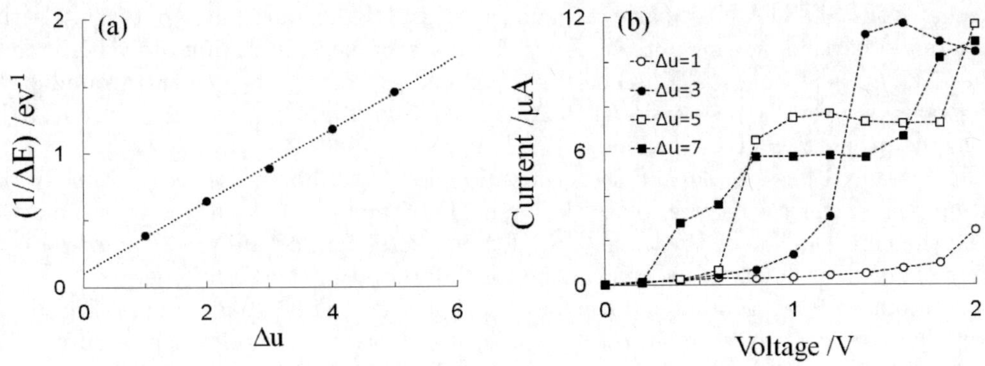

Figure 3. (a) Relation of the reciprocal of the gap between the peaks of the transmission function around the Fermi energy and the interval between h-BN arrays in 11-AGRN/h-BN. (b) *I-V* characteristics of 11-AGNR/h-BN with two h-BN arrays $\Delta u = 1 \sim 7$U

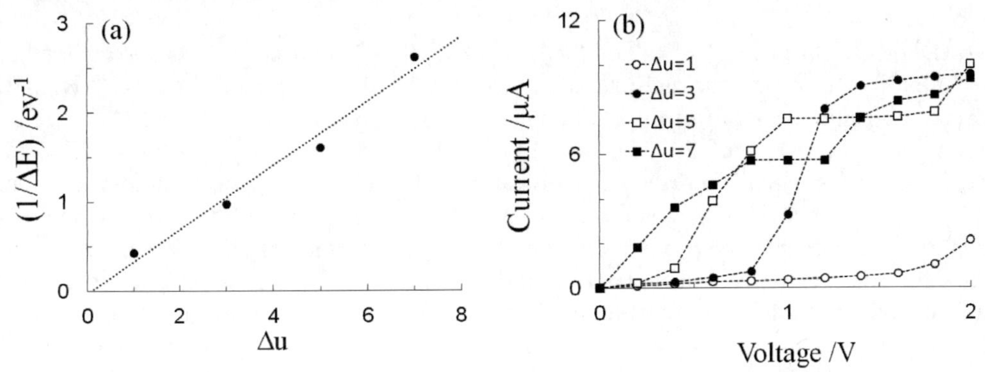

Figure 4. (a) Relation of the reciprocal of the gap between the peaks of the transmission function around the Fermi energy and the interval between h-BN arrays in 14-AGRN/h-BN. (b) *I-V* characteristics of 14-AGNR/h-BN with two h-BN arrays $\Delta u = 1 \sim 7$U

energy with NEGF, several spiky peaks are orderly lined up symmetrically on both sides of the Fermi energy. The number of peaks also increases, when the interval between the walls becomes broader and the peaks get narrower as the TF[24]. These features are also well reproduced by $\mathcal{T}$.

However, the Fermi velocity of the electron has to be rectified and set to a slightly lower value, i.e., $v_F = c/350$(c is the velocity of light) than the widely accepted value of $v_F = c/300$. The velocity of the electron that we have chosen would be some kind of ensemble throughout the first Brillouin zone. Actually, the original v_F is the estimation of the tight binding model at the Dirac cone structure where an electron must act as a massless fermion. With arbitrary momentum, there is an electron band gap and effectively massive dispersion relation. Hence, the averaged velocity should be slightly lower.

The *I-V* characteristics suggests that the resonant tunneling, which allows electrons to penetrate the double barriers system, must occur when the energy of electrons gets close to the level of the pseudo-eigenstates inside the double barrier well. Remarkably, the TP of our Dirac fermion model nicely reproduces the TF of the (3n-1)-AGNR/h-BN double barrier system

IOP Conf. Series: Journal of Physics: Conf. Series **1290** (2019) 012010 doi:10.1088/1742-6596/1290/1/012010

in NEGF results from SIESTA. It clearly explains why ΔE is proportional to $1/\Delta u$. If the height of the double barriers as the potentials, or the mass of electron, is infinitely high, then the eigenvalues are $E_n = \frac{\pi}{\Delta u}(n + \frac{1}{2})$[35]. The threshold energy corresponds to the ground state $n = 0$, and its value is inversely proportional to Δu. The actual finite mass would cause a slight deviation from the proportionality relation.

Thinner (3n-1)-AGNR has simpler TF structure and has less number of other peaks than the orderly lined-up peaks, which can be explained by our 1D Dirac model. It means that the other peaks must be the consequence of the transverse degree of freedom of AGNR. Our model gets more precise in the thinner AGNR, at least in the neighborhood of the Fermi energy.

Finally, on adopting the Landauer formula, Eq.(5), the spiky peaks in TFs(Figure 2) and the Fermi distribution function at room temperature form the moderately step-wise *I*-*V* characteristics. There is no bottom of the conduction band in metallic GNR. When the bias voltage is increased, more peaks just allow electrons to pass through the double barriers. Higher bias voltage results in a wider energy window: $f_L(E) - F_R(E)$ in the integrand of the formula. Thus, the *I*-*V* characteristics become step-wise, and the negative resistance of the double barrier system made of semiconductors does not appear in our simulations.

6. Conclusion

We calculated the electrical properties of the (3n-1)-AGNR/h-BN double barrier systems, which are assumed to exhibit resonant tunneling. They are made of the two in-plate AGNR/h-BN heterojunctions. The pure (3n-1)-family is known to be a conductor like the GS[17, 18]. The TF and the *I*-*V* characteristics of 8-, 11-, and 14-AGNR produced similar results. The TFs had sharp peaks near the Fermi energy and the interval between peaks was almost inversely proportional to the distance between the two transverse h-BN arrays. The orderly lined-up peaks in the TFs exhibited the step-wise shape of their *I*-*V* characteristics. Then, increasing the bias voltage for AGNR/h-BN double barriers increased the number of contributing channels inside the energy window, which were actually spiky peaks in TF. The conduction band overlaps with the valence band in (3n-1)-family of AGNR. Therefore no channel enters the band gap and negative resistance cannot exit. The threshold voltage of the *I*-*V* characteristics was detemined using the energy gap value ΔE in the TF and was propotional to $1/\Delta u$. This implies that the threshold voltage is adjustable to a certain extent.

We also introduced the 1D Dirac equation model[24]. The model works significantly well for the (3n-1)-AGNR. The TF of the quantum tunneling phenomenon can be well simulated as the TP by our 1D Dirac model. The Dirac model showed more similar results for thinner AGNR. The model confirmed the quantum resonant tunneling in (3n-1)-AGNR/h-BN in-plate heterojunctions. Lastly, our study suggests the possibilities that the electronic properties of GNR can be tuned to manufacture the next generation nanodevices, by inserting h-BN arrays at GNR in varying positions.

References

[1] Cui T, Lv R, Huang Z-H, Chen S, Zhang Z, Gan X, Jia X, Wang K, Wu D, and Kang F 2013 *J. Mater. Chem. A* **1**, 5736
[2] Nepal N, Wheeler V D, Anderson T J, Kub F J, Mastro M A, Myeys-Ward R L, Qadri S B, Freitas J A, Hernandez S C, Nykiti, Walton S G, Gaskill K, and Eddy C R Jr. 2013 *J. Appl. Phys. Express* **6**, 061003
[3] Lin F, Chen S-H, Meng J, Tse G, Fu X-W, Xu F-J, Shen B, Liao Z-M, and Yu D-P 2014 *Appl. Phys. Lett.* **106**, 073103
[4] Cuo G-C, Wang D, Wei X-L, Zhang Q, Liu H, Lau W-M, and Liu L-M 2015 *J. Phys. Chem. Lett.* **6**, 5002
[5] Cai Y, Zhang G, and Zhang Y-W 2015 *J. Phys. Chem. C* **119**, 13929
[6] Padilha J E, Fazzio A, and da Silva A J R 2015 *Phys. Rev. Lett.* **114**, 066803
[7] Hu W, Lin L, Yang C, Dai J, and Yang J 2016 *Nano Lett.* **16**, 1675
[8] Zhang Y-Y, Pei Q-X, Mai Y-W, and Lai S-K 2016 *J. Phys. D* **49**, 465301
[9] Kumar A, Kashid R, Ghosh A, Kumar V, and Singh R 2016 *ACS Matter. Interfaces* **8**, 8213

[10] Yang Q, Tan C-J, Meng R-S, Jiang J-K, Liang Q-H, Sun X, Yang D-G, and Chen X-P 2016 *IEEE Electron Device Lett.* **38**, 145
[11] Zhu J, Zhang J, and Hao Y 2016 *Jpn. J. Appl. Phys.* **55**, 080306
[12] Kalita G, Shaarin M D, Paudei B, Mahyavanshi R, and Tanemura M 2017 *Appl. Phys. Lett.* **111**, 013504
[13] Hu X-R, Zheng J-M, Ren Z-Y 2018 *Front. Phys.* **13**, 137302
[14] Berger C, Song Z, Li T, Li X, Ogbazghi A Y, Feng R, Dai Z, Marchenkov A N, Conrad E H, First P N, and de Heer W A 2004 *J. Chem. Phys. B* **108** 19912
[15] Han M Y, Özyilmaz B, Zhang Y, and Kim P, 2007 *Phys. Rev. Lett.* **98** 206805
[16] Berger C, Song Z, Li X, Wu X, Brown N, Naud C, Mayou D, Li T, Hass J, Marchenkov A N, Conrad E H, First P N, and de Heer W A 2006 *Science* **312** 1191
[17] Yang L, Park C -H, Son Y -W, Cohen M L, and Louie S G 2007 *Phys. Rev. Lett* **99** 186801
[18] Neto A H C, Guinea F, Peres N M R, Novoselov K S, and Geim A K 2009 *Rev. Mod. Phys.* **81** 109
[19] Behera H, and Mukhopadhhyay G 2012 *J. Phys. Chem. Solids* **73** 818
[20] Liu L, Feng Y P, and Shen Z X 2003 *Phys. Rev. B* **68**, 104102
[21] Liu L, Park J, Siegel D A, McCarty K F, Clark K W, Deng W, Basile L, Idrobo J C, Li A -P, and Gu G 2014 *Science* **343** 163
[22] Jin C, Lin F, Suenaga K, and Iijima S 2009 *Phys. Rev. Lett.* **102** 195505
[23] Tsu R and Esaki L 1973 *Appl. Phys. Lett.* **70**, 562
[24] Wakai T, Sakamoto S, Tomiya M, 2018 *J. Phys. Condens. Matter* **30** 265302
[25] Wallace P R 1947 *Phys. Rev.* **71**, 622
[26] Chang L L, Esaki L, and Tsu R 1974 *Appl. Phys. Lett.* **24** 593
[27] Sánchez-Portal D, Ordejón P, Artacho E, and Soler J M 1997 *Int. J. Quantum Chem.* **65** 453
[28] Soler J M, Artacho E, Gale J D, García A, Junquera J, Ordejón P, and Sánchez-Portal D 2002 *J. Phys.: Condens. Matter* **14** 2745
[29] Brandbyge M, Mozos J -L, Ordejón P, Taylor J, and Stokbro K 2002 *Phys. Rev. B* **65** 165401
[30] Kleinman L, and Bylander D M 1982 *Phys. Rev. Lett.* **48** 1425
[31] Perdew J P, Burke K, and Ernzerhof M 1996 *Phys. Rev. Lett.* **77** 3865
[32] Büttiker M, Imry Y, Landauer R, and Pinhas S 1985 *Phys. Rev. B* **31** 6207
[33] Haas P, Tran F, Blaha P, and Schwarz K 2011 *Phys. Rev. B* **83** 205117
[34] Brandbyge M, Mozos J -L, Ordejón P, Taylor J, and Stokbro K 2002 *Phys. Rev. B* **65** 165401
[35] Berry M V and Mondragon R J 1987 *Proc. R. Soc. Lond. A* **412** 53

Electronic properties of binary compounds with high fidelity and high throughput

Protik Das, Timur Bazhirov
Exabyte Inc., San Francisco, California 94103, USA

Abstract. We present example applications of an approach to high-throughput first-principles calculations of the electronic properties of materials implemented within the Exabyte.io platform[1, 2]. We deploy computational techniques based on the Density Functional Theory with both Generalized Gradient Approximation (GGA) and Hybrid Screened Exchange (HSE) in order to extract the electronic band gaps and band structures for a set of 775 binary compounds. We find that for HSE, the average relative error fits within 22%, whereas for GGA it is 49%. We find the average calculation time on an up-to-date server centrally available from a public cloud provider to fit within 1.2 and 36 hours for GGA and HSE, respectively. The results and the associated data, including the materials and simulation workflows, are standardized and made available online in an accessible, repeatable and extensible setting.

1. Introduction

First-principles modeling for the purpose of materials design and discovery by means of high-throughput calculations received much attention with multiple success stories reported and multiple large-scale efforts deployed to date[3, 4, 5, 6, 7]. Such efforts establish the data-centric approach to materials science and new collaborative inter-disciplinary work. The electronic properties of materials, such as the electronic band structures and band gaps, represent a natural target for the first principles modeling and have been successfully extracted earlier. Despite the attention, however, reaching high fidelity of the resulting predictions for the electronic structural properties is often given a lower priority during the high-throughput screening due to the complexity of the computational implementation of the underlying techniques.

Among the initiatives that assembled large datasets of electronic properties of materials are the Materials Project [3], AFLOW[4], the Open Quantum Materials Database[5], AIIDA[6] which also provides a set of building blocks for the construction of the simulation workflows, NOMAD[7] with an open access data repository and data analytics tools, and the Computational 2D Materials Database[8, 9]. All except the latter have primarily focused on obtaining results within the most robust, but less accurate implementation of Density Functional Theory within the Generalized Gradient Approximation, which has a number of well-known drawbacks when applied to the extraction of the electronic properties of materials[10].

We present the approach implemented in Exabyte.io[1] and earlier described in more details in our prior publication [2]. The approach is able to deliver high fidelity in a repeatable way transferable from one material to another and able to facilitate high throughput as well. We apply this approach to extract the electronic properties of binary semiconducting compounds. We employ Density Functional Theory in the plane-wave pseudopotential formalism[11, 12] and obtain the electronic band structures and band gaps for a diverse set of 775 binary compounds, further called EBSC-775. We extract the results within the Generalized Gradient Approximation [10] for the full set, and within the Hybrid Screened Exchange[13] for a subset of 270 materials. We compare the results with the available experimental data and present the assessment of the accuracy levels for each model.

This manuscript is structured as follows. We first explain the categorization of materials studied with respect to the modeling workflows and discuss the methodology and the parameters used using during the calculations. Next, we present the results for all the materials, compare them with the available experimental and computational data, and review the outliers. Finally, we discuss and analyze the results more in depth and suggest the pathways toward further improved accuracy. This work presents all the following: the results, the tools that generated the results, all associated data, and an easy-to-access way to reproduce and further improve upon our work.[14]

2. Methodology

2.1. General logic

We demonstrate the general execution flow employed in this work in a prior publication [2]. As it is mentioned therein, our general routine includes the creation of the modeling workflows, followed by the ingestion of the structural data about materials and its conversion to database entries, followed by the execution of simulation jobs and further analysis of the results. In this work we employ the simulation workflows within the Generalized Gradient Approximation (GGA) and the Hybrid Screened Exchange (HSE) approach with a similar logic as previously explained. The users of Exabyte platform can clone the associated entities (eg. materials, workflows, jobs) - and re-create our calculations in order to reproduce or further improve the results.

2.2. Materials

All materials studied in this work constitute the EBSC-775 set and are divided in 7 categories according to the difficulty levels of the corresponding simulation workflows. The details about the materials studied, including the corresponding counts per categories are given in Table 1. Our selection is based initially on the binary semiconducting compounds (i.e. compounds with non-zero band gap) available from the Materials Project (MP) database[3]. We optimized the choice of compounds according to the size of the crystal unit cell and the associated difficulty levels for the workflows as explained below. For the HSE calculations we used a subset containing 270 total compounds.

2.3. Workflows

In order to organize the information about the simulation workflows we employ the categorization illustrated in Table 1. This categorization was explained in more details in a prior publication [2]. The categorization depends on: (a) the inclusion of the semi-core electronic states in the pseudopotentials, (b) the treatment of spin-orbit coupling within the calculation, and (c) the treatment of magnetic interactions. As it can be seen from the table, we prioritize compounds with lower difficulty levels where the computational burden is lower and focus our attention on the first 5 categories.

2.4. Computational setup

2.4.1. Model and precision All calculations were performed using Density Functional Theory[15, 16] in the plane-wave pseudopotential projector augmented wave (PAW)[17] formalism using the Vienna Ab initio Simulation Package (VASP)[18, 19]. Within the generalized gradient approximation the exchange-correlation effects were modeled using the Perdew-Berke-Ernzerhof (PBE)[10] functional, for hybrid screened exchange the Heyd-Scuseria-Ernzerhof

IOP Conf. Series: Journal of Physics: Conf. Series **1290** (2019) 012011 doi:10.1088/1742-6596/1290/1/012011

Difficulty	Semi-core	SOC	Magnetism	N_{mat}^{GGA}	N_{mat}^{HSE}	N_{at}^{max}
1	no	no	no	159	129	24
2	yes	no	no	126	71	20
3	no	yes	no	303	43	20
4	yes	yes	no	123	8	20
5	no	no	yes	60	19	20
6	no	no	yes	2	0	16
7	yes	yes	yes	3	0	16

Table 1. Summary of the simulation workflows categorization employed in this work. "Semi-core" indicates that the pseudopotentials with semi-core states were used, "SOC" stands for the inclusion of the spin-orbit coupling, and "Magnetism" is used to denote the inclusion of collinear magnetic moments, except for the difficulty 7 when spin-orbit coupling and magnetism are included both, which lead to the treatment of non-collinear magnetic interactions. "GGA" indicates number of materials calculated with GGA level of theory. Similarly "HSE" indicates band gaps calculated with HSE level of theory.

(HSE06) implementation[20] is used. We sample in the reciprocal cell based on the k-points per reciprocal atom (KPPRA) with a uniform unshifted grid. The KPPRA value of 2,000 is used. The band structures are calculated using the default paths per each crystal lattice type as defined in the AFLOW methodology[4]. The electronic density of states calculations were performed on a denser grid with KPPRA of 16,000 using tetrahedron interpolation as implemented in VASP[18].

2.4.2. Method and parameters used Atomic positions for all structures studied were optimized until the forces acting on each atom are less than 0.01 eV/Å. The kinetic energy cutoff was set to 520 eV. We used the Gaussian smearing algorithm, and the blocked Davidson iteration scheme[21] as the initial minimization algorithm for the calculation of the electronic subsystem, with the ionic positions updated using the conjugated gradient algorithm. A smearing value of 50 meV was employed. The semi-empirical Grimme-D2 correction was used[22]. The HSE calculations were performed with the default 25% mixing parameter for the short-range Hartree-Fock exchange[13]. The screening parameter μ is set to 0.2 Å^{-1}.

2.4.3. Hardware and related We performed all calculations described in this work using the high-performance computing hardware readily available from Microsoft Azure cloud computing service[23] in the same manner as described in a prior publication[2] using H16r and H16mr instances. Computational resources were provisioned and assembled on-demand by the software available within the Exabyte platform[1]. The calculations were executed within a two-week period from June 6th to June 20th of 2018. The peak size of the computational infrastructure used during this work was administratively limited to 750 nodes or over 10,000 total computing cores. In this manuscript we omit the discussion of the infrastructure capable of supporting the large-scale calculations deployed during this work, however, we note that this work serves as another proof point demonstrating the viability and readiness of cloud computing for the high-performance computing workloads involved in the first-principles computational materials modeling[24].

IOP Conf. Series: Journal of Physics: Conf. Series **1290** (2019) 012011 doi:10.1088/1742-6596/1290/1/012011

2.5. Data access and repeatability

The data about the materials, workflows, batch jobs for each material, and the associated properties, including the files for each step of the simulation workflows are all available online.[14, 25]. Readers may freely access the results. Alternatively, in order to reproduce the results, the readers may create an account, copy materials and/or workflows to their account collection, and recreate the simulations for these materials. Optionally, the readers may also adjust the model parameters, and improve upon our results. For further assistance on how to do that the readers may consult the online documentation[26].

3. Results

We present the comparison of all calculated band gaps within GGA and HSE with their experimental values[27, 28] for the materials where the experimental data is available in Fig. 3. We also include the results of Materials Project[3] (further referred to as MP) calculated within GGA and GGA+U approaches for reference (further referred to as MP-GGA and MP-GGA+U, respectively). We organize and make all data available online for the readers at this link[25]. As expected, the GGA underestimates the band gaps, and HSE significantly improves the results. We fit the results with a linear regression fit, as shown in Fig. 2. It can be seen that for a simple $y = kx + b$ relation the resulting values for the model-wise errors based on the coefficient of proportionality k are: GGA - 37%, HSE - 22%. We proceed to discuss the results per each difficulty level below and focus on the "outlier" cases where the results deviate significantly from their expected values.

3.1. Difficulty 1 (D1)

Fig. 3 shows the band gaps of materials for difficulty level 1. Overall, our results are in good agreement with MP for this difficulty level, with minor differences related to the inclusion of semi-core pseudopotentials in our case. The average error for GGA calculations is 46.2%. HSE calculations improve the band gap and reduce the average error to 13%. For MgF_2, with the experimentally found gap of 10.8 eV[27] is an outlier case. Both GGA and MP-GGA underestimate the value by 34.5% and 36.8%, respectively, and HSE reduces the band gap error to 8%. At the same time, the lattice constant of MgF_2 matches within 0.04% of the experimental values[29].

3.2. Difficulty 2 (D2)

Fig. 4 compares the calculated band gaps of D2 materials with their experimental values. Here we see good agreement with MP results as well. Within GGA the average error is 43.4%. For HSE the average error is reduced to 27.6%. SrSe is an important outlier case. The experimental gap of SrSe is 4.45 eV[27]. The lattice constants of the relaxed structure match experiment within 0.1% [30]. Both GGA and MP-GGA underestimate the band gap by 51.5% and 48.3%, respectively. On the contrary, HSE calculation with the default 25% exact exchange overestimates the band gap by 48.3%. This suggests that the mixing parameter needs to be adjusted in order to predict the gap correctly.

For Ca_2Si, both GGA and HSE calculations predict zero band gap (with a semi-metallic nature of the band structure) while MP predicts a band gap of 0.29 eV. We include semi-core electronic states for both elements in our calculations (titled "pv" for Ca and "sv" for Si), MP calculations do so only for Ca (titled "sv"). In addition, the lattice constants b and c of our relaxed structure are 9.9% and 3.4% smaller compared to the MP result. All of this contribute to the reduced value of the band gap in our case.

3.3. Difficulty 3 (D3)

The results for the D3 category are shown in Fig. 5. For most of the materials the GGA band gaps agree well with the MP-GGA values. We concentrate on discussing the outliers below.

PtS_2 and $PtSe_2$ are layered two-dimensional (L2D) materials with experimental band gap values of 0.75 eV and 0.10 eV, respectively. The MP-GGA band gaps for PtS_2 and $PtSe_2$ are overestimated by 105.7% and 727.4% compared to the experimental values. As the interaction between the layers of L2D materials is dominated by the van-der-Waals (vdW) forces, the inclusion of vdW correction is necessary to calculate the structural parameters accurately. It is our understanding that due to the absence of vdW correction within MP, the lattice constant c is overestimated by 24% and 20.4%[31], respectively, which leads to the overly large band gaps. Our GGA calculations predict PtS_2 and $PtSe_2$ to be metallic. The lattice constant c within our calculation is underestimated compared to the experimental value by 13.5% and 8.56% only, which we believe closes the band gaps in both. We suspect that the DFT-D2 correction incorporated in our calculation does not fully capture the vdW interaction in these materials[32, 33].

HgS has an experimental gap of 0.70 eV[27]. GGA and MP-GGA both overestimate it by 16.1% and 143.8%, respectively. The lattice constant a within GGA is 8.9% smaller than MP-GGA value which can account for the band gap difference. Ag_2S has an experimental band gap of 1.03 eV. Our calculations predict Ag_2S to be metallic and MP-GGA predicts a band gap of 1.36 eV. The kinetic energy cutoff parameters used in our work are same as in MP calculations. The final relaxed structure in our case, however, is different than that of MP, perhaps due to the inclusion of spin-orbit coupling and vdW correction. The final total energy of the structure is 0.29 eV/atom lower in our case, which might indicate a different stable phase.

Pr_2O_3 and CdSe have experimental band gaps of 0.84 eV and 1.87 eV[27], respectively. MP-GGA overestimate the values by 345.9% and 92%. While MP used pseudopoential titled "Pr 3" which has 11 electronic states in valence, we used the default pseudopotential (titled "Pr default") with 13 states in valence and smaller cutoff radius. We include a further discussion on this in section 4). The band gap within MP is calculated on a coarse grid of k-points which might also contribute to the error. For CdSe, the lattice constant in the calculation matches the experiment within 1% [34]. MP shows the band gap as 3.58 eV while the band structure plot suggests a gap value of about 0.5 eV at the moment of this writing.

Te_2Ru, P_2Pd and As_2Pt have experimental band gaps of 0.25 eV, 0.65 eV and 0.55 eV, respectively. HSE predictions for these materials are 0.92 eV, 1.11 eV and 0.70 eV, and hence are overestimated by 267.3%, 70.6% and 27.3%. It appears that in order to get accurate gap values for these materials within HSE, the tuning of the exact exchange mixing parameter is needed.

3.4. Difficulty 4 (D4)

Fig. 6 compares the calculated band gaps with their experimental values for the D4 category. Our results agree well with that of MP except for the following cases. PbS has an experimentaly found gap of 0.28 eV. MP-GGA result for this material is 0.47 eV, whereas our GGA calculation predicts PbS to be semi-metallic with no bands crossing. The lattice constant of the relaxed structure of PbS is 1.5% smaller compared to MP-GGA which might lead to no gap in our case. CsF has an experimental gap of 10 eV. The band gap predicted by GGA and MP-GGA are 5.03 eV and 5.26 eV, respectively which are significantly underestimated compared to the experimental value by 49.7% and 47.4%.

3.5. Difficulty 5 (D5)

Fig. 7 has the band gaps for category D5 materials. FeO and CoO have experimental band gaps of 2.10 eV and 2.8 eV[35], correspondingly. In our GGA calculations, these materials are predicted to be metallic. Inclusion of GGA+U within MP leads to non-zero band gaps. For CoO, for example, the MP-GGA+U result leads to an error of 74%. Within our HSE calculation, the error is reduced to 13.2%.

For MnF_2 the experimental band gap is 9.90 eV[27]. Our GGA calculation predicts 2.26 eV with an error of 77.2%. The MP-GGA+U value is close to ours at 2.87 eV. HSE produces the estimate of 3.6 eV, still with a 63% error. The large difference between the experimental and calculated band gaps indicate that the experimental value might be extracted for a different polymorph of MnF_2. For instance, MP has three polymorphs of this material and the values for the energy above Hull are within 47 meV of each other for all.

4. Discussion

We meant this study as a data-centric benchmark of the ability of the current generation of pseudopotential density functional theory (DFT) to predict the electronic properties of materials. We also focused our attention on how it can be applied in an accessible way with minimal additional computational setup (i.e. no specialized hardware or compilation routines). We elaborate on the results of our prior work[2] and report the results for a significantly larger number of compounds.

4.1. Fidelity and error analysis

4.1.1. Comparison with experimental results When comparing with the available experimental data we point out some important conditions used within our approach that are known to affect the calculation results. Similar to our prior study[2], we introduce a van-der-Waals (vdW) correction as implemented in VASP[36, 22], and conduct the structural relaxation within the GGA and subsequently use the resulting structure for HSE calculations. Fig. 9 has the data about the average errors per each category. The width of each column is proportional to the number or materials.

Unlike the prior study, where we manually assembled the experimental results, here we programmatically collect the data using an online platform[37]. At the moment of this writing this platform only contains the stoichiometric data and lacks the structural parameters enough to uniquely identify the materials. We therefore apply filters and remove from consideration some data points as described further. We start with 231 total experimental data entries[27]. We completely remove from comparison the 23 entries where GGA results are over 25% larger than the experimental values. In addition, when calculating the averages, we remove the 2 cases where the HSE values are over 150% larger than experimental: Sb_2Os (0.2 eV), Te_2Ru (0.25 eV). The resulting average values for the errors are 49% and 22% for GGA and HSE over the sets of 215 and 75 total experimental entries, correspondingly.

4.1.2. Comparison with MP data We conduct a comparison with the results of MP for all calculated values within GGA (and exclude the MP-GGA+U). The top and bottom parts of Fig. 4 illustrate the absolute and relative band gap differences. The absolute difference E^{BG}_{abs} is defined as:

$$E^{BG}_{abs} = sign(E_{bg} - E_{MP}) \times min(E_{bg} - E_{MP}, 1.0), \tag{1}$$

where we intentionally limit the maximum difference to 1.0 eV for the visualization purposes. E_{bg} is the band gap calculated in this work, E_{MP} is the value from MP, and $sign$ and min are the sign and minimum functions.

As can be seen from E^{BG}_{abs}, generally our values match closely with MP-GGA gaps. The cases when the values are different were discussed in the previous section and are generally attribute to one of the following: (a) the difference in pseudopotentials used, eg. inclusion of the semi-core states in this work, which leads to the somewhat better GGA results for band gaps (this is very evident for Ge, for example, in our prior study[2]); (b) inclusion of the van-der-Waals correction in our calculations, which leads to the shortened structural parameters, and, therefore, smaller band gaps. Due to the inclusion of vdW correction and spin-orbit coupling the relaxed structures in our calculations often have smaller lattice constants compared to that of MP-GGA.

The gap value correction also occurs for materials containing lanthanides Nd, Pr, Eu, Gd, Ho, Dy. For these compounds our calculations predict metallicity whereas MP suggests the presence of a gap. Such a discrepancy leads to the relative difference of 100%. It appears as a common practice to use pseudopotentials with a reduced number of states in valence for the elements in the Ce-Lu row for VASP calculations in order to reduce the computation time, and MP is following this convention, which results in non-zero band gaps. In our calculations the default pseudopotentials are used where all f electrons are considered, and this fact is leading to the metallic behavior. We demonstrated it clearly for the HoN, for example (and included the comparison in the data available online): our calculations reproduce the MP gap value of 0.087 eV (also notably rather small compared to the smearing employed for the electronic occupations) when a "Ho_3" pseudopotential is used, for the "default" Ho pseudopotential we get no gap.

Another important note can be derived from KF_2 and RbF_2 that are extracted as metals in our calculation, although both are semiconducting in MP results. The Fermi level for both compounds is calculated to be close to the top of the conduction band, however, and we believe that both compounds should indeed be perceived as wide-gap semiconductors with a gap value in the 5-7 eV range. MP results suggest values below 2 eV and, perhaps erroneously, ferromagnetic ordering in these compounds. We attribute the difference in the gap values to the absence of magnetic treatment in our case, and to the pseudopotentials used for alkali metals with "p" and "s/p" states in valence in our and MP case, respectively.

4.2. Further improvements to accuracy

One way to improve the accuracy of the HSE calculations would be to use an adjustment scheme where first the improved value for the mixing parameter is calculated per material based on a statistical model, and then a single HSE calculation is executed. This work suggests that within the HSE the band gaps are significantly overestimated for the following materials (experimental values in braces): Sb_2Os (0.2 eV), Te_2Ru (0.25 eV), As_2Pt (0.55 eV), P_2Pd (0.65 eV), TiS_3 (1.30 eV), FeP_2 (1.60 eV), Mg_3As_2 (1.60 eV), $SiSe_2$ (2.40 eV), and SrSe (4.45 eV). Finding a robust way to classify the outliers with respect to a set of descriptors based on the atomistic data would provide the next step in building such a scheme.

Alternatively, another way to improve the accuracy of the results would be to use a dynamically self-consistently adjustable value for the HSE mixing parameter similar to how it is done in [38]. This approach would be more computationally intensive as it requires the convergence of the static dielectric constant with respect to the mixing parameter to be achieved during the calculation. As we demonstrated earlier[2], the GW approximation, both in the G0W0 and in the self-consistent implementation can be used efficiently in order to further improve the accuracy of band gap predictions.

4.3. Computational time and cost

Fig. 9 has the data about the average calculation runtime per each category for both GGA and HSE. Similar to our prior consideration from [2], in order to provide an insight into the feasibility of improved accuracy approaches, we construct a simple logarithmic regression using the data obtained for the GGA and the HSE results. We assume that the average simulation lifetime increases exponentially as the average error is dropping. As can be seen from Table 4.3, within this logic one would need to run a simulation for about 30 days on average in order to produce an exact result.

Our motivation for the above is to provide a metric of the extent to which the physics-based first-principles modeling can augment the trial-and-error experimental approach when compared with respect to the capital and time investments required. We suggest that for the equivalent of one month of calculation time (human time) on a commodity compute server readily available from a cloud provider it is possible to obtain results that are accurate well within 20% and potentially within 1-5% range for the properties that we study in the current work.

Model	Avg. Err, (%)	Avg. time	Cost ($)	Note
Exact	0	30 days	5,000	extrapolated
HSE	24	35 hrs	200	factual
GGA	46	1.2 hrs	10	factual

Table 2. Average errors and the associated average calculation time (human time) for the HSE and GGA cases studied in this work. *Exact* and *Zero* values are constructed through a simple logarithmic fit of the HSE/GGA data for the (hypothetic) models that would produce exact and zero-fidelity results correspondingly. NOTE: only the subset of materials where HSE values are also available is used to calculate the average error for GGA case shown above.

4.4. Future outlook

Computational materials design is rapidly evolving toward a data-driven science where the modeling results are aggregated and classified by their precision/accuracy, as the recent work from NOMAD demonstrates[7]. Major improvements in the way computational materials science is used would be possible when increased veracity of this data also becomes a norm. The approach described in this work can assist with achieving this goal.

Modeling workflows accessible in a standardized and repeatable way let the field evolve away from the "medieval artisan-like" model[6] still prevalent nowadays. This work serves as a proof that precision within 20%, perhaps only for the electronic materials at this moment, is readily achievable using existing first-principles modeling techniques.

Our intent is to welcome collaborative contributions in order to, firstly, further grow the online repository of high fidelity results; secondly, allow contributions from other high-fidelity modeling techniques beyond studied here; and, finally, facilitate the creation of statistical (machine learning) models based on the available data.

5. Conclusions

We present the applications of a novel approach to materials modeling from nanoscale implemented within the Exabyte platform[1] and capable of delivering both high fidelity and high throughput in an accessible and data-centric manner to a set of 775 binary semiconducting compounds. We report the results for the electronic band gaps obtained within the Generalized Gradient Approximation (GGA) for the full dataset and with Hybrid Screened Exchange (HSE) for a subset of 270 materials. We analyze the level of fidelity for the predictions delivered by each of the models used, compare the results with experimental data and prior similar calculation attempts, when available, and discuss the corresponding computational costs and pathways to further improved accuracy.

We find the average relative error in the estimates for the electronic band gaps obtained in this work to be 22% for HSE and 49% for GGA, respectively. We further find the average calculation time on a current up-to-date compute server centrally available from a public cloud provider to fit within 1.2 and 36 hours for GGA and HSE, correspondingly. We present not only the results and the associated data, but also an easy-to-access way to reproduce and extend the results by means of the Exabyte platform.[14] Our work provides an accessible and repeatable practical recipe for performing high-fidelity first-principles calculations of the electronic structural properties of materials in a high-throughput manner.

6. References

[1] 2018 *Exabyte.io: materials discovery cloud* URL `https://exabyte.io/`
[2] Das P, Mohammadi M and Bazhirov T 2018 *arxiv.org/abs/1807.05623*
[3] Jain A, Ong S P, Hautier G, Chen W, Richards W D, Dacek S, Cholia S, Gunter D, Skinner D, Ceder G *et al.* 2013 *Apl Materials* **1** 011002
[4] Curtarolo S, Setyawan W, Wang S, Xue J, Yang K, Taylor R H, Nelson L J, Hart G L, Sanvito S, Buongiorno-Nardelli M *et al.* 2012 *Computational Materials Science* **58** 227–235
[5] Saal J E, Kirklin S, Aykol M, Meredig B and Wolverton C 2013 *Jom* **65** 1501–1509
[6] Pizzi G, Cepellotti A, Sabatini R, Marzari N and Kozinsky B 2016 *Computational Materials Science* **111** 218–230
[7] 2018 *NOMAD laboratory: A European Centre of Excellence* URL `https://www.nomad-coe.eu/`
[8] Haastrup S, Strange M, Pandey M, Deilmann T, Schmidt P S, Hinsche N F, Gjerding M N, Torelli D, Larsen P M, Riis-Jensen A C *et al.* 2018 *arXiv preprint arXiv:1806.03173*
[9] Rasmussen F A and Thygesen K S 2015 *The Journal of Physical Chemistry C* **119** 13169–13183
[10] Perdew J P, Burke K and Ernzerhof M 1996 *Physical review letters* **77** 3865
[11] Hohenberg P and Kohn W 1964 *Phys. Rev.* **136** B864–B871
[12] Ihm J, Zunger A and Cohen M 1979 *Journal of Physics C: Solid State Physics* **12** 4409
[13] Heyd J, Scuseria G E and Ernzerhof M 2003 *The Journal of chemical physics* **118** 8207–8215
[14] 2018 *Exabyte platform: project URL with data about simulations* URL `https://platform.exabyte.io/exabyte-io/projects/exabyte-io-2018-bg-study-phase-iii/`
[15] Kresse G and Furthmüller J 1996 *Physical review B* **54** 11169
[16] Kohn W and Sham L J 1965 *Physical review* **140** A1133
[17] Blöchl P E 1994 *Physical review B* **50** 17953
[18] Kresse G 1996 *Phys. Rev. B* **54** 169
[19] Hacene M, Anciaux-Sedrakian A, Rozanska X, Klahr D, Guignon T and Fleurat-Lessard P 2012 *Journal of computational chemistry* **33** 2581–2589
[20] Krukau A V, Vydrov O A, Izmaylov A F and Scuseria G E 2006 *The Journal of Chemical Physics* **125** 224106 (*Preprint* `https://doi.org/10.1063/1.2404663`) URL `https://doi.org/10.1063/1.2404663`
[21] Johnson S G and Joannopoulos J D 2001 *Optics express* **8** 173–190
[22] Grimme S 2006 *Journal of computational chemistry* **27** 1787–1799 URL `http://www.ncbi.nlm.nih.gov/pubmed/16955487`
[23] *Microsoft Azure Cloud Computing platform: web page.* URL `https://docs.microsoft.com/en-us/azure/virtual-machines/virtual-machines-linux-sizes`
[24] Mohammadi M and Bazhirov T 2018 Comparative benchmarking of cloud computing vendors with high performance linpack *Proceedings of the 2Nd International Conference on High Performance Compilation, Computing and Communications* HP3C (New York, NY, USA: ACM) pp 1–5 ISBN 978-1-4503-6337-2 URL `http://doi.acm.org/10.1145/3195612.3195613`
[25] 2018 *Exabyte platform: data sheet all band gap values discussed in this manuscript* URL `https://exabyte.io/data/bg-study-2018-ph-iii`
[26] 2018 *Exabyte.io: online documentation* URL `https://docs.exabyte.io/`
[27] Strehlow W and Cook E 1973 *Journal of Physical and Chemical Reference Data* **2** 163–200
[28] Deml A M, Holder A M, OHayre R P, Musgrave C B and Stevanovi V 2015 *The Journal of Physical Chemistry Letters* **6** 1948–1953 pMID: 26263275 (*Preprint* `https://doi.org/10.1021/acs.jpclett.5b00710`) URL `https://doi.org/10.1021/acs.jpclett.5b00710`
[29] Vidal-Valat G, Vidal J P, Zeyen C M and Kurki-Suonio K 1979 *Acta Crystallographica Section B* **35** 1584–1590
[30] Primak W, Kaufman H and Ward R 1948 *Journal of the American Chemical Society* **70** 2043–2046
[31] Furuseth S, Selte K and Kjekshus A 1965 Redetermined crystal structures of nite2 pdte2 pts2 ptse2 and ptte2
[32] Zhao Y, Qiao J, Yu P, Hu Z, Lin Z, Lau S P, Liu Z, Ji W and Chai Y 2016 *Advanced Materials* **28** 2399–2407
[33] Zhang W, Qin J, Huang Z and Zhang W 2017 *Journal of Applied Physics* **122** 205701
[34] Landolt-Bornstein S 1987 111, vol. 17a intrinsrc properties of group iv elements, iii-v, ii-vi, and i-vil compounds

[35] Deml A M, Holder A M, OHayre R P, Musgrave C B and Stevanovic V 2015 *The journal of physical chemistry letters* **6** 1948–1953
[36] Harl J, Schimka L and Kresse G 2010 *Phys. Rev. B* **81**(11) 115126 URL https://link.aps.org/doi/10.1103/PhysRevB.81.115126
[37] 2018 *Citrine Informatics: Materials Data Platform* URL https://citrination.com/
[38] Skone J H, Govoni M and Galli G 2014 *Phys. Rev. B* **89**(19) 195112 URL https://link.aps.org/doi/10.1103/PhysRevB.89.195112

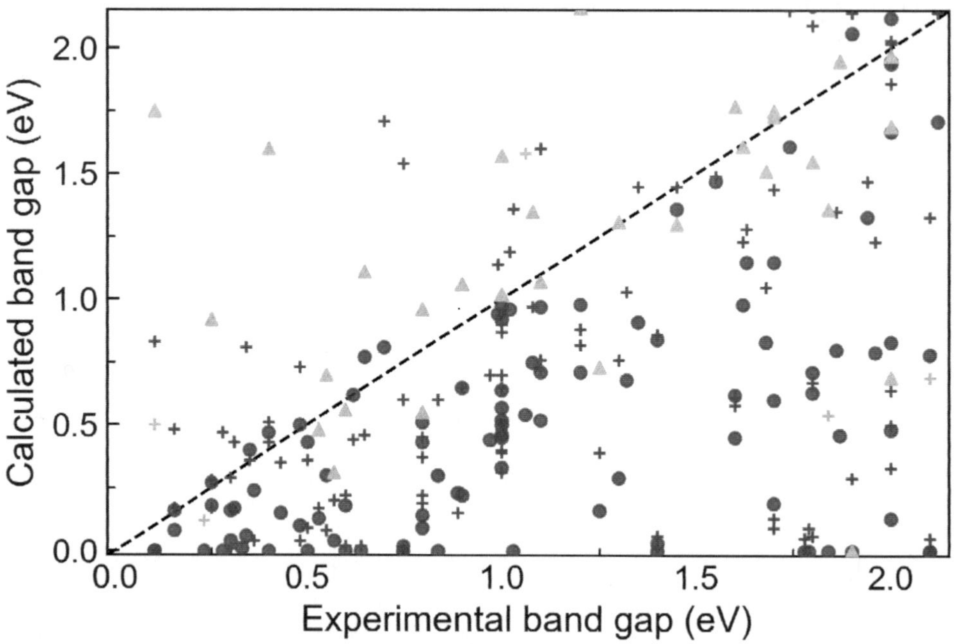

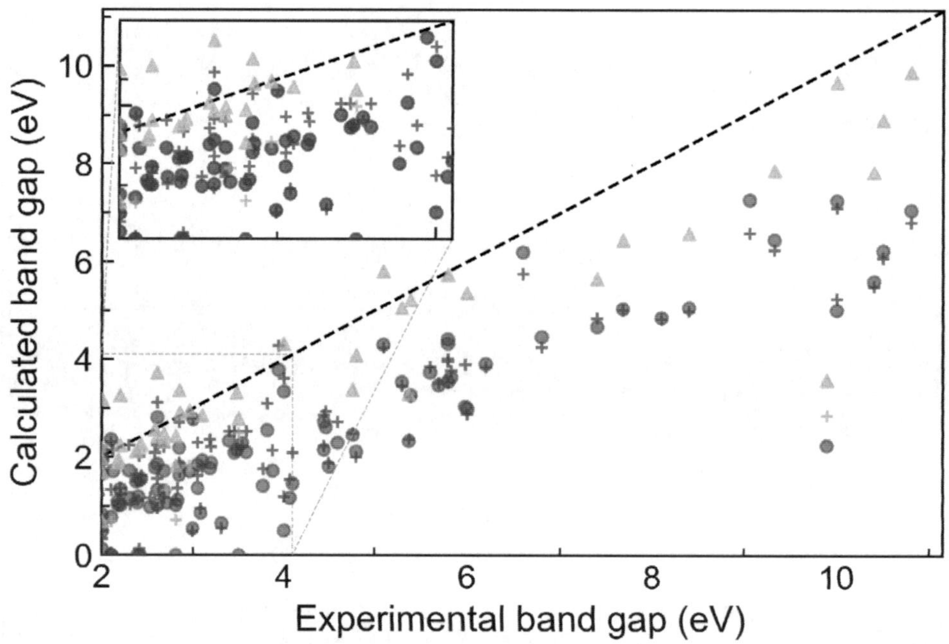

Figure 1. (Legend same as in Fig. 3) Comparative plot of the calculated and experimentally available values for the electronic band gaps obtained in the current work. Legend: GGA and HSE denote the results of this work for the corresponding level of theory. MP-GGA and MP-GGA+U denote the results of Materials Project[3] available at the moment of this writing and calculated within the GGA and GGA+U approaches, respectively. The legend is the same as Fig. 3. We separate the materials with the experimental gaps in the 0-2 eV range into the top sub-figure. The compounds with gaps larger that 2 eV are shown in the bottom sub-figure. The latter also has an inset where the 2-4 eV region is expanded for better visibility.

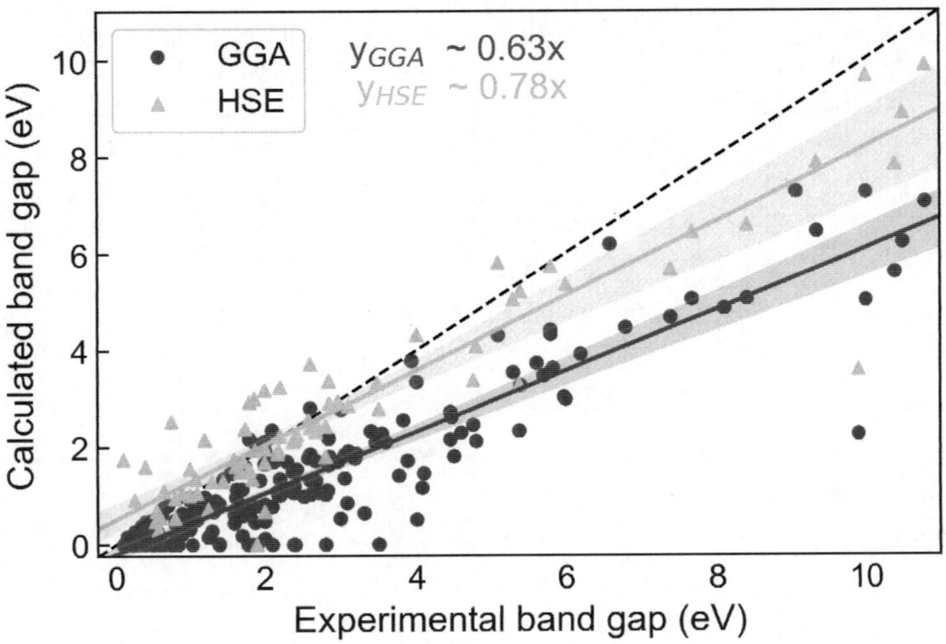

Figure 2. Comparative plot of the calculated and experimentally available values for the electronic band gaps obtained in the current work We include a linear $y = kx + b$ fit to data per each model (GGA, HSE). The legend is same as in Fig.3. The proportionality coefficients for each of the linear fits are shown in the figure.

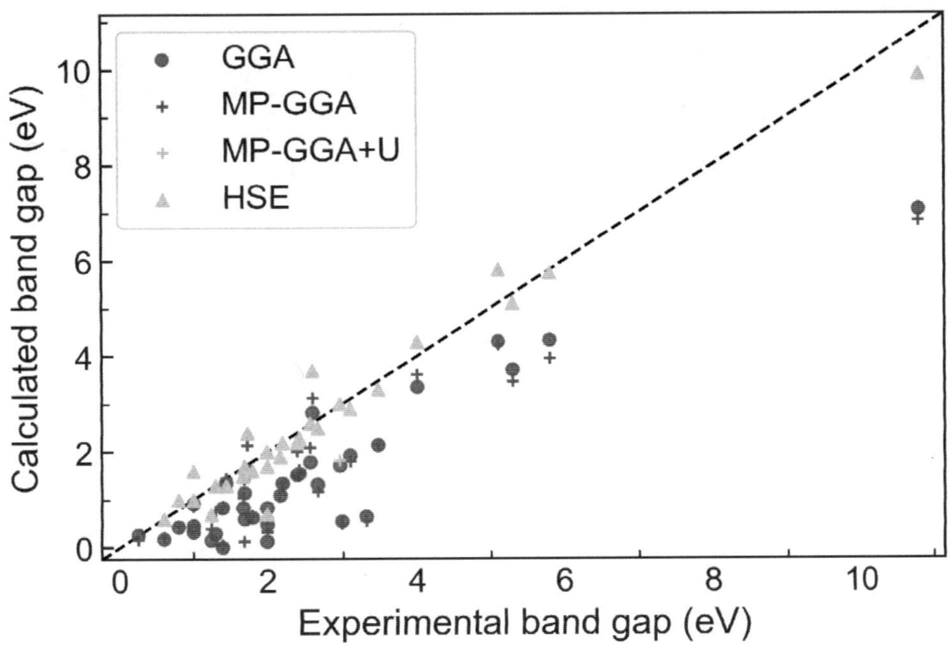

Figure 3. Comparative plot of the calculated and experimentally available band gap values for the materials in the D1 category.

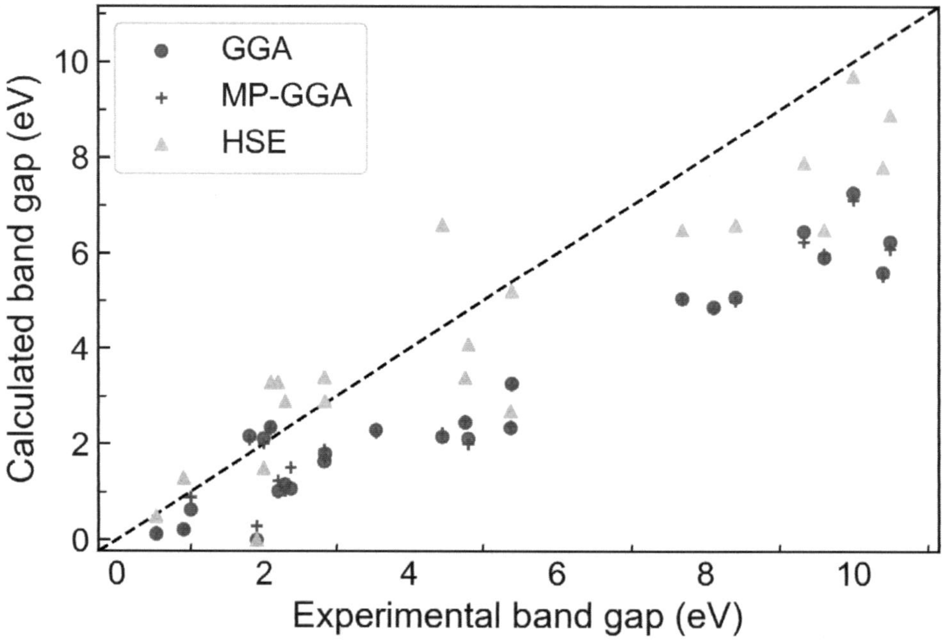

Figure 4. Same as Fig. 3 for the D2.

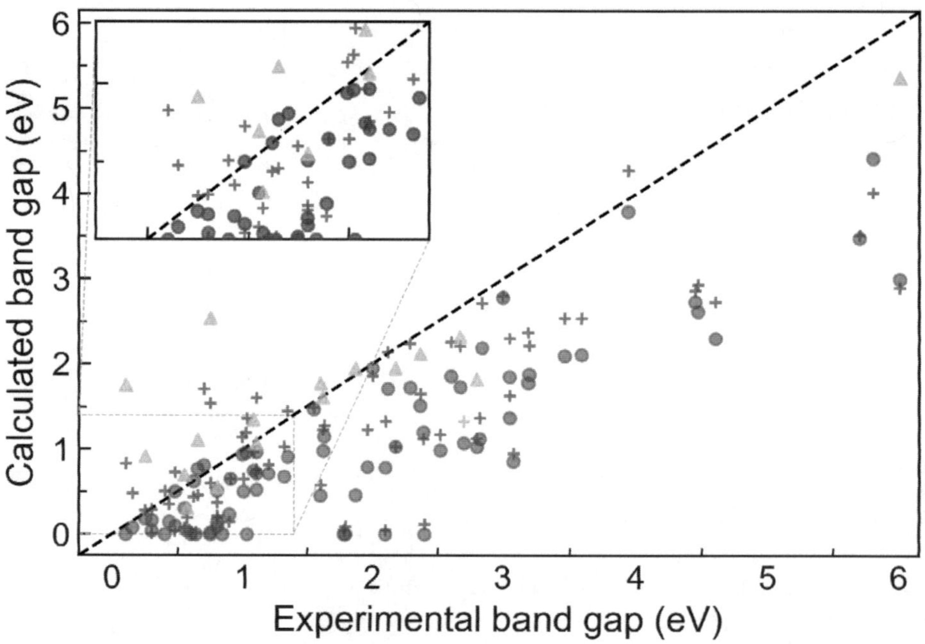

Figure 5. Same as Fig. 3 for the D3. Legend follows Fig. 3.

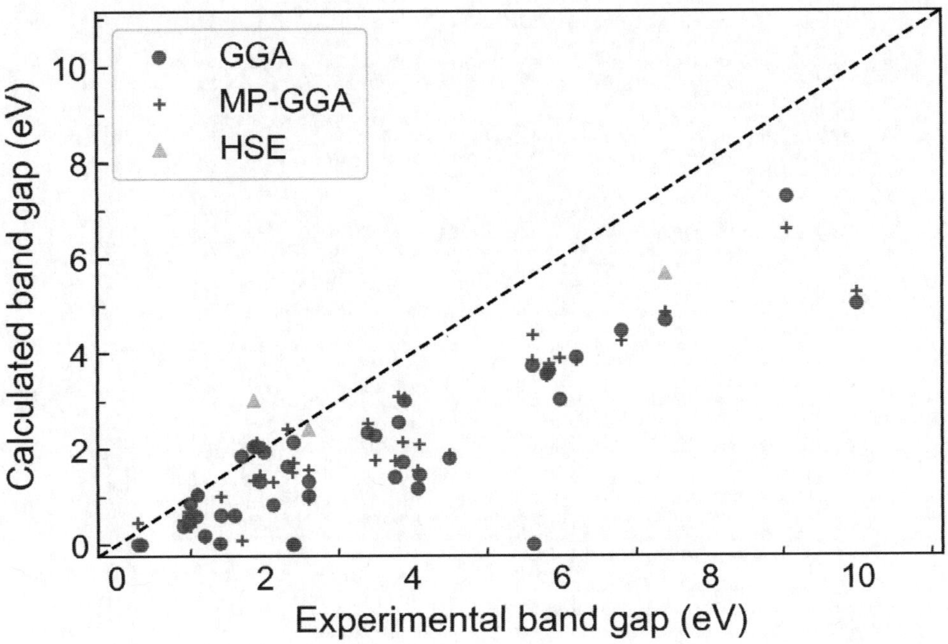

Figure 6. Same as Fig. 3 for the D4.

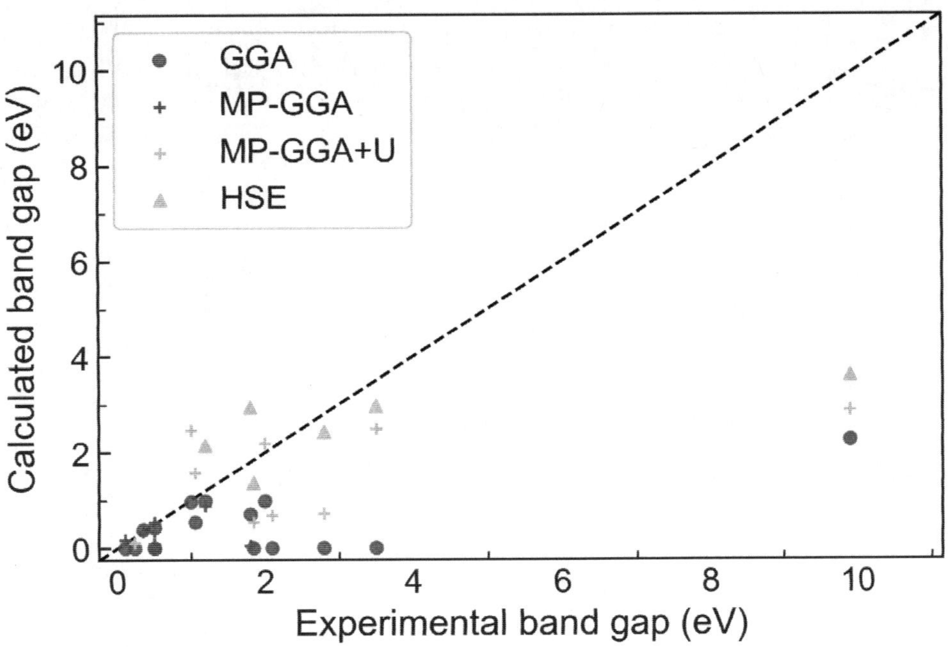

Figure 7. Same as Fig. 3 for the D5.

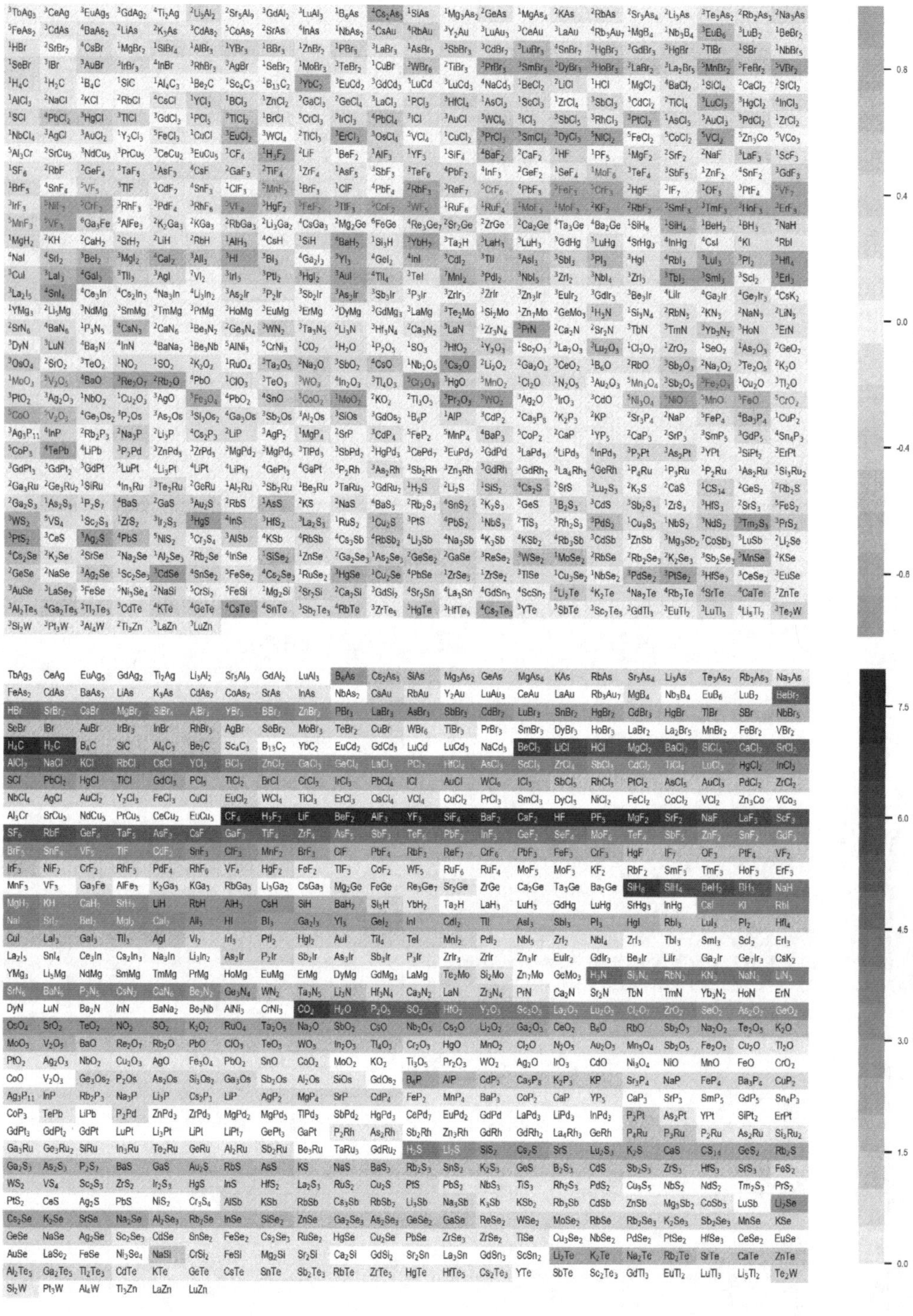

Figure 8. (TOP) Heatmap plot for E_{abs}^{BG}, the absolute difference in the band gap values calculated within GGA in this work and MP[3], as defined in Eq. 1. Chemical formulas for compounds where MP results use GGA+U are highlighted in violet. (BOTTOM) Same for the GGA band gap results of this work. Both plots use same compounds order sorted by the second element in formula and the gap value starting from the top right corner. Superscripts denote the corresponding difficulty levels as defined in the methodolody section.

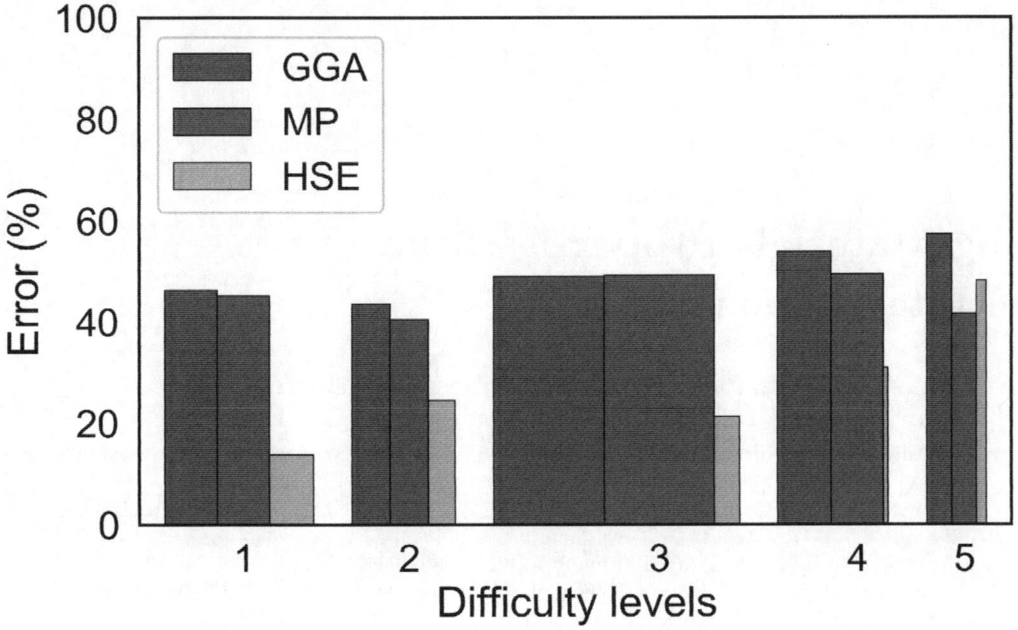

Figure 9. Difficulty-wise average errors. The width of the bars are proportional to the number of materials in category.

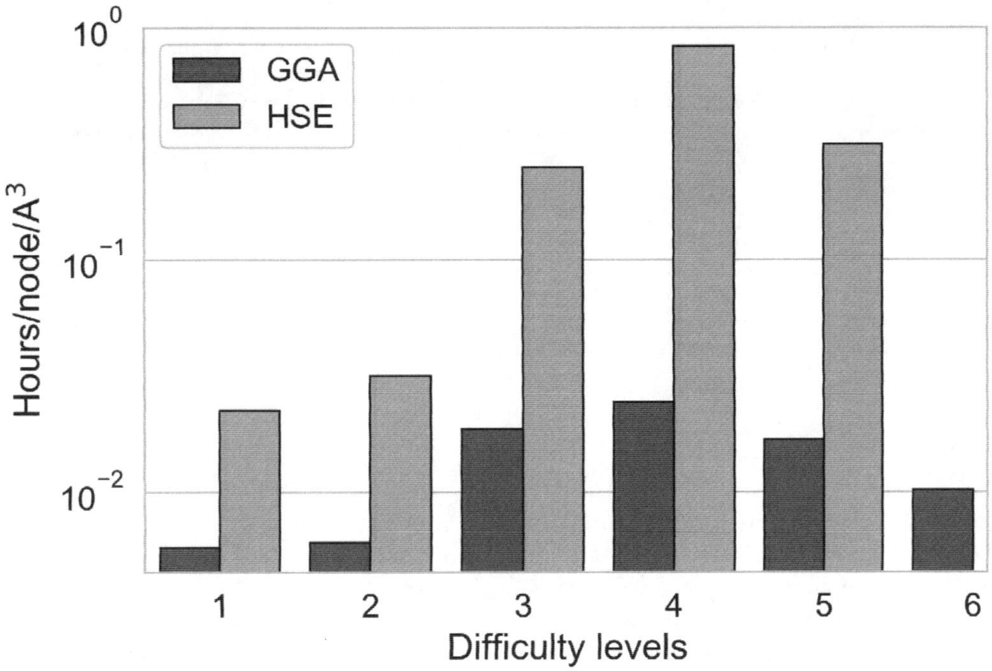

Figure 10. Calculation time per each difficulty level (as defined in section 2). The time is normalized per one compute node and unit cell volume (Å^3).

IOP Conf. Series: Journal of Physics: Conf. Series **1290** (2019) 012012 doi:10.1088/1742-6596/1290/1/012012

Learning Materials Properties from Orbital Interactions

Tien-Lam Pham[1,2,3], Tran-Thai Dang[1], Van-Doan Nguyen[2], Hiori Kino[2,4], Takashi Miyake[2,3,4], and Hieu-Chi Dam[1,2,5]

[1]Japan Advanced Institute of Science and Technology, 1-1 Asahidai, Nomi, Ishikawa 923-1292, Japan

[2]Center for Materials Research by Information Integration, NIMS, 1-2-1 Sengen, Tsukuba, Ibaraki 305-0047, Japan

[3]ESICMM, NIMS 1-2-1 Sengen, Tsukuba, Ibaraki 305-0047, Japan

[4]CD-FMat, AIST, 1-1-1 Umezono, Tsukuba 305-8568, Japan

[5]JST, PRESTO, 4-1-8 Honcho, Kawaguchi, Saitama 332-0012, Japan

E-mail: `phamtlam@jaist.ac.jp`

Abstract. Orbital field matrix (OFM) descriptors were developed with an emphasis on atomic orbitals for representing material structures in datasets of multi-element compounds. The descriptors were based on atomic valence shell electrons and their coordination. In addition to original OFM and OFM1 which is OFM with a column representing information on the center atom, in this work, we present another version, named OFM0, which is OFM1 without information on atomic distances, for predicting the properties of unoptimized structures. We focus on formation energy and phase stability of crystalline systems, while the atomization energy is examined for molecules. With the emphasis on the ability to identify materials with similar properties, here, the applicabilities of OFM, OFM1, and OFM0 are systematically examined with decision tree (DT) regression, random forest (RF) regression, and kernel ridge regression (KRR). We show that the family of OFM descriptors are highly capable to build predictive models for the properties of solids and molecules. The accuracy of a DT and a forest of trees (RF) is comparable to that of the KRR models. The KRR with a Laplacian kernel estimated by OFM1 yields the most accurate predictions, with the formation energy, phase stability, and atomization energy having mean absolute errors (MAEs) of 0.072 eV/atom, 0.059 eV/atom, and 6.74 kcal/mol, respectively. The OFM0 without atomic distances also yields acceptable predictions with respective MAEs of 0.090 eV/atom, 0.069 eV/atom, and 7.77 kcal/mol. The results imply that our descriptors are highly useful to find similar materials.

1. Introduction

In the last 50 years, there has been an exponential increase in the amount of materials science data, including both experimental and quantum-computational data. Together with the development of machine-learning techniques, this has created a new research field of materials informatics, which is an integration of machine-learning, or artificial intelligence, and materials science. Machine-learning algorithms have been demonstrated to automatically extract knowledge of materials, such as patterns and chemical and physical rules, using both first-principles-calculated and experimental data [1, 2, 3, 4, 5, 6, 7, 8, 9]. Materials science aims to discover new materials for specific applications and to provide insight into their properties by first searching for hidden variables or correlations (including linear and non-linear correlation), then building theoretical models to predict future outcomes. Machine-learning algorithms, which search for latent structure, hidden correlation, predictive models, etc, can thus enable the discovery of hidden knowledge from a material dataset.

In a material dataset, a material is identified by a set of atoms with their coordinates and periodic unit-cell vectors which are required for crystalline systems. However, this primitive structural information depends on the reference frame and the number of atoms. This is not feasible for machine-learning algorithms. To apply well-established machine-learning methods, such as predictive and descriptive learning, an appropriate transformation from the primitive structural information to vectors or matrices (i.e., material descriptors), is required. Various methods for encoding materials have been developed in so far. Behler et al. [10, 11, 12, 13, 14, 15] utilized atom-distribution-based symmetry functions to represent the local chemical environment of atoms, and employed a multilayer perceptron to map this representation to atomic energy. The arrangement of structural fragments has also been used to encode materials to predict physical properties of molecular and crystalline systems [16, 5]. Isayev used band structure and density-of-states fingerprint vectors as a representation to visualize the material space [5]. Rupps et al. developed the Coulomb matrix descriptor to predict atomization energies and formation energies [17, 18, 19].

Recently, we have proposed a family of descriptors named orbital field matrix (OFM) which incorporate the valence electron configuration to describe both the local and the entire material structure. It was based on the consideration that certain essential aspects of electronic structures could be deduced from a simple description of the valence electrons surrounding a central atom [20]. An extended version of OFM, named OFM1, [9], was proposed, which contained a column explicitly representing information on center atoms. To extend the capability of our descriptors to new (unoptimized) structure, here, we present another version of OFM1 (OFM0) that does not include information on atomic distances. In this paper, we extend our previous studies [9, 20] and focus on the systematic investigation of the applicabilities of OFM, OFM1, and OFM0 descriptors for the prediction of formation energies and phase stabilities of solid systems, and atomization energies of molecules with the emphasis on identifying materials with similar properties (the formation energy, phase stability, or atomization energy). We studied OFM, OFM1, and OFM0 to predict the properties by using kernel ridge regression, decision tree regression, and random forest regression. With the simple decision tree, an accuracy was obtained that was comparable to that for KRR with a Gaussian kernel; whereas, the random forest regression exhibited a slight improvement. The KRR with a Laplacian kernel estimated by OFM1 yielded the most accurate prediction for formation energy, phase stability, and atomization energy with mean absolute errors (MAEs) of 0.072 eV/atom, 0.059 eV/atom, and 6.74 kcal/mol, respectively. OFM0 without distances of the center atom to neighboring atoms also gave acceptable accuracy with respective MAEs of 0.090 eV/atom, 0.069 eV/atom, and 7.77 kcal/mol. Hence, this descriptor is expected to be highly useful for predicting a hypothesized structure that has not been optimized. The accurate predictions of our descriptors implies that they are highly useful in finding similar materials base on distance measurements and decision trees.

2. Methodology

2.1. Encoding Materials: Orbital Field Matrix

To apply machine-learning methods to solve the issues of materials science, we developed the chemistry-oriented descriptors OFM and OFM1 [20, 9] for representing a multi-element material adapted to a machine-learning algorithm. In the method, a material (molecule or solid) was decomposed into local structures, which included a center atom and neighboring atoms, by a Voronoi diagram analysis [21]. A matrix, OFM, was designed to represent a local structure whose element OFM_{ij} conveyed information on the number of an atomic orbital j, coordinated to a central atomic orbital i. Here, i and j were in the dictionary comprising the valence subshell orbitals, $D = \{s^1, s^2, p^1, p^2, ..., p^6, d^1, d^2, ..., d^{10}, f^1, f^2, ..., f^{14}\}$. A one-hot row vector $\vec{O}_{atom}$ was used with the dictionary of D to encode the electron configuration of an atom. The OMF was built by using the one-hot row vector representation for the central atom $\vec{O}_{central}$ and the neighboring atoms $\vec{O}_k$, where k was the index of the neighboring atoms. The vector of the coordinating atoms was then summed to form a vector that represented the environment surrounding a central $\vec{O}_{env}$:

$$\vec{O}_{env} = \sum_k^K w_k \vec{O}_k \tag{1}$$

where the weight w_k measured the contribution of the kth neighboring atom, and K was the number of neighboring atoms. The OFM of a local structure now was:

$$OFM = \vec{O}^T_{central} \times \vec{O}_{env}, \tag{2}$$

where the column vector $\vec{O}^T_{central}$ was the transpose of $\vec{O}_{central}$.

The definition of neighboring atoms by O'Keeffe [21] was adopted; it utilized the solid angles θ_k, determined by faces of the Voronoi polyhedra. The weight w_k was determined by the solid angles and the distances to the center atom: $w_k = \frac{\theta_k}{\theta_{max}} \zeta(r_k)$, where θ_k was the solid angle determined by the face of the Voronoi polyhedra between the central atom and the neighboring atom k and θ_{max} was the maximum solid angle among those between the central atom and the neighbor atoms. To incorporate information on valence orbital sizes, the distance r_k between the central atom and the kth atom was included in $\zeta(r_k)$. The inverse of the distance was used as the distance-dependent weight function: $\zeta(r_k) = 1/r_k$.

To explicitly incorporate information on the central atom, $\vec{O}^T_{central}$ was concatenated to the OFM matrix as a new column. Finally, the form for representing a local structure was given by:

$$OFM1 = \vec{O}^T_{central} \times \left(1.0, \vec{O}_{env}\right). \tag{3}$$

To apply a machine-learning model to the new (unoptimized) material, predictive models were built for mapping the chemical composition and the prototype structure to the formation energy. Also used was the OFM1, Eq. 3, version without $\zeta(r)$. This was OFM0:

$$OFM0 = \vec{O}^T_{central} \times \left(1.0, \vec{O}^\circ_{env}\right), \tag{4}$$

where $\vec{O}^\circ_{env} = \sum_k^K \frac{\theta_k}{\theta_{max}} \vec{O}_k$. By the name of OFM0, the absence of the atomic distances was emphasized.

Composing the descriptor for a structure (molecule or crystal) from its local structure representation required careful consideration. From a data science viewpoint, the composed descriptors should include as much information as possible. However, from a materials science viewpoint, the descriptors should appropriately reflect the nature of the target physical properties. Here, for the formation energy per atom of a crystal, the mean was used over the local structure descriptors as the descriptor for the entire structure:

$$x_i = \frac{1}{N_i} \sum_j^{N_i} x_{ij}, \tag{5}$$

where i was the index of a structure, j was the index of a local structure in the structure i, x_{ij} was the OFM, OFM1, or OFM0 representation of the local structure j, and N_i was the number of atoms in the unit cell; x_i was the OFM1 representing the entire crystalline material. For the atomization energy of a molecule, which is proportional to the size of the systems and can be considered an accumulation of contributions of constituent local structures, the descriptors of molecules could also be constructed by the sum of the OFM, OFM1, or OFM0 of all atoms:

$$x_i = \sum_{j}^{N_i} x_{ij}. \tag{6}$$

2.2. Machine-learning Modeling

2.2.1. Kernel Ridge Regression

Here, the formation energy, the phase stability, and the formation energy of material were modeled by a kernel ridge regression (KRR)[22], which was a combination of the kernel method and ridge regression that has been used successfully in materials science and chemistry. In the KRR algorithm, the property of a material can be given by the weighted kernel function:

$$y = f(x, c) = \sum_{k \in D_{ref}} c_k K(x, x_k), \tag{7}$$

where k ran over all the reference data (D_{ref}), and the $K(x_i, x_j)$ kernel function measured the dissimilarity between material i and j. Here, the Gaussian and Laplacian kernels for the KRR were examined. To minimize the prediction risk, the coefficients c_k were determined by minimizing the total square error regularized by the L2 norm (ridge regression):

$$\arg\max_{c} \left(\sum_{i} [f(x_i) - y_i]^2 + \lambda \sum_{k} c_k^2 \right). \tag{8}$$

2.2.2. Decision Tree Regression

Decision trees (DTs) are a non-parametric, supervised learning method used for classification and regression. They learn simple decision rules on data features (independent variables or predictors) to predict the values of dependent variables (target variables). A tree is built to divide the feature space into J distinct and non-overlapping regions $R_1, R_2, ..., R_J$, which were defined as the leaves of the tree. For every observation that fell into the region R_j, the same prediction was made, which was simply the mean of the response values for the training observations in R_j. The tree was built to minimize the sum of square residuals $\sum_j^J \sum_{x_i \in R_j} (y_i - \hat{y}_j)^2$. Recursive binary splitting, which is a specific decision tree for numerical data, was used to build the decision tree, which was implemented in scikit-learn library [23]. After training, the DTs divide the data space into subspaces (the leaves), and the materials in the same leaf are considered as similar materials. Therefore, the accuracy of DT regression in a reasonable indication of how our descriptors can be utilized to find similar materials.

Because averaging a set of observations reduces variance, a natural way to reduce the variance and thus increase the prediction accuracy was to build a set of models trained from many training sets. The prediction of a new outcome was thus the average of predictions obtained by the models. A forest of many trees was used, i.e., the random forest algorithm, to enhance the predictability of the decision tree. K subsets of the training set were randomly selected and K DTs were trained to predict the target variables. The prediction of new data was the average of the outcomes of the K trees.

3. Learning from Crystalline Material Data: Prediction of Formation Energy and Phase Stability

The applicability of OFM, OFM1, and OFM0 for predicting the formation energy and phase stability of crystalline materials was analyzed. The focus was magnetic materials consisting of lanthanide (LA), transition metal (T), and light (X) atoms (LATX). The materials were formed by 1 or 2 LA, 1 or 2 transition T and one X. Selected were LA in {Y, La, Ce, Pr, Nd, Pm, Sm, Eu, Gd, Tb, Dy, Ho, Er, Tm, Yb, Lu}, T in {Ti, V, Cr, Mn, Fe, Co, Ni, Cu, Zn, Y, Zr, Nb, Mo, Tc, Ru, Rh, Pd, Ag, Cd, Hf, Ta, W, Re, Os, Ir, Pt, Au, Hg}, and X in {H, B, C, N, O}. 5967 LATX materials were collected with formation energies from OQMD [24, 25]. This dataset was named as the LATXb dataset. Local structures of each material in LATXb were represented by OFM, OFM1, or OFM0 descriptors calculated by using Eq. 2, 3, and 4, respectively. The mean of all the descriptors of a material was used to represent it, as in Eq. 5. The dataset was divided into the training set with 4773 materials (80 %) and the test set with 1194 materials (20 %).

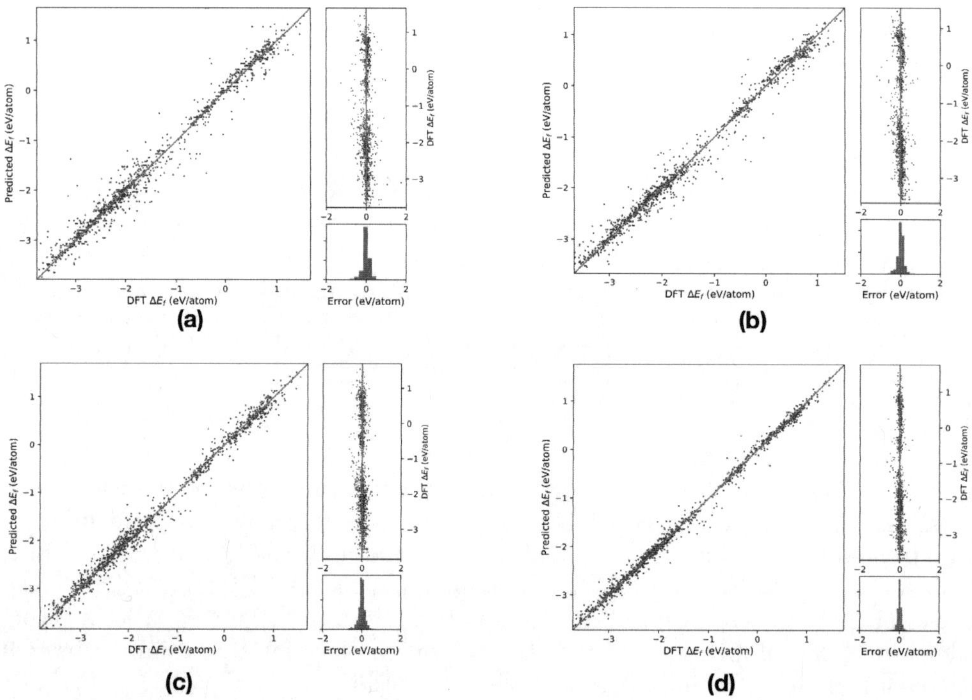

Figure 1. Comparison of machine-learning and DFT-predicted formation energy ΔE_f. Decision tree regression (a), random forest regression (b), and kernel ridge regression with a Gaussian kernel (c) and a Laplacian kernel (d) by using OFM1.

First, the KRR was examined, Eq. 7, where the x were the material descriptors calculated by OFM, OFM1, or OF0 to represent the formation energy of materials in LATXb. We investigated the Gaussian kernel with Euclidean distance:

$$K(x, x_k) = e^{-\gamma \sum_p (x_{ip} - x_{jp})^2}, \tag{9}$$

and the Laplacian kernel with Manhattan distance:

$$K(x, x_k) = e^{-\gamma \sum_p |x_{ip} - x_{jp}|}. \tag{10}$$

These kernels place a higher impact on the closer reference data x_k to the interest data, x, which are measured by the Euclidean and Manhattan distances. The accuracy of the prediction is considered as a measurement of the capability of identifying similar materials.

Table 1. The accuracy of the formation energy: RMSE (eV/atom), MAE (eV/atom), and R^2 estimated on the test set of LATXb dataset obtained using OFM, OFM1, and OFM0 descriptors.

Descriptor		DT	RF	KRR-Gaus	KRR-Lapl
	MAE	0.134	0.100	0.129	0.075
OFM	RMSE	0.250	0.186	0.188	0.148
	R^2	0.963	0.979	0.979	0.987
	MAE	0.127	0.101	0.113	0.072
OFM1	RMSE	0.242	0.162	0.160	0.109
	R^2	0.970	0.986	0.986	0.994
	MAE	0.153	0.117	0.122	0.090
OFM0	RMSE	0.239	0.189	0.171	0.137
	R^2	0.969	0.981	0.984	0.990

The hyper-parameters γ and λ were determined by a ten-fold cross-validation-grid search on the training set. The train set with 4773 materials was divided into the validation set (10%) and the sub training set (90 %) for finding model parameters c_k by minimizing the cost function, as seen in Eq. 8. The values of the hyper-parameters were selected to minimize the estimated root mean square error (RMSE) on the validation set. The accuracy of the model, the RMSE, the MAEs, and the R^2 score were estimated on the test set.

Table 1 shows that the Laplacian kernel with Manhattan distance gave a more accurate prediction than that obtained by the Gaussian kernel with Euclidean distance. Fig. 1 shows a comparison between density functional theory (DFT) and machine-learning predicted formation energy. OFM1 had the most accurate predictions, indicating the importance of the information of the center atom that was added to OFM. By using OFM1 and the Laplacian kernel, RMSE was 0.109 eV/atom, MAE was 0.072 eV/atom, and R^2 was 0.994; whereas, that of the Gaussian kernel was 0.160 eV/atom, 0.113 eV/atom, and 0.996, respectively. Interestingly, OFM0 without the distance weight, which was only determined by the shape of the material structures, also yielded a reasonably good result with a RMSE of 0.137 eV/atom, a MAE of 0.090 eV/atom, and a R^2 of 0.990. This might be a good descriptor for estimating the stability of un-optimized structures for new materials searched by virtual screening or Bayesian optimization [26].

The OFM descriptors were sparse representations with the one-hot vector encoding the electron configuration. Therefore, the pairwise distances between materials, such as Euclidean and Manhattan distances, were difficult to estimate. Hence, the descriptors may not be useful for distance-based algorithms like KRR. The tree-based methods such as decision tree and random forest regression did not require calculation of the distance between materials. In the DT regression, the critical step was to build a tree which divided the feature space into subspaces (leaves). Therefore, the tree-based methods using the descriptors were also expected to yield high-accuracy predictions.

The DT regression was investigated for the prediction of formation energy of the LATXb dataset. The depth of the tree and the number of data points per leaf were important parameters to prevent the model from over-fitting. The maximum depth of the tree (MaxD) and the minimum number of data points (MinP) in a leaf were determined with a grid search on the training set using 10-fold cross-validation. For formation energy, the optimum values of MaxD and the MinP were 40 and 5, respectively. With this simple DT regression and OFM1, a reasonable accuracy was obtained with RMSE of 0.242 eV/atom, MAE of 0.127 eV/atom, and R^2 of 0.970. This prediction was comparable to the KRR using a Gaussian kernel. Similar to

IOP Conf. Series: Journal of Physics: Conf. Series **1290** (2019) 012012 doi:10.1088/1742-6596/1290/1/012012

Table 2. The prediction accuracy of the phase stability: RMSE (eV/atom), MAE (eV/atom), and R^2 estimated on the test set of the LATXb dataset using the KRR with the Laplacian kernel (KRR-Lapl), the KRR with the Gaussian kernel (KRR-Gauss), and the DT and RF regressions. The values in parentheses were obtained by the machine models of formation energy; whereas, values outside the parentheses were given by machine-learning models directly representing the phase stability.

Descriptor		DT	RF	KRR-Gaus	KRR-Lapl
OFM	MAE	0.106 (0.129)	0.082 (0.114)	0.110 (0.117)	0.065 (0.077)
	RMSE	0.199 (0.225)	0.140 (0.181)	0.152 (0.166)	0.100 (0.116)
	R^2	0.849 (0.819)	0.913 (0.868)	0.907 (0.889)	0.961 (0.946)
OFM1	MAE	0.089 (0.133)	0.081 (0.118)	0.099 (0.113)	0.059 (0.072)
	RMSE	0.172 (0.221)	0.140 (0.179)	0.139 (0.160)	0.092 (0.109)
	R^2	0.882 (0.822)	0.915 (0.869)	0.921 (0.897)	0.966 (0.952)
OFM0	MAE	0.105 (0.154)	0.090 (0.135)	0.105 (0.122)	0.069 (0.090)
	RMSE	0.201 (0.238)	0.158 (0.206)	0.148 (0.171)	0.111 (0.137)
	R^2	0.846 (0.792)	0.894 (0.825)	0.911 (0.883)	0.950 (0.925)

the k-nearest neighbor (KNN) regression, the DT regression also relied on the local behavior of the data. In the KNN, the inference of a new observation was determined by its k nearest neighbors. Whereas, the DT partitions the training data into the leaves and the inference for a new observation was based on the leaf information that was assigned to the new observation. This DT prediction accuracy was significantly better than that obtained by using KNN regression with Manhattan distance, with RMSE of 0.301, MAE of 0.160, and R^2 of 0.960.

To enhance its predictive ability, the random forest algorithm was applied with a forest of 1000 trees trained on the subset of the training set. The same MaxD of 40 and MinP of 5 was used for all the trees. The prediction of new data was given by the average of the output of all the trees. A better prediction accuracy was obtained, with RMSE of 0.162 eV/atom, MAE of 0.101 eV/atom, and R^2 of 0.986.

The applicability of the descriptors to the prediction of the phase stability of a material was then investigated. The center issues in finding new materials are stability calculations of a hypothesized candidate. First-principles calculations can accurately estimate the formation energy and build phase diagrams for the systems of interest. Hence, the phase stability of a material is obtained by a convex hull (CH) analysis of the phase diagram and the decomposition of the materials into other phases [27]. OQMD [25, 24] was used for the analysis, and the decomposition energy of a material was estimated by:

$$\Delta E = \Delta E_f - E_H \tag{11}$$

where ΔE_f was the formation energy, and E_H was the CH energy corresponding to the chemical composition. The ΔE was essentially the degree of thermodynamic phase stability. A material lying on the CH surface, $\Delta E = 0$, was stable; whereas, one slightly above the CH surface was in a metastable phase. Hereafter, ΔE was referred to as the material stability.

It was straightforward to build the machine-learning models to estimate ΔE_f^{ML}, then use Eq. 11 to estimate ΔE. There was also another approach that developed the machine-learning models to directly represent the phase stability. Following both methods, OFM, OFM1, OFM0 were used to build the machine-learning model for ΔE. Table 2 summarizes the prediction accuracy estimated on the test set. Fig. 2 shows the comparison of DFT and machine-learning

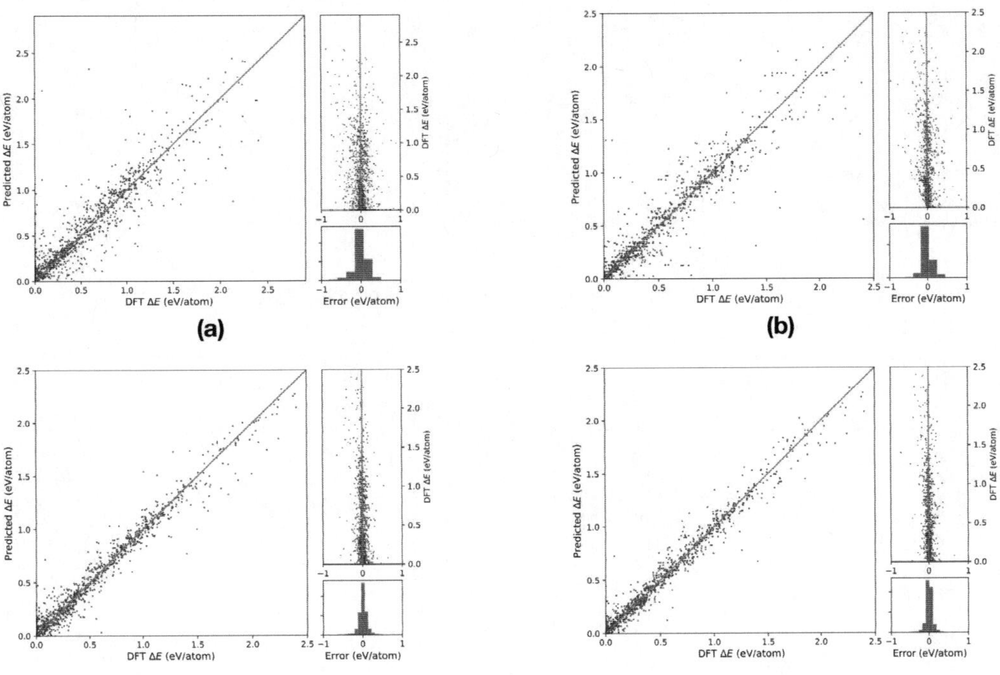

Figure 2. Comparison of machine-learning and DFT-predicted phase stability $\Delta E = \Delta E_f - E_H$: DT regression for formation energy ΔE_f (a) and direct DT regression for stability (b), KRR with Laplacian kernel for formation energy (c), and direct KRR for stability (d).

prediction of the phase stability. In Table 2, the values in parentheses were the results of the former method, and the values outside were obtained by the models that directly represented the phase stability. The latter method exhibited significantly more accuracy than the former. Once again, the OFM1 prediction was the most accurate. OFM0 also yielded acceptable accuracy, and gave a reasonable prediction accuracy with MAE of 0.069 eV/atom, RMSE of 0.092 eV/atom, and R^2 of 0.950 using the KRR with the Laplacian kernel. DT and RF regressions also yielded reasonable accuracy relative to KRR with the Gaussian kernel.

4. Learning from Molecular Data: Prediction of Atomization Energy

Atomization energy predictions via OFM, OFM1, and OFM0 were investigated. The atomization energies of organic molecules were examined with the QM7 dataset with 6915 organic molecules [17, 28]. (Originally, the QM7 dataset had 7195 molecules, but more than 100 were removed because of a technical problem in determining Voronoi polyhedra for flat structures). Local structures of atoms were represented by OFM, OFM1, or OFM0. Because the atomization energy of a molecule depends on the number of atoms, the sum of the descriptors of the local structures was used as the molecule descriptor. This dataset was also divided randomly into the test set of 1383 molecules (20%) and the training set of 5532 molecules (80%).

Descriptors for the molecules were calculated by using Eq. 6 and OFM, OFM1, or OFM0. The machine-learning models were built for the atomization energy based on KRR with the Laplacian kernel (KRR-Lapl), KRR with the Gaussian kernel (KRR-Gauss), DT, or RF regressions. The evaluation of the prediction accuracy is summarized in Table 3. As with the solid systems,

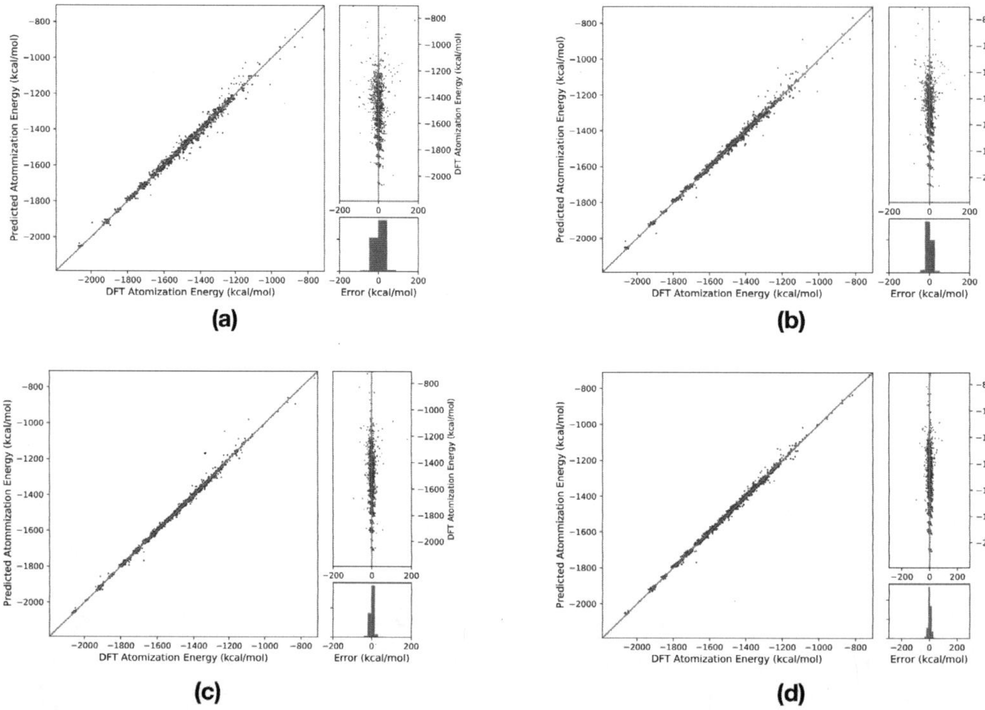

Figure 3. Comparison of machine-learning and DFT-predicted formation energy, ΔE_f. Decision tree regression (a), random forest regression (b), and kernel ridge regression with Gaussian kernel (c) and Laplacian kernel (d).

Table 3. Prediction accuracy of the atomization energy for molecules. RMSE (kcal/mol), cross-validated MAE (kcal/mol), and R^2 were estimated on the test set of QM7 using KRR with the Laplacian kernal (KRR-Lapl), KRR with the Gaussian kernel (KRR-Gauss), and DT and RF regressions.

Descriptor		DT	RF	KRR-Gaus	KRR-Lapl
OFM	MAE	23.84	19.95	22.72	14.97
	RMSE	43.39	34.88	33.87	26.58
	R^2	0.960	0.973	0.975	0.985
OFM1	MAE	12.16	10.79	8.14	6.74
	RMSE	22.62	19.34	11.90	11.69
	R^2	0.989	0.992	0.997	0.997
OFM0	MAE	11.91	10.38	8.78	7.77
	RMSE	21.07	17.41	12.62	11.72
	R^2	0.990	0.993	0.996	0.997

the OFM1 results were the most accurate compared with those of OFM and OFM0. OFM0, which did not include information on the distance from a center atom to its neighboring atoms, exhibited better predictions than those of OFM. These results again indicated the importance of the explicit appearance of information for the center atoms in the OFM. The information on the center atoms had bigger impact than the distances between atoms because the difference between OFM1 and OFM was small. Again, the prediction given by OFM1 was the most accurate with RMSE of 11.69 kcal/mol, MAE of 6.74 kcal/mol, and R^2 of 0.997 by using KRR with Laplacian kernel. Furthermore, a good prediction could be achieved with OFM0.

5. Conclusion

We present a systematic study of the application of OFM, 0FM1, and OFM0 descriptors, which were developed with an emphasis on atomic orbital interactions, for representing material structures in datasets of multi-element materials. We showed that the descriptors are highly applicable in predicting physical properties of materials, including solids and molecules. For crystalline systems, the focus was on the formation energy and phase stability, while, for molecules, the atomization energy was studied. The prediction capability of OFM, OFM1, and OFM0 for properties was examined with kernel ridge regression, decision tree regression, and random forest regression with the emphasis on the identification of similar materials. The accuracy of the simple decision tree was comparable to that of the KRR with Gaussian kernel; whereas, the random forest regression exhibited a slight improvement. The KRR with Laplacian kernel estimated by OFM1 yielded the most accurate predictions. The formation energy, the phase stability, and atomization energy had MAEs of 0.072 (eV/atom), 0.059 eV/atom, and 6.74 (kcal/mol), respectively. The OFM0 without distances of the center atom to the neighboring atoms also had acceptable accuracy. It provided a reasonable prediction accuracy with MAEs of 0.090 eV/atom, 0.069 eV/atom, and 7.77 kcal/mol for the formation energy, the phase stability, and the atomization energy, respectively. Therefore, it would be highly useful for predicting a hypothesized structure that is not yet optimized. The prediction ability of our descriptors implies that they are highly useful in finding similar materials base on distance measurements and decision trees.

Acknowledgments

This work was supported by Precursory Research for Embryonic Science and Technology from the Japan Science and Technology Agency (JST), the Elements Strategy Initiative Project under the auspices of MEXT, the 'Materials Research by Information Integration' Initiative (MI^2I) project of the Support Program for Starting Up Innovation Hub from JST, and MEXT as a social and scientific priority issue employing the post-K computer (creation of new functional devices and high-performance materials to support next generation industries; CDMSI). We thank the Edanz Group (www.edanzediting.com/ac) for editing this manuscript.

References

[1] Yousef S, Da G, Thanh N, Scotty B, R C J and Wanda A 2012 Phys. Rev. B **85** 104104
[2] Yang S, Lach-hab M, Vaisman I I and Blaisten-Barojas E 2009 Phys. Chem. C **113** 21721
[3] Hautier G, Fischer C C, Jain A, Mueller T and Ceder G 2010 Chem. Mater **22** 3762
[4] Snyder J C, Rupp M, Hansen K, Müller K R and Burke K 2012 Phys. Rev. Lett **108** 253002
[5] Isayev O, Fourches D, Muratov E N, Oses C, Rasch K, Tropsha A and Curtarolo S 2015 Chem. Mater **27** 735
[6] Ghiringhelli L M, Vybiral J, Levchenko S V, Draxl C and Scheffler M 2015 Phys. Rev. Lett. **114**(10) 105503
[7] Dam H C, Pham T L, Ho T B, Nguyen A T and Nguyen V C 2014 J. Chem. Phys. **140** 044101
[8] Pham T L, Kino H, Terakura K, Miyake T and Dam H C 2016 J. Chem. Phys. **145** 154103
[9] Pham T L, Nguyen N D, Nguyen V D, Kino H, Miyake T and Dam H C 2018 J. Chem. Phys **148** 204106
[10] Behler J and Parrinello M 2007 Phys. Rev. Lett. **98** 146401
[11] Behler J 2011 J. Phys. Chem **134** 074106
[12] Eshet H, Khaliullin R Z, Kuhne T D, Behler J and Parrinello M 2010 Phys. Rev. B **81** 184107
[13] Eshet H, Khaliullin R Z, Kuhne T D, Behler J and Parrinello M 2012 Phys. Rev. Lett **108** 115701
[14] Artrith N, Morawietz T and Behler J 2011 Phys. Rev. B **83** 153101

IOP Conf. Series: Journal of Physics: Conf. Series **1290** (2019) 012012 doi:10.1088/1742-6596/1290/1/012012

[15] Artrith N and Behler J 2012 Phys. Rev. B **85** 045439
[16] Pilania G, Wang C, Jiang X, Rajasekaran S and Ramprasad R 2013 Scientific Reports **3** 2810
[17] Rupp M, Tkatchenko A, Muller K R and von Lilienfeld O A 2012 Phys. Rev. Lett. **108**(5) 058301
[18] Faber F, Lindmaa A, von Lilienfeld O A and Armiento R 2015 Int. J. Quantum Chem. **115** 1094–1101
[19] Rupp M 2015 Int. J. Quantum Chem. **115** 1058–1073
[20] Lam Pham T, Kino H, Terakura K, Miyake T, Tsuda K, Takigawa I and Chi Dam H 2017 Sci Technol Adv Mater **18** 756–765
[21] O'Keeffe M 1979 Acta. Cryst **A35** 772–775
[22] Murphy K P 2012 Machine Learning: A Probabilistic Perspective (MIT Press)
[23] Pedregosa F, Varoquaux G, Gramfort A, Michel V, Thirion B, Grisel O, Blondel M, Prettenhofer P, Weiss R, Dubourg V, Vanderplas J, Passos A, Cournapeau D, Brucher M, Perrot M and Duchesnay E 2011 Journal of Machine Learning Research **12** 2825–2830
[24] Kirklin S, Saal J E, Meredig B, Thompson A, Doak J W, Aykol M, Rühl S and Wolverton C 2015 Npj Computational Materials **1** 15010 EP –
[25] Saal J E, Kirklin S, Aykol M, Meredig B and Wolverton C 2013 JOM **65** 1501–1509
[26] Yamashita T, Sato N, Kino H, Miyake T, Tsuda K and Oguchi T 2018 Phys. Rev. Materials **2**(1) 013803
[27] Sun W, Dacek S T, Ong S P, Hautier G, Jain A, Richards W D, Gamst A C, Persson K A and Ceder G 2016 Science Advances **2**
[28] Blum L C and Reymond J L 2009 J. Am. Chem. Soc. **131** 8732

IOP Conf. Series: Journal of Physics: Conf. Series **1290** (2019) 012013 doi:10.1088/1742-6596/1290/1/012013

A Comparison of Medium-Sized Basis Sets for the Prediction of Geometries, Vibrational Frequencies, Infrared Intensities and Raman Activities for Water

Brian C. Ferrari

Department of Physics, University of Central Florida, 4111 Libra Drive, Orlando FL 32816

E-mail: `Brian.Ferrari@ucf.edu`

Chris J. Bennett

Department of Physics, University of Central Florida, 4111 Libra Drive, Orlando FL 32816
Department of Chemistry, University of Central Florida, 4111 Libra Drive, Orlando FL 32816
Florida Space Institute, 12354 Research Parkway, Partnership 1 Building, Suite 214, Orlando FL 32826
Department of Chemistry & Biochemistry, Georgia Institute of Technology, 901 Atlantic Drive, Atlanta, GA 30332

E-mail: `Christopher.Bennett@ucf.edu`

Abstract.
Optimized geometries, vibrational frequencies, as well as infrared intensities and Raman activities were calculated for water (H_2O) utilizing popular quantum mechanical approaches. Here, density functional theory (DFT) calculations were performed using the B3LYP (Becke, three-parameter, Lee-Yang-Parr) functional, as well as *ab initio* calculations using second-order Møller-Plesset (MP2) perturbation theory and coupled-cluster with single, double and perturbative triple excitations [CCSD(T)] levels of theory were used. We assess and benchmark the performance of 69 different atomic orbital basis sets including various popular families of medium-sized basis sets typically of two to four zeta quality and differing levels of augmentation by polar and diffuse functions. The basis sets range from the commonly adopted Pople-style (6-31G & 6-311G), Dunning's correlation consistent (cc-pV(n+d)Z & aug-cc-pV(n+d)Z, as well as Truhlar's calendar variations, Jensen's polarization consistent (pc-n & aug-pc-n), Ahlrichs (def2-...), Sapporo's and Karlsruches as well as atomic natural orbitals (ANOs) such as NASA Ames (ANOn), Neese-style, and Roos-style. We also compare several basis sets specifically designed to calculate vibrational and electronic properties, including the Sadlej-pVTZ (and LPol-X families), as well as SNS families of Barone. The results are compared to experimental values where available, or calculations performed with 5 or 6 zeta-level (e.g., cc-pV6Z). The performance of each family of basis sets is discussed in terms of their accuracy (and pitfalls), as well as computational resource scaling and efficiency. The Def2 basis family performs very well overall, yielding more accurate results with lower runtimes than traditional basis sets. 'May' basis sets also provide accurate predictions of vibrational frequencies at significantly lower costs. Raman activities can be accurately calculated using MP2 under harmonic approximation with several 'spectroscopic' families performing well.

1. Introduction

Predictions of molecular properties, such as geometries, vibrational frequencies, infrared intensities and Raman activities have become powerful tools for chemists and physicists performing spectroscopic analysis of substances. Particularly involving the characterization

IOP Conf. Series: Journal of Physics: Conf. Series **1290** (2019) 012013 doi:10.1088/1742-6596/1290/1/012013

or quantification of unstable species, or radicals which are produced *in situ* during experiments, where experimental values are often not available [1, 2]. Commonly employed quantum mechanical approaches to calculation these properties include density functional theory (DFT) and *ab initio* methods, of which extensive studies have been done investigating accuracy of their predictions [3–11]. However, only a few studies exist systematically comparing the performance of a limited number of differing atomic orbital basis sets to accurately calculate these properties [12]. Moreover, careful selection from the plethora of basis set choices now easily accessible (for example through the EMSL database, https://bse.pnl.gov/bse/portal) can lead to optimization of the approach implemented based upon computational resources and required accuracy levels [12–14]. Since computational resources are often constrained, Pople-style basis sets are often employed since they are often not demanding, and widely available in commercial packages, however, they are often out-performed by other basis sets at similar computational cost. On the other hand, correlation consistent basis sets are widely available and - although computationally heavy - they do systematically converge on accurate results when either 5/6 zeta basis sets are used, or complete basis set extrapolation schemes are incorporated (where each zeta, ζ, typically approximately doubles the number of atomic orbitals). Since computational costs for these calculations scale with N^4 to N^7, the runtimes for even medium-sized molecules under 5/6 zeta become unfeasible. However, these large calculations don't necessarily produce the most accurate predictions. The results from several basis set families can be extrapolated to complete basis set (CBS) limit via a series of calculations performed at lower zeta level [15–17]. This study compares the results of several medium-sized basis sets across three levels of theory, in order to identify which basis sets can be reliably utilized to predict molecular properties (such as geometries, frequencies, IR intensities and Raman activities) maximizing the precision/accuracy for a given set of computational resources. Here, we only focus on preliminary results for water, but the findings are in line with those observed for other molecules we have investigated. Both DFT and *ab initio* calculations were perfomed, using Becke's three parameter exchange funtional [18] along with Lee-Yang-Parr's correlation functional [19] (B3LYP), second-order Møller-Plesset (MP2) perturbation theory [20] and coupled-cluster with single, double and perturbative triple excitations [CCSD(T)] [21, 22]. For each level of theory 69 different atomic orbital basis sets were used varying from Pople-style (6-31G & 6-311G) [23], Dunning's correlation consistent (cc-pV(n+d)Z & aug-cc-pV(n+d)Z) [24], as well as Truhlar's calendar variations [25–30], Jensen's polarization consistent (pc-n & aug-pc-n) [31–33], Alrich's (def2-...) [34], Sapporo's and Karlsruhe's, as well as atomic natural orbitals (ANOs) such as NASA Ames (ANOn), Neese-style [35], and Roos-style. We also compare several basis sets specifically designed for calculating vibrational and electronic properties, including the Sadlej-pVTZ (and LPol-X families) [36,37], as well as SNS & NO7 families of Barone [38,39] and a new generation of Thakkar DZ basis sets, NLO-1 [40]. Calculated values are compared to both experimental values [41–43] where available, as well as various convergences found either running large zeta calculations or through extrapolation. Though many medium-sized basis sets may converge on the solutions provided by extrapolation schemes and high-zeta results, this does not guarantee, and in fact rarely results in the prediction of molecular properties being accurately calculated due to inherent flaws in the level of theory taken. Therefore, minor improvements at the cost of significantly larger runtimes are rarely worthwhile. It is of note that this study does not account or attempt to correct for errors arising from inadequacies in the levels of theory described here, although it is possible to do so (e.g., by correcting for the truncation of the theoretical approaches, core-correlation effects, or relativistic effects which need to be taken into account to reproduce frequencies to within 1 cm^{-1}). However, due to fortuitous cancellation of errors, the CBS limits, or cc-pV6Z results are often within 5-10 cm^{-1} of the corrected values [44]. Here, we report on which basis sets can be reliably utilized to predict molecular properties (such as geometries, frequencies, IR intensities and Raman activities) maximizing the precision/accuracy for a given set of computational resources. The need for high zeta basis sets, or extrapolated results can often be avoided, if the basis sets are more carefully considered for the property being evaluated, and more accurate results reliably obtained at a lower computational cost. We briefly address the benefits of using anharmonic corrections to determine vibrational frequencies versus the use of scaling factors for a few select cases.

2. Methods

Calculations were done using the General Atomic and Molecular Electronic Structure System, GAMESS(US) [45–47], using the 14 FEB 2018 (R1) version of Gamess(US). All calculations were run using spherical harmonics; Pople-style basis sets were typically ran using cartesian functionals unless they incorporated f-type polarization functions, where spherical harmonics were used. The segmented versions of polarization consistent basis sets were used, it should also be noted that within the Def2 family QZVPD is equivalent to QVZPPD and that QZVP is equivalent to QVZPP. Tight convergence criteria were used throughout (e.g., integral cutoff 10^{-12}, primitive cutoff 10^{-25}, 10^{-7} SCF convergence, gradient convergence 10^{-6} for geometries). Hessian and Raman calculations were run with semi-numerical (B3LYP and MP2) or fully-numerical gradients [CCSD(T)] under the double harmonic approximation with projected frequencies and two displacements in each cartesian direction during force field computations. The frozen-core approximation was used for both MP2 and CCSD(T) calculations. B3LYP calculations were run using an "army grade" pruned grid (JANS=2) with roughly 71,000 grid points and 155 radial points per atom. In general calculations were run on four processors, with only a few exceptions. Larger basis sets such cc-pV5Z, cc-pV6Z and Pcseg-4 were run with fully numerical gradients for all theory methods, and required calculations to be run on a single processor. For a few special cases, the anharmonic frequencies were calculated using Vibrational self-consistent field (VSCF) [48]. These calculations were run with third order mode couplings, ten grid points, with second order degenerate perturbation theory corrections. Extrapolated complete basis set (CBS) limits were obtained using the following two-point linear extrapolation formula generated by Halkier *et al* [49].

$$E_{X,Y} = \frac{E_X^{corr} X^3 - E_Y^{corr} Y^3}{X^3 - Y^3} \quad (1)$$

Where X and Y are the basis set's zeta value, with $Y = X - 1$, CBS limits were found using the highest zeta basis sets calculated per basis "family". Error analysis was done by taking the average error between the calculated values and either experimental data or CBS limits.

3. Results and Discussion

Table 1: Molecular Geometry Predictions,
Bond Lengths are in Angstroms (Å), Bond Angles are in Degrees (°), {1} is fastest time rank

	Optimization	Bond Length{Time Rank}			Bond Angle		
	Experimental [42]	0.9578			104.4776		
#	Basis Set	B3LYP	MP2	CCSD(T)	B3LYP	MP2	CCSD(T)
1	cc-pV6Z	0.9606{67}	0.9582{68}	0.9582{69}	105.120	104.338	104.422
2	cc-pV5Z	0.9605{63}	0.9580{63}	0.9580{61}	105.091	104.286	104.373
3	Aug-cc-pVQZ	0.9608{55}	0.9589{55}	0.9589{55}	105.110	104.269	104.365
4	Jul-cc-pVQZ	0.9608{53}	0.9588{52}	0.9589{52}	105.109	104.263	104.360
5	Jun-cc-pVQZ	0.9608{50}	0.9586{48}	0.9587{49}	105.114	104.209	104.301
6	May-cc-pVQZ	0.9608{48}	0.9584{45}	0.9585{45}	105.122	104.219	104.306
7	cc-pVQZ	0.9605{45}	0.9577{43}	0.9579{43}	104.877	104.018	104.116
8	Aug-cc-pVTZ	0.9621{41}	0.9614{39}	0.9616{39}	105.082	104.109	104.180
9	Jul-cc-pVTZ	0.9620{39}	0.9614{37}	0.9616{37}	105.100	104.125	104.188
10	Jun-cc-pVTZ	0.9620{35}	0.9609{34}	0.9611{35}	105.110	104.063	104.118
11	May-cc-pVTZ	0.9620{32}	0.9603{31}	0.9604{31}	105.052	103.893	103.920
12	cc-pVTZ	0.9616{29}	0.9591{27}	0.9594{29}	104.517	103.518	103.582
13	Aug-cc-pVDZ	0.9651{21}	0.9659{20}	0.9665{21}	104.741	103.873	103.937
14	Jul-cc-pVDZ	0.9650{14}	0.9653{14}	0.9658{14}	105.035	104.232	104.256
15	Jun-cc-pVDZ	0.9686{8}	0.9665{8}	0.9672{8}	104.045	102.835	102.798
16	May-cc-pVDZ	0.9686{10}	0.9665{8}	0.9672{9}	104.045	102.835	102.798

17	cc-pVDZ	0.9689{4}	0.9649{5}	0.9663{6}	102.720	101.902	101.912
18	6-311++G(3df,3pd)	0.9614{34}	0.9589{33}	0.9592{34}	105.040	104.037	104.086
19	6-311G(2df,2pd)	0.9610{30}	0.9569{26}	0.9575{23}	103.936	103.109	103.211
20	6-311++G(2d,2p)	0.9612{14}	0.9583{17}	0.9584{20}	105.138	104.279	104.272
21	6-311G(2d,2p)	0.9617{13}	0.9573{13}	0.9578{12}	103.902	103.246	103.313
22	6-311++G(d,p)	0.9624{11}	0.9600{8}	0.9607{13}	105.043	103.358	103.299
23	6-311+G(d,p)	0.9624{6}	0.9600{5}	0.9607{10}	105.047	103.374	103.314
24	6-311G(d,p)	0.9624{6}	0.9586{5}	0.9596{4}	103.752	102.300	102.306
25	6-31+G(d,p)	0.9653{4}	0.9632{1}	0.9633{5}	105.744	105.421	105.338
26	pc4	0.9606{64}	0.9576{66}	0.9577{63}	105.128	104.347	104.430
27	aug-pc3	0.9606{57}	0.9581{57}	0.9583{59}	105.137	104.370	104.449
28	pc3	0.9606{49}	0.9578{49}	0.9580{48}	105.139	104.341	104.415
29	aug-pc2	0.9610{40}	0.9590{38}	0.9593{38}	105.145	104.306	104.340
30	pc2	0.9606{23}	0.9580{30}	0.9582{30}	105.149	104.177	104.165
31	aug-pc1	0.9661{19}	0.9662{17}	0.9662{18}	104.831	103.922	104.000
32	pc1	0.9686{3}	0.9650{1}	0.9654{3}	103.865	103.553	103.518
33	Def2-QZVPD	0.9607{53}	0.9585{51}	0.9586{47}	105.121	104.237	104.321
34	Def2-QZVP	0.9604{46}	0.9580{46}	0.9581{44}	105.099	104.223	104.301
35	Def2-TZVPPD	0.9613{37}	0.9598{36}	0.9600{34}	105.127	104.070	104.118
36	Def2-TZVPP	0.9610{26}	0.9586{27}	0.9589{26}	104.893	103.823	103.859
37	Def2-TZVPD	0.9636{31}	0.9644{31}	0.9650{32}	105.111	104.377	104.401
38	Def2-TZVP	0.9630{17}	0.9622{21}	0.9630{19}	105.236	104.597	104.548
39	Def2-SVPD	0.9644{18}	0.9643{14}	0.9649{15}	105.370	104.444	104.476
40	Def2-SVP	0.9670{2}	0.9625{1}	0.9638{2}	103.078	102.438	102.465
41	SPK-AQZP	0.9607{56}	0.9585{56}	0.9586{57}	105.157	104.351	104.440
42	SPK-QZP	0.9604{47}	0.9580{47}	0.9582{46}	105.082	104.234	104.321
43	SPK-ATZP	0.9612{42}	0.9598{40}	0.9602{40}	105.102	104.227	104.266
44	SPK-TZP	0.9612{33}	0.9588{35}	0.9593{33}	104.355	103.493	103.565
45	SPK-ADZP	0.9645{22}	0.9640{23}	0.9645{22}	104.790	103.996	104.047
46	SPK-DZP	0.9674{8}	0.9637{8}	0.9647{7}	102.772	102.267	102.287
47	KTZVPP	0.9610{24}	0.9586{27}	0.9589{26}	104.893	103.823	103.859
48	KTZV	0.9750{1}	0.9757{1}	0.9781{1}	109.965	110.567	110.162
49	Sadlej-LPolX-fl	0.9616{69}	—[a]	0.9621{69}	105.038	—[a]	104.364
50	Sadlej-LPolX-fs	0.9613{60}	0.9615{60}	0.9622{60}	105.048	104.431	104.489
51	Sadlej-LPolX-dl	0.9622{54}	0.9607{54}	0.9611{54}	105.055	104.420	104.454
52	Sadlej-LPolX-ds	0.9616{44}	0.9614{50}	0.9621{51}	105.038	104.265	104.296
53	Sadlej-pVTZ	0.9675{28}	0.9689{25}	0.9697{28}	104.592	103.541	103.610
54	N07T	0.9644{27}	0.9665{22}	0.9676{25}	104.754	103.609	103.632
55	N07D	0.9644{12}	0.9641{8}	0.9645{11}	104.643	103.679	103.733
56	SNST	0.9649{24}	0.9653{24}	0.9659{24}	104.702	103.853	103.896
57	SNSD	0.9644{20}	0.9643{17}	0.9648{17}	104.662	103.748	103.802
58	NLO-1	0.9751{16}	0.9840{14}	0.9847{16}	106.514	105.024	105.088
59	Roos-Aug-TZ-ANO	0.9607{61}	0.9581{61}	0.9583{62}	105.128	104.251	104.310
60	Roos-Aug-DZ-ANO	0.9616{43}	0.9609{44}	0.9613{50}	105.054	104.333	104.371
61	NASA-Ames-ANO2	0.9606{65}	0.9587{64}	0.9588{65}	105.042	104.249	104.340
62	NASA-Ames-ANO1	0.9629{49}	0.9614{58}	0.9617{56}	104.923	103.928	104.000
63	NASA-Ames-ANO0	0.9647{36}	0.9638{41}	0.9642{41}	104.393	103.662	103.602
64	Neese-ANO-AVQZ	0.9606{68}	0.9588{67}	0.9588{68}	105.144	104.330	104.414
65	Neese-ANO-VQZ	0.9608{66}	0.9589{65}	0.9589{66}	105.127	104.303	104.391
66	Neese-ANO-AVTZ	0.9609{62}	0.9581{62}	0.9583{64}	105.085	104.213	104.286
67	Neese-ANO-VTZ	0.9628{59}	0.9617{59}	0.9619{58}	105.087	104.069	104.130
68	Neese-ANO-AVDZ	0.9633{52}	0.9605{53}	0.9606{53}	105.053	104.275	104.335
69	Neese-ANO-VDZ	0.9648{38}	0.9645{42}	0.9646{42}	104.584	103.861	103.794

[a] Removed due to convergence issues

Table 2: Harmonic Vibrational Frequencies in Wavenumbers (cm^{-1})

Hes	ω_1{Time Rank}			ω_2			ω_3		
Exp [5]	3832.17			1648.47			3942.53		
The [44]	3835.89			1649.39			3946.65		
#	B3LYP	MP2	CCSD(T)	"	"	"	"	"	"
1	3806.0{67}	3845.0{68}	3837.3{}[a]	1631.4	1633.0	1651.3[a]	3907.4	3971.2	3947.2[a]
2	3805.6{62}	3848.9{62}	3840.1{}[a]	1630.0	1635.6	1653.4[a]	3906.9	3974.3	3949.3[a]
3	3801.6{55}	3839.8{55}	3831.2{54}	1628.4	1632.2	1649.6	3903.3	3965.6	3940.8
4	3802.2{53}	3840.5{52}	3831.8{51}	1628.8	1632.5	1649.9	3903.8	3966.4	3941.5
5	3801.7{51}	3843.4{50}	3834.4{49}	1628.4	1635.6	1652.5	3903.3	3968.8	3943.5
6	3801.9{47}	3845.0{47}	3836.1{45}	1628.3	1636.5	1653.4	3903.5	3970.3	3945.1
7	3803.1{45}	3855.2{43}	3844.5{42}	1633.9	1642.5	1658.8	3903.4	3977.8	3951.5
8	3793.8{40}	3822.0{39}	3810.9{38}	1626.4	1628.1	1645.4	3896.4	3947.8	3920.0
9	3798.2{38}	3824.6{37}	3813.7{36}	1627.3	1629.9	1647.4	3901.1	3952.0	3924.4
10	3799.6{34}	3832.6{33}	3821.9{32}	1626.1	1633.8	1650.8	3903.3	3960.0	3932.7
11	3799.9{33}	3843.4{32}	3833.6{31}	1625.9	1642.0	1661.0	3902.8	3967.0	3940.1
12	3797.8{27}	3855.5{29}	3841.0{29}	1638.8	1651.9	1668.4	3898.0	3975.8	3945.6
13	3791.7{19}	3803.4{19}	3787.0{20}	1617.8	1622.1	1637.7	3901.4	3937.6	3904.9
14	3794.9{14}	3818.9{13}	3804.2{13}	1624.2	1632.4	1648.8	3904.6	3952.7	3922.1
15	3772.7{8}	3842.6{8}	3824.1{8}	1617.5	1638.4	1655.8	3875.2	3964.2	3930.4
16	3772.7{8}	3842.6{8}	3824.1{9}	1617.5	1638.4	1655.8	3875.2	3964.2	3930.4
17	3748.4{5}	3851.8{5}	3821.7{5}	1658.1	1677.6	1689.8	3849.8	3971.1	3927.6
18	3810.8{39}	3869.8{34}	3855.9{34}	1625.7	1623.7	1640.5	3909.7	3989.1	3959.7
19	3805.3{27}	3890.6{26}	3871.5{25}	1655.0	1670.3	1685.6	3901.1	4005.0	3971.4
20	3816.4{21}	3861.2{19}	3848.9{21}	1637.6	1659.7	1678.6	3917.8	3982.7	3953.9
21	3803.7{16}	3872.5{16}	3852.7{15}	1666.0	1686.1	1701.1	3900.1	3988.6	3953.7
22	3813.3{13}	3880.1{11}	3860.8{12}	1601.6	1629.7	1648.0	3918.7	3999.6	3965.1
23	3813.2{11}	3879.6{7}	3860.2{11}	1601.8	1629.7	1647.9	3918.7	3999.4	3964.9
24	3804.3{7}	3896.6{5}	3868.9{6}	1636.8	1667.2	1681.7	3902.3	4007.4	3966.7
25	3806.8{4}	3867.1{3}	3855.8{4}	1602.5	1622.8	1639.4	3928.7	4013.9	3987.5
26	3805.5{64}	—[b]{64}	—[b]{61}	1630.8	1654.2	1677.8	3907.1	3968.6	3945.0
27	3802.6{68}	3822.4{58}	3811.5{58}	1628.0	1632.9	1649.5	3904.4	3948.2	3921.1

28	3802.9{49}	3822.7{49}	3811.5{46}	1627.6	1635.1	1651.7	3904.6	3948.7	3921.2
29	3805.3{41}	3846.2{38}	3832.8{37}	1626.3	1632.5	1649.8	3908.9	3972.2	3942.6
30	3806.6{27}	3859.3{28}	3846.8{30}	1623.5	1637.7	1656.3	3910.9	3982.8	3954.2
31	3780.2{18}	3805.5{18}	3797.2{18}	1615.1	1630.7	1648.5	3887.6	3940.3	3916.1
32	3762.0{7}	3854.7{2}	3840.2{3}	1627.5	1648.7	1664.7	3873.5	3991.1	3962.4
33	3805.0{50}	3849.3{51}	3840.0{48}	1628.4	1635.3	1652.2	3906.8	3975.6	3949.9
34	3807.2{46}	3855.9{45}	3846.5{43}	1629.6	1637.0	1654.1	3908.3	3980.7	3955.0
35	3808.2{37}	3838.1{35}	3826.0{35}	1627.2	1636.3	1653.3	3911.2	3963.5	3935.2
36	3807.3{31}	3853.7{29}	3841.1{28}	1631.9	1646.5	1664.1	3910.2	3976.3	3947.7
37	3783.4{26}	3798.0{27}	3783.1{26}	1622.9	1620.4	1636.9	3884.8	3925.2	3893.5
38	3782.9{19}	3819.0{21}	3801.3{19}	1616.2	1613.3	1629.4	3888.1	3946.1	3912.4
39	3816.8{15}	3843.4{14}	3828.0{14}	1584.2	1596.4	1612.3	3923.1	3974.9	3944.5
40	3788.9{3}	3895.3{3}	3867.3{2}	1638.2	1654.5	1666.9	3884.0	4011.3	3971.1
41	3805.6{57}	3845.2{56}	3836.5{56}	1627.5	1631.4	1648.7	3907.1	3971.4	3946.5
42	3807.1{47}	3852.9{46}	3843.2{44}	1629.8	1635.4	1652.5	3908.7	3978.2	3952.5
43	3799.9{42}	3835.5{42}	3821.3{41}	1625.4	1626.4	1643.6	3901.5	3961.3	3930.8
44	3799.1{36}	3852.4{36}	3835.5{33}	1648.0	1655.2	1670.8	3895.5	3971.9	3939.3
45	3778.4{22}	3797.3{22}	3780.4{22}	1604.9	1617.9	1634.3	3888.1	3931.8	3899.3
46	3759.5{6}	3844.6{12}	3817.3{7}	1653.6	1665.2	1679.3	3864.5	3972.5	3932.2
47	3807.3{30}	3853.7{29}	3841.1{27}	1631.9	1646.5	1664.1	3910.2	3976.3	3947.7
48	3610.8{1}	3598.2{1}	3551.4{1}	1567.0	1601.4	1610.4	3775.9	3787.2	3725.3
49	3798.6{69}	—[c]	4031.4{62}	1634.4	—[c]	2126.1	3898.0	—[c]	4126.8
50	3803.0{60}	3811.3{60}	3793.1{59}	1638.6	1642.1	1658.6	3904.4	3943.8	3910.5
51	3796.7{54}	3820.2{54}	3805.0{53}	1630.4	1629.3	1645.5	3899.0	3949.7	3918.3
52	3799.1{44}	3811.9{48}	3793.4{49}	1629.8	1640.1	1655.9	3900.3	3937.6	3902.9
53	3794.7{24}	3797.9{25}	3779.9{24}	1633.8	1642.9	1658.3	3908.4	3935.8	3900.8
54	3770.5{25}	3775.4{24}	—[b]	1631.4	1622.9	—[b]	3871.7	3901.6	—[b]
55	3812.6{8}	3841.5{8}	3829.4{10}	1640.9	1650.4	1665.3	3921.1	3975.9	3948.0
56	3785.8{23}	3810.4{23}	3794.3{23}	1629.8	1632.6	1647.6	3896.4	3946.3	3914.3
57	3808.6{17}	3834.6{17}	3820.6{17}	1633.6	1640.4	1655.4	3918.0	3968.3	3938.1
58	3716.6{12}	3620.4{15}	3597.6{16}	1600.3	1605.1	1623.5	3862.4	3790.4	3750.2
59	3807.0{61}	3850.1{61}	3838.1{60}	1627.6	1639.3	1656.2	3906.6	3974.4	3946.0
60	3807.9{43}	3828.9{44}	3812.6{47}	1635.6	1644.9	1660.5	3908.3	3954.9	3921.7
61	3803.7{65}	3843.7{65}	3835.1{65}	1631.6	1636.0	1652.7	3906.6	3972.5	3947.5
62	3798.1{56}	3838.6{57}	3827.2{55}	1629.5	1637.9	1653.3	3901.7	3969.1	3942.3
63	3810.2{32}	3854.4{40}	3835.0{39}	1638.1	1642.0	1659.1	3927.6	3991.6	3958.2
64	3804.4{68}	3840.9{67}	3832.9{64}	1628.6	1631.3	1649.0	3904.8	3967.5	3943.2
65	3802.1{66}	3842.2{66}	—[b]	1628.5	1634.2	—[b]	3907.0	3973.2	—[b]
66	3803.6{63}	3851.9{63}	3842.0{63}	1629.3	1640.0	1656.8	3904.8	3977.6	3951.6
67	3800.2{59}	3833.9{59}	3823.8{57}	1624.3	1633.2	1649.3	3904.2	3965.2	3939.0
68	3796.7{52}	3845.0{53}	3832.9{52}	1628.8	1637.9	1653.1	3900.3	3972.3	3944.1
69	3811.5{35}	3847.8{41}	3833.0{40}	1633.8	1635.3	1654.0	3931.3	3989.1	3959.2

[a] These results were taken from Tew *et al.* (2007)

[b] An error in the calculation of this vibrational frequency caused this particular result to be invalid

[c] Removed due to convergence issues

Table 3: Infrared Intensities in $Km \cdot mol^{-1}$ and Raman Activities in $\AA^4 amu^{-1}$

—	Infrared Intensities						Raman Intensities					
Exp [5]	2.93		62.5		41.7		111 ± 12		0.9 ± 0.2		19 ± 2	
—	IR_1		IR_2		IR_3		R_1		R_2		R_3	
#	B3LYP	MP2	"	"	"	"	"	"	"	"	"	"
3	4.63{57}	6.09{57}	76.10	73.17	63.27	78.36	98.26	109.50	0.81	0.83	25.4	22.1
4	5.20{52}	6.68{51}	76.19	73.04	63.65	77.85	87.63	97.56	0.71	0.75	25.0	21.9
5	5.21{55}	6.80{54}	76.03	73.50	63.67	78.52	87.75	96.98	0.72	0.75	25.0	21.8
6	5.16{49}	6.69{48}	76.56	72.64	63.16	77.23	88.55	97.72	0.71	0.79	25.4	22.2
7	3.89{45}	6.55{43}	72.79	69.11	51.46	68.30	78.51	78.28	2.75	2.78	26.0	23.5
8	4.59{41}	5.56{39}	75.71	71.68	62.83	75.46	96.82	107.62	1.02	1.09	25.6	22.8
9	5.86{36}	7.44{33}	77.62	72.50	62.87	74.21	80.43	87.64	0.86	0.97	26.1	23.6
10	5.91{39}	7.61{37}	77.20	73.81	63.08	75.77	79.33	87.53	0.86	0.91	25.8	23.1
11	6.53{34}	9.57{32}	69.58	63.43	60.65	73.11	78.84	81.66	3.79	4.00	28.5	26.2
12	3.16{29}	5.74{29}	69.46	64.52	40.54	55.20	71.40	69.55	4.34	4.25	26.1	24.8
13	4.00{19}	4.14{19}	71.28	67.46	60.41	67.02	96.28	104.28	1.86	1.96	25.5	24.2
14	11.75{8}	16.28{9}	65.74	60.84	59.66	66.22	80.40	81.55	4.52	4.82	37.5	36.3
15	6.33{14}	9.13{13}	81.06	79.23	55.36	65.70	72.34	74.73	1.35	1.39	29.9	27.6
16	11.75{9}	16.28{9}	65.74	60.84	59.66	66.22	80.40	81.55	4.52	4.82	37.5	36.3
17	2.84{4}	6.56{5}	55.64	56.98	19.70	32.64	73.03	68.32	6.24	5.83	33.3	33.8
18	4.45{37}	6.42{34}	72.10	67.09	59.30	72.70	94.11	96.06	2.20	2.28	26.5	23.5
19	2.72{26}	5.12{26}	68.08	64.04	33.76	49.29	69.05	65.45	4.11	4.13	23.8	22.8
20	7.23{22}	9.84{19}	70.95	65.33	61.76	73.54	82.31	82.95	4.54	4.74	28.3	26.1
21	2.06{17}	3.60{15}	65.23	61.36	29.89	43.09	73.50	69.74	4.71	4.75	24.5	23.8
22	9.01{13}	12.66{11}	66.60	56.65	56.58	61.67	83.54	81.66	7.45	7.94	34.7	34.1
23	9.48{11}	12.96{7}	67.05	57.13	57.07	61.97	80.38	78.95	7.77	8.09	34.2	33.8
24	3.41{7}	5.82{6}	57.32	50.17	24.14	32.17	71.51	66.12	7.90	7.92	30.6	31.1
25	6.34{5}	10.08{3}	91.20	92.87	57.07	66.68	80.34	78.73	3.36	3.48	35.1	34.0
27	4.64{60}	6.09{60}	76.26	72.91	63.30	78.90	97.55	111.47	0.67	0.75	25.5	21.8
28	5.07{50}	7.04{50}	74.88	71.04	62.08	78.03	86.39	95.65	1.63	1.76	26.4	23.0
29	4.66{40}	5.60{38}	75.92	71.88	63.41	75.58	97.96	109.88	0.74	0.81	25.7	22.4
30	5.76{28}	9.42{28}	71.38	66.09	54.89	68.70	76.92	76.53	4.31	4.38	29.9	27.7
31	4.57{16}	4.79{18}	75.72	72.47	65.99	73.73	96.78	104.35	0.69	0.65	25.3	22.3
32	7.35{2}	13.81{3}	74.19	78.31	42.98	63.56	76.17	70.14	5.08	4.57	37.1	35.7

IOP Conf. Series: Journal of Physics: Conf. Series **1290** (2019) 012013 doi:10.1088/1742-6596/1290/1/012013

33	4.61{54}	6.08{52}	76.07	71.96	63.09	77.54	97.95	111.63	0.77	0.84	25.4	22.1
34	5.47{46}	8.14{45}	74.45	70.25	58.55	74.74	77.32	82.65	2.49	2.57	26.6	23.7
35	4.66{38}	5.59{36}	75.81	70.19	62.89	74.20	96.54	107.74	0.78	0.89	25.5	22.6
36	4.46{30}	7.23{29}	69.75	64.39	48.72	62.46	76.52	75.31	4.51	4.44	27.8	26.0
37	4.08{32}	5.13{27}	77.76	76.03	61.03	74.42	98.89	108.40	0.66	0.73	26.1	23.3
38	4.38{21}	8.83{21}	80.64	80.97	43.76	60.96	78.80	74.43	5.09	4.98	32.9	31.5
39	7.05{15}	7.22{14}	77.96	73.99	75.28	82.02	93.91	103.84	0.92	0.90	24.5	22.5
40	4.79{2}	9.87{2}	55.38	58.08	26.65	42.65	74.23	69.19	7.01	6.44	32.7	33.1
41	4.74{59}	6.25{58}	76.38	73.41	63.66	79.15	94.50	109.18	0.81	0.83	25.3	22.0
42	4.97{48}	7.96{47}	74.87	71.14	56.06	73.35	78.04	80.52	2.58	2.61	26.8	23.9
43	4.73{42}	5.95{42}	75.66	71.85	63.25	76.44	90.30	110.88	0.87	0.89	25.6	22.4
44	3.08{35}	4.94{35}	68.75	64.64	44.39	59.51	75.62	78.28	3.78	3.88	25.1	23.2
45	4.03{20}	4.36{22}	71.41	68.02	60.28	68.26	92.34	103.12	1.28	1.27	25.8	23.6
46	4.40{6}	8.40{11}	60.07	62.63	28.94	44.97	72.22	69.20	5.38	5.10	31.1	31.4
47	4.46{31}	7.23{31}	69.75	64.39	48.72	62.46	76.52	75.31	4.51	4.44	27.8	26.0
48	6.91{1}	3.27{1}	94.34	103.74	19.53	29.17	98.06	94.75	9.97	9.17	37.2	39.0
50	4.55{62}	5.15{62}	76.26	72.75	62.69	75.00	106.73	121.62	0.64	0.83	25.7	22.6
51	4.57{56}	4.93{56}	76.56	72.06	63.43	73.14	103.55	106.04	0.85	0.95	26.0	23.2
52	4.79{44}	5.13{49}	76.05	71.78	63.30	72.97	95.17	118.33	0.76	0.91	25.9	23.1
53	4.66{23}	4.68{25}	73.71	69.16	61.79	68.04	102.57	110.91	1.08	1.00	24.8	22.8
54	4.63{24}	4.49{24}	75.96	69.42	60.95	67.08	98.22	109.62	0.86	0.95	25.7	23.8
55	5.65{10}	5.39{7}	78.46	74.25	62.45	69.30	72.52	76.56	0.87	0.81	21.1	20.3
56	5.09{25}	4.78{23}	76.43	71.70	62.76	69.52	95.51	108.23	0.88	0.92	25.5	23.4
57	4.56{17}	4.53{17}	74.88	69.79	61.51	67.60	97.11	105.17	0.81	0.83	25.7	23.8
58	3.95{11}	3.73{15}	81.39	73.79	69.13	70.28	92.47	99.53	2.86	3.26	24.2	24.6
59	4.65{63}	5.57{63}	76.27	72.09	63.68	76.90	90.80	111.12	0.96	0.96	25.3	22.1
60	3.99{43}	4.48{44}	73.68	69.35	59.44	69.17	95.38	113.06	1.06	1.30	24.7	22.5
61	4.36{65}	6.95{65}	73.82	71.56	56.02	73.41	72.29	86.85	2.44	2.24	24.4	22.8
62	4.51{58}	6.24{59}	73.82	69.50	47.62	61.96	66.56	72.05	3.74	3.51	21.8	21.2
63	7.78{27}	10.17{40}	77.07	76.05	47.81	61.06	68.46	65.61	5.11	4.96	25.9	27.3
64	6.37{67}	8.55{67}	75.87	73.08	63.32	78.99	73.72	87.14	2.25	2.17	25.1	22.6
65	5.00{66}	7.89{66}	74.08	71.86	58.17	76.13	74.70	85.16	2.57	2.38	25.0	23.2
66	5.12{64}	7.51{64}	73.25	70.42	59.13	74.38	78.62	83.35	2.71	2.71	25.6	23.4
67	5.10{61}	7.16{61}	74.30	69.80	51.45	65.63	65.97	73.44	3.69	3.49	22.0	21.5
68	6.32{51}	7.70{55}	71.70	67.09	53.11	64.38	68.66	73.80	4.15	3.93	23.5	23.0
69	9.09{33}	12.02{41}	78.73	77.57	51.62	65.64	65.21	67.87	5.25	5.15	26.7	27.9

Table 4: VSCF Calculations
for Anharmonic Frequencies in cm^{-1} and Infrared Intensities in $Km \cdot mol^{-1}$

VSCF	ω_1	ω_2	ω_3	IR_1	IR_2	IR_3
Exp [50]	3657	1595	3756	2.93	62.5	41.7
B3LYP/cc-pV6Z	3639.4	1546.2	3713.5	4.39	76.68	61.01
B3LYP/cc-pV5Z	3639.8	1546.9	3713.7	4.17	76.16	58.07
B3LYP/Aug-cc-pVQZ	3637.1	1545.8	3711.3	3.92	77.46	61.74
B3LYP/May-cc-pVTZ	3635.2	1542.4	3710.7	5.61	71.47	59.03
B3LYP/aug-pc3	3638.5	1545.5	3712.9	3.92	77.63	61.76
B3LYP/Def2-QZVPD	3639.5	1545.8	3713.8	3.90	77.46	61.57
MP2/cc-pV5Z	3680.3	1553.0	3779.6	6.53	72.86	74.81
MP2/Aug-cc-pVQZ	3671.5	1550.1	3771.2	5.32	74.24	76.47
MP2/May-cc-pVTZ	3677.3	1559.5	3775.3	8.49	65.02	71.19
MP2/aug-pc3	3669.3	1551.0	3769.4	5.33	74.05	76.98
MP2/Def2-QZVPD	3679.6	1552.5	3779.6	5.30	73.16	75.67
SCS-MP2/Def2-QZVPD	3673.7	1569.2	3768.0	5.29	73.31	76.07
CCSD(T)/Def2-QZVPD	3664.6	1569.4	3746.8	13.03	93.92	79.31
CCSD(T)/Def2-TZVPPD	3650.2	1569.5	3731.9	13.11	93.55	78.27
CCSD$(2)_T$/Def2-QZVPD	3670.9	1571.1	3752.2	3.64	73.08	58.12
CCSD$(2)_T$/Def2-TZVPPD	3656.4	1571.0	3737.2	3.04	70.65	54.36

3.1. Calculations Removed from Study

After careful consideration, certain combinations of basis set and theory methods were dropped from the study due to issues discussed in this section. The version of Gamess(US) used for this study, does not support basis sets with high angular momentum i-functions (5/6 zeta basis sets) to be utilizied for predicting IR intensities or Raman activities. Zeta 5/6 Hessian calculations for CCSD(T) were too computationally intensive and did not provide significant improvements on molecular property predictions to justify their extreme runtimes. Sadlej-LPolX-fl was also

IOP Conf. Series: Journal of Physics: Conf. Series **1290** (2019) 012013 doi:10.1088/1742-6596/1290/1/012013

dropped from most calculations (MP2 within optimization and hessian calculations, across all theories within Raman calculations) due to convergence issues.

3.2. Molecular Geometries

The floating integer values within Figures 1-4 throughout the following sections are the zeta values for the basis sets. As expected, CCSD(T) performs best for calculating molecular geometries yielding consistent results with few errors. B3LYP converges faster than MP2 and CCSD(T), with the latter providing only slightly more accurate bond lengths and angles, for water.

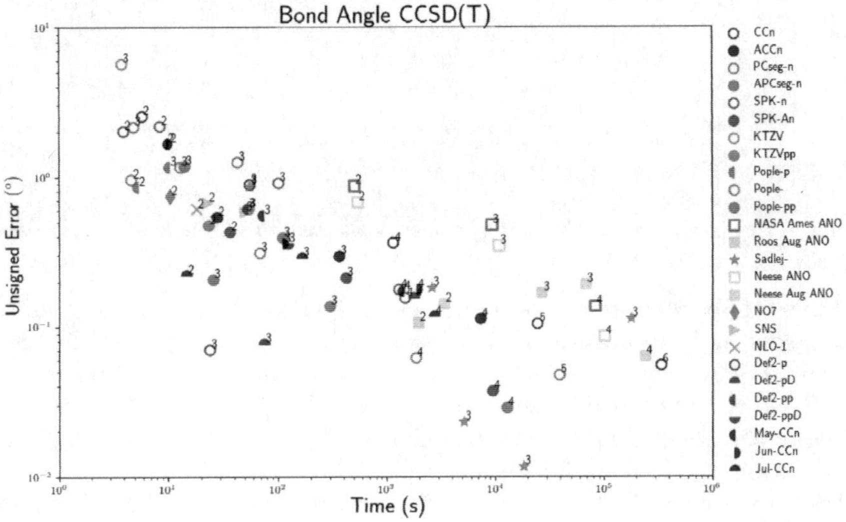

Figure 1: Unsigned Error of CCSD(T) Bond Angle Calculations Compared to Experimental Data

It is notable to mention that the Pople-style basis sets provide reasonably accurate geometries, and it requires considerably high zeta basis sets to converge on the accurate geometry; particularly for the bond angle of water. While bond length converges quickly, for all levels of theory studied here, the bond angle converges more slowly, as shown for CCSD(T) in Figure 1. In general, most basis families show linear trend lines of slower convergence as either diffuseness or zeta values increase; making DFT with pople basis sets particularly well-suited for optimizing the geometries of large molecules. The Def2 family performs well in both runtime and accuracy, similarly both the PCseg and APCseg yield high accuracy predictions with lower runtimes, particularly when compared to their CC counterparts.

3.3. Vibrational Frequencies

When comparing vibrational frequencies to experimental data CCSD(T) calculations yielded the most accurate results, as expected. Interestingly, the NO7 and SNS basis set families yield results within ten wavenumbers (here, we consider this a reasonable threshold for the characterization or discrimination of most molecular species) with minimal runtimes [11].

IOP Conf. Series: Journal of Physics: Conf. Series **1290** (2019) 012013 doi:10.1088/1742-6596/1290/1/012013

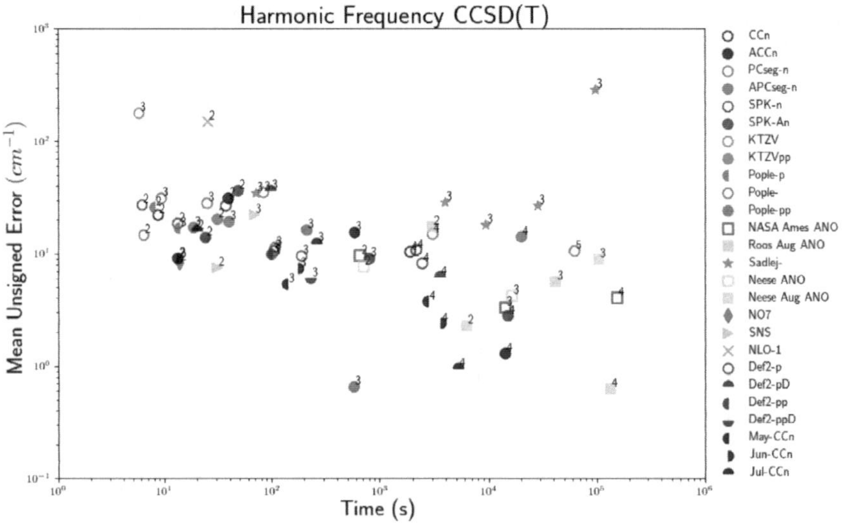

Figure 2: Mean Unsigned Error of CCSD(T) Frequency Calculations Compared to Experimental Data

The Def2 basis family consistently outperforms both correlation consistent and polarization consistent basis sets, at significantly reduced computational expense. Both the augmented and non-augmented SPK- basis sets also perform very well in calculating frequencies, producing similar accuracy and runtimes as Def2 sets. While, for B3LYP, the commonly used Pople family, converges above the 10 cm^{-1} threshold with similar runtimes. It is noteworthy to mention that fortuitous cancellation of errors often plays a large role here, particularly with smaller basis sets. However, if these errors consistently off-set a deficiency in the level of theory encountered at the basis set limit, or that is not overcome unless much larger basis sets are used, this may prove useful. As an example, for the prediction of vibrational frequencies, the 'may' calendar basis sets consistently out-perform even their fully augmented counterparts at only a minor additional computational expense compared to the CCn basis at all levels of theory studied. B3LYP is known to over-estimate bond-lengths, and under-estimate harmonic frequencies. Often, a scaling factor is applied to reduce the errors in predicting anhamonic values from calculations performed under the harmonic approximation [51–53]. In Table 4 we can see clearly that the anharmonic frequencies for B3YLP are all below the experimental values. Again in Table 4, the anharmonic frequencies are shown for a few select levels of theory and basis sets. Here, we can see that B3LYP consistently underestimates the experimental frequencies, and that even the may-cc-pVTZ basis set values lie within 5 cm^{-1} of the cc-pV6Z values. Similarly, we see for MP2 that the def2, may and pc families provide excellent agreement with the cc-pV5Z values. For each level of theory, the def2-QZVPD basis set approaches the accuracy of cc-pV5Z/cc-pV6Z much faster than most other basis sets of similar quality. Since anharmonic calculations are computationally expensive, CCSD(T) calculations were only performed using two of the def2 basis sets, where it could be seen that although an improvement over MP2 is observed, there is still not overall agreement with the experimental harmonic frequencies. This is likely because MP2 and CCSD(T) do not perform as well when used to calculate energetics at geometries far from the equilibrium position. To emphasize this point, calculations were additionally ran using the SCS-MP2 and CCSD(2)$_T$ levels of theory, which in each case show a marked improvement in the agreement with experimental anharmonic frequencies. All of the levels of theory considered here, struggle to reproduce the ν_2/ω_2 frequency. We note that the CCSD(2)$_T$/def2-QVZPD level of theory only has a MUE of $\approx$13 cm^{-1}.

3.4. Infrared Intensities

From Figure 3, we can see that though the Pople and other small basis sets can outperform larger basis sets at predicting some infrared intensities, however, caution is urged since this is not always the case. Both the B3LYP and MP2 levels of theory have significant errors when calculating infrared intensities under the harmonic approximation, even when extrapolated to their CBS limits. The B3LYP theory level convergence line appears to lie just above the ten percent error margin, some basis sets - particularly small basis sets, such as Pople - deviate from this producing more accurate results. In some cases where small basis sets are utilized, the errors arising actually compensates for the deficiencies in B3LYP, producing seemingly more accurate results. However, these results may occur sporadically, and caution is urged using these results, which could easily be taken out of context [54].

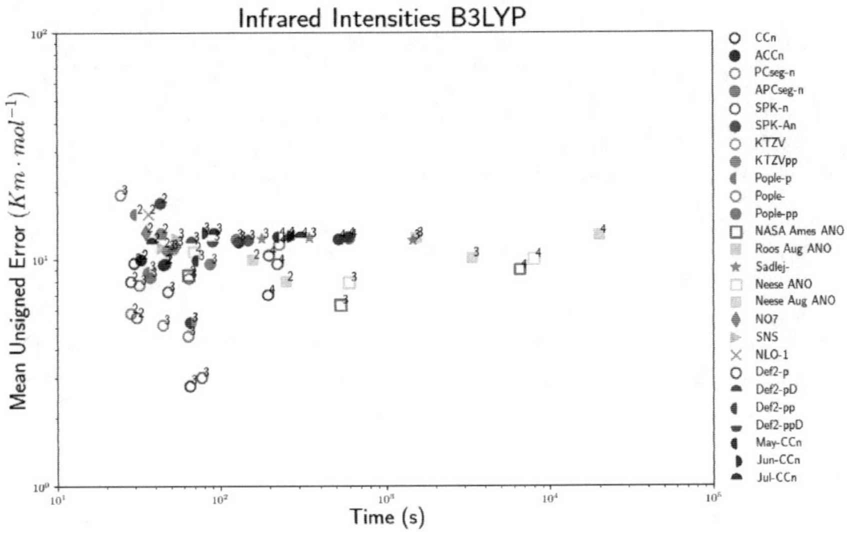

Figure 3: Mean Unsigned Error of B3LYP Infrared Intensity Calculations Compared to Experimental Data

Both the harmonic and anharmonic calculation of the infrared frequencies is subject to the accuracy of the initial Hessian used to project the geometries, as well as the accuracy of the level of theory and basis set used to calculate the energetics and dipole moment. As explained previously, the MP2 and CCSD(T) levels of theory are not as reliable at geometries required for anharmonic, or even harmonic calculations. As a result, infrared intensities are not as dependent upon diffuse and polar functions as Raman activities, but rather how accurately the relative energies are calculated at each geometry, which explains why some basis sets that do not incorporate large polarization and diffuse functions are only necessary to the extent that the energetics and dipole moment are calculated with sufficient accuracy. Table 4 shows that by utilizing SCS-MP2 [55] or CCSD$(2)_T$ [56] levels of theory (the former at no additional cost, the latter at twice the cost of regular CCSD(T) calculations), further improvements in the frequencies and IR intensities can be obtained.

3.5. Raman Activities

In contrast to the IR intensities, the Raman activities can be very accurately calculated under the harmonic approximation, in particular when the MP2 level of theory is used (in agreement with previous studies [3, 4]) and are extremely sensitive to the degree of polarization and diffusivity incorporated into the basis sets.

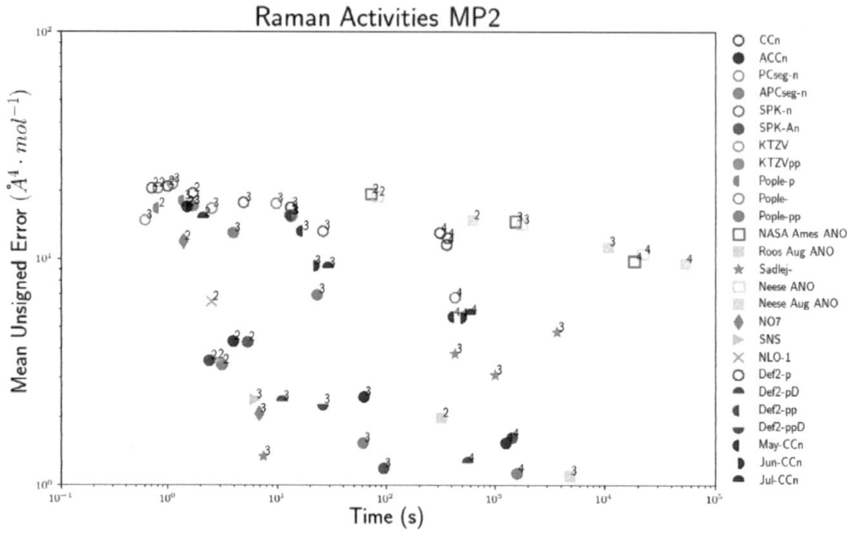

Figure 4: Mean Unsigned Error of MP2 Raman Activities Calculations Compared to Experimental Data

It is clear from Figure 4 that while increasing the zeta-level or overall size of basis set is helpful, those basis sets that incorporate higher polarization levels and more diffuse functions are generally more accurate at reproducing Raman activities. In particular, the spectroscopic basis sets all perform well here (e.g., Sadlej, N07, SNS). The (PPD versions of) Def2 and (aug) Roos basis sets perform very well here also. The NLO-1 basis set also performs fairly well in calculating Raman activities, comparable to many quadruple-zeta basis sets. The dependence on the level of diffusivity is well demonstrated from the correlation consistent and calendar basis sets (cc→ may→ jun→ jul→ aug).

4. Conclusion

From our investigation we have identified various trends for the predictions of water's molecular properties, of which the most notable is the Def2-n basis set family's overall performance. Pople-style basis sets tend to be very consistent with their predictions, typically producing minimal changes between each individual basis set's calculation. Pople basis sets are a reasonable choice for exploration, but there is little systematic improvement within their family, and are not recommended when results require high accuracy or quantitative values. Although they are occasionally somewhat accurate for infrared intensities when used with the B3LYP level of theory, they are not recommended to predict Raman activities. The Aug-CCn family is good for predicting Raman activities, particularly with the MP2 level of theory, however, they are computationally heavy, and outperformed by less computationally heavy options, such as aug-PCn, Def2-PD, and SPK augmented versions, as well as the majority of spectroscopic basis sets in this area. These basis sets have been shown to produce results comparable to six zeta basis sets, with minimal runtimes for the calculation of many optical properties. Another notable trend is the NO7 and SNS basis families success with calculating vibrational frequencies with harmonic approximation methods. Regarding the ANO series of basis sets, the NASA and Neese series offer high accuracy for geometries, as well as vibrational frequencies and infrared intensities, but lack sufficient diffuse functions to be able to accurately determine Raman activities (even the Aug-Neese sets). Only the Roos augmented ANO basis sets were able to accurately reproduce Raman activities, and represent an excellent all-round choice. We noticed that for the calculation of Raman activities basis sets which incorporated large numbers of diffuse functions always provided superior accuracy, however our scope was limited to the water molecule and further

IOP Conf. Series: Journal of Physics: Conf. Series **1290** (2019) 012013 doi:10.1088/1742-6596/1290/1/012013

investigations need to be conducted.We are currently increasing the scope of this study to include various other molecular species in the hopes of validating trends found in this study.

References

[1] Bennett C J, Ennis C P and Kaiser R I 2014 *The Astrophysical Journal* **782** 63
[2] Glaser R, Hodgen B, Farrelly D and McKee E 2007 *Astrobiology* **7** 455–470
[3] Zvereva E E, Shagidullin A R and Katsyuba S A 2010 *The Journal of Physical Chemistry A* **115** 63–69
[4] Jiménez-Hoyos C A, Janesko B G and Scuseria G E 2008 *Physical Chemistry Chemical Physics* **10** 6621–6629
[5] Huber K P 2013 *Molecular spectra and molecular structure: IV. Constants of diatomic molecules* (Springer Science & Business Media)
[6] Helgaker T, Jorgensen P and Olsen J 2014 *Molecular electronic-structure theory* (John Wiley & Sons)
[7] Halls M D and Schlegel H B 1998 *The Journal of chemical physics* **109** 10587–10593
[8] Halls M D and Schlegel H B 1999 *The Journal of chemical physics* **111** 8819–8824
[9] Scott A P and Radom L 1996 *The Journal of Physical Chemistry* **100** 16502–16513
[10] Buczek A, Kupka T and Broda M A 2011 *Journal of molecular modeling* **17** 2029–2040
[11] Galabov B, Yamaguchi Y, Remington R B and Schaefer H F 2002 *The Journal of Physical Chemistry A* **106** 819–832
[12] Hill J G 2013 *International Journal of Quantum Chemistry* **113** 21–34
[13] Jensen F 2013 *Wiley Interdisciplinary Reviews: Computational Molecular Science* **3** 273–295
[14] Kupka T and Lim C 2007 *The Journal of Physical Chemistry A* **111** 1927–1932
[15] Helgaker T, Klopper W, Koch H and Noga J 1997 *The Journal of chemical physics* **106** 9639–9646
[16] Truhlar D G 1998 *Chemical Physics Letters* **294** 45–48
[17] Petersson G and Al-Laham M A 1991 *The Journal of chemical physics* **94** 6081–6090
[18] Becke A D 1993 *The Journal of chemical physics* **98** 5648–5652
[19] Lee C, Yang W and Parr R G 1988 *Physical review B* **37** 785
[20] Møller C and Plesset M S 1934 *Physical Review* **46** 618
[21] Piecuch P, Kucharski S A, Kowalski K and Musiał M 2002 *Computer Physics Communications* **149** 71–96
[22] Olson R M, Bentz J L, Kendall R A, Schmidt M W and Gordon M S 2007 *Journal of Chemical Theory and Computation* **3** 1312–1328
[23] Hariharan P C and Pople J A 1973 *Theoretica chimica acta* **28** 213–222
[24] Woon D E and Dunning Jr T H 1993 *The Journal of chemical physics* **99** 3730–3737
[25] Papajak E and Truhlar D G 2010 *Journal of chemical theory and computation* **7** 10–18
[26] Dunning Jr T H 1989 *The Journal of chemical physics* **90** 1007–1023
[27] Woon D E and Dunning Jr T H 1994 *The Journal of chemical physics* **100** 2975–2988
[28] Kendall R A, Dunning Jr T H and Harrison R J 1992 *The Journal of chemical physics* **96** 6796–6806
[29] Papajak E and Truhlar D G 2010 *Journal of chemical theory and computation* **6** 597–601
[30] Papajak E, Zheng J, Xu X, Leverentz H R and Truhlar D G 2011 *Journal of chemical theory and computation* **7** 3027–3034
[31] Jensen F 2001 *The Journal of Chemical Physics* **115** 9113–9125
[32] Jensen F 2002 *The Journal of chemical physics* **116** 7372–7379
[33] Jensen F 2014 *Journal of chemical theory and computation* **10** 1074–1085
[34] Weigend F and Ahlrichs R 2005 *Physical Chemistry Chemical Physics* **7** 3297–3305
[35] Neese F and Valeev E F 2010 *Journal of chemical theory and computation* **7** 33–43
[36] Baranowska A and Sadlej A J 2010 *Journal of computational chemistry* **31** 552–560
[37] Benkova Z, Sadlej A J, Oakes R E and Bell S E 2005 *Journal of computational chemistry* **26** 145–153
[38] Barone V, Cimino P and Stendardo E 2008 *Journal of chemical theory and computation* **4** 751–764
[39] Barone V, Biczysko M and Bloino J 2014 *Physical Chemistry Chemical Physics* **16** 1759–1787
[40] Paschoal D, Costa M F and Dos Santos H F 2014 *International Journal of Quantum Chemistry* **114** 796–804
[41] Shimanouchi T 1977 *Journal of physical and chemical reference data* **6** 993–1102
[42] Hoy A and Bunker P R 1979 *Journal of Molecular Spectroscopy* **74** 1–8
[43] Culot F and Liévin J 1992 *Physica Scripta* **46** 502
[44] Tew D P, Klopper W, Heckert M and Gauss J 2007 *The Journal of Physical Chemistry A* **111** 11242–11248
[45] Gordon M S and Schmidt M W 2005 *Advances in electronic structure theory: GAMESS a decade later* (Amsterdam: Elsevier) pp 1167–1189
[46] Sayvetz A 1939 *The Journal of Chemical Physics* **7** 383–389
[47] Ishimura K, Acc S N T C, McMurchie L, Davidson E and Rys J 1993 *J. Comput. Chem* **14** 1347–1363
[48] Krasnoshchekov S V, Schutski R S, Craig N C, Sibaev M and Crittenden D L 2018 *The Journal of chemical physics* **148** 084102

IOP Conf. Series: Journal of Physics: Conf. Series **1290** (2019) 012013 doi:10.1088/1742-6596/1290/1/012013

[49] Halkier A, Helgaker T, Jørgensen P, Klopper W, Koch H, Olsen J and Wilson A K 1998 *Chemical Physics Letters* **286** 243–252
[50] Johnson B G, Gill P M and Pople J A 1993 *The Journal of chemical physics* **98** 5612–5626
[51] Merrick J P, Moran D and Radom L 2007 *The Journal of Physical Chemistry A* **111** 11683–11700
[52] Andersson M P and Uvdal P 2005 *The Journal of Physical Chemistry A* **109** 2937–2941
[53] Halls M D, Velkovski J and Schlegel H B 2001 *Theoretical Chemistry Accounts* **105** 413–421
[54] Hudson R L, Gerakines P A and Ferrante R F 2018 *Spectrochimica Acta Part A: Molecular and Biomolecular Spectroscopy* **193** 33–39
[55] Grimme S 2003 *The Journal of chemical physics* **118** 9095–9102
[56] Piecuch P, Włoch M, Gour J R and Kinal A 2006 *Chemical physics letters* **418** 467–474

IOP Conf. Series: Journal of Physics: Conf. Series **1290** (2019) 012014 doi:10.1088/1742-6596/1290/1/012014

A Wannier orbital based method for resonant inelastic x-ray scattering simulation

Chunjing Jia
Stanford Institute for Materials and Energy Sciences, SLAC National Accelerator Laboratory and Stanford University, Menlo Park, California 94025, USA
E-mail: chunjing@stanford.edu

Abstract. We report an algorithm for simulating oxygen K-edge RIXS for weakly correlated systems, using maximally localized Wannier functions as the basis set. The N-electron wavefunctions are formulated using single Slater determinants, and many-body effects are treated explicitly at the dipole matrix element level. The simulated results for oxygen K-edge RIXS from solid state Li_2CO_3 matches well with the experimental data. Aside from being efficient and reasonably accurate, this algorithm also shows potential to extend to more complex RIXS problems.

1. Introduction

Since its discovery by Wilhelm Roentgen more than a hundred years ago, x-rays have not only improved the quality of life of human beings via chemical imaging and radiotherapy etc., but also provided scientists a powerful tool to directly investigate the microscopic scales of the world. Take transition-metal oxides as an example. X-rays with high photon energy above 5-10 keV, usually called hard x-rays, can be tuned to coincide with the energy difference between specific transition-metal $1s$ core-levels and the valence shell, thus selectively exciting the electronic state of the transition metal in the system. X-rays with lower photon energies, or soft x-rays, can be used to excite transition-metal $2p$ electrons or oxygen $1s$ electrons in a similar fashion. The energy and edge selectivity and their large cross-section are among the main advantages of x-ray spectroscopic techniques. Thanks to the recent development of high-throughput, high-resolution synchrotron light sources and detection instruments, x-ray spectroscopies have become increasingly important in many fields. In physics, x-ray absorption spectroscopy (XAS) and resonant inelastic x-ray scattering (RIXS) have been employed extensively for understanding electronic structure, generally, and collective excitations, specifically, in strongly correlated systems [1, 2, 3, 4]. In chemistry, x-ray spectroscopy has been used to fingerprint electronic structure and look into static and transient states of matter [5]. Recently, XAS and RIXS also have been used to search for better battery materials by "fingerprinting" the redox process [6, 7].

While x-ray spectroscopies have been used successfully in variety of fields, theoretical understanding remains a great challenge. We lack a universal theoretical scheme to treat different systems and different edges. For strongly correlated systems, correlation effects and localized physics dominate, while for more itinerant systems band dispersion matters more. For many other systems lying between the two extremes, both effects remain non-negligible to some extent. To speak of different edges, the commonly measured ones are transition-metal K, L, and M

edges and the oxygen K edge. Dipole selection rules guarantee that specific valence orbitals are excited for a certain edge, requiring that one employ the right method to describe the physics of the electrons in those valence orbitals. One of the main theoretical difficulties lies in the choice of basis – how the wavefunction is formulated should be treated very differently between weakly and strongly correlated electronic systems. See Table 1 for a comparison between the single-particle and the many-body methods of constructing the Hilbert space. To treat strongly correlated electronic systems, one has to use the full many-body basis as the starting point. The size of the Hilbert space scales exponentially with the number of available single-particle states, which could represent the number of orbitals for atomic systems, the number of real space lattice sites for single band model Hamiltonians, or a combination of both for a realistic multi-orbital/multi-site system. This exponential scaling leads to a severe limitation of the largest systems one can attempt to simulate. Usually, full band dispersion, which comes along with the inclusion of hoppings at intermediate and long range, cannot be addressed with this construction. On the other hand, for weakly correlated electronic systems, the many-body electronic wavefunction can be expressed as a single Slater determinant of the single-particle states, thus the size of the Hilbert space scales only linearly with the number of single-particle states. Using this basis set and the corresponding construction, we more easily can address the band dispersion by including intermediate and long range hoppings in a supercell with hundreds of atoms. However, many-body effects are "intentionally" missing, as the single Slater determinant wavefunction construction cannot account for the complex response of the remaining $N-1$ electrons when one electron has been perturbed, as in the XAS or RIXS process. For certain systems, such as strongly correlated electronic systems, this missing part proves crucial and not including it usually proves fatal when trying to describe material characteristics. In a word, with current computing power and current algorithms, it has not been possible to address both strong correlations and the intermediate/long range hoppings precisely at the same time.

Instead of directly tackling this hard problem, and trying to find a universal theory for understanding x-ray spectroscopies of both strongly and weakly correlated systems, scientists in the theoretical x-ray spectroscopy community have taken another route: looking at each problem carefully and developing a number of different methods, each suitable for a specific set of problems. *ab initio* based methods such as FEFF [8] and OCEAN [9] have been developed to target weakly correlated systems and oxygen K-edge spectra. Many-body techniques such as exact diagonalization have been used for strongly correlated systems and transition-metal K/L-edges [10, 11, 12].

In this study, we will report on a newly developed method that adds to the existing algorithm "zoo" for x-ray spectroscopy simulations. We target the oxygen K-edge for weakly correlated systems or at least "not too" strongly correlated systems. We will use a single Slater determinant to construct the N electron wavefunction, and treat many-body effects explicitly on the matrix element level. The main advantage of this method is that instead of using a plane-wave basis, a typical choice in the *ab initio* methods for spectroscopy simulations of solid state materials, we use Wannier orbitals as the single-particle basis. As shown in Table 1, for the single-particle construction there is a one-to-one correspondence between each building block (Wannier orbital in our method or plane-wave for solid state DFT simulations) and Hilbert space basis vector, thus we will use the two terms, Wannier orbitals and basis set, interchangeably in the discussion of our algorithm. One can demonstrate that the most compact basis set for valence electrons is a basis constructed from Wannier orbitals, thus largely accelerating the algorithm and enhancing the numerical accuracy. Moreover, since Wannier orbitals are very similar to atomic orbitals in the solid state systems, by construction it is easier to extend the current algorithm to include correlation effects. Aside from these advantages, we will demonstrate the effectiveness of this algorithm by showing numerical results for RIXS and comparison with experiments on an example material. The numerical simulations show reasonably good agreement and even catch

	single-particle construction	many-body construction
orbitals		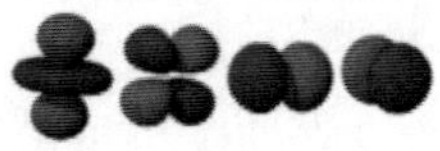
Hilbert space basis set	b_1→ 1 0 0 0 b_2→ 0 1 0 0 b_3→ 0 0 1 0 b_4→ 0 0 0 1	b_1→ 1 1 0 0 b_2→ 1 0 1 0 b_3→ 1 0 0 1 b_4→ 0 1 1 0 b_5→ 0 1 0 1 b_6→ 0 0 1 1 (2-electron subspace)
Hilbert space dimension	$O(N)$	$O(2^N)$
wavefunction	single Slater determinant of the single-particle states	linear combination of the basis vectors

Table 1. Comparison of single-particle and many-body constructions for a spinless four-orbital system. In both cases, each Hilbert space basis vector can be written as a 4-bit binary number. In the single-particle construction, each Hilbert space basis vector has a one-to-one correspondence to the orbital. Thus, the size of the Hilbert space scales linearly with the number of orbitals. In the many-body construction, each bit can be either 0 or 1. This gives a total Hilbert size of 2^N, where N is the number of orbitals. If we confine the total number of electrons to be 2, then we have 6 different basis vectors as shown in the table, where 6 is given by C_4^2.

fine experimental details for the example material Li_2CO_3 at the oxygen K-edge.

The remainder of the manuscript is organized as follows: In section 2, we will talk about the methods by showing the derivation of the algorithm for XAS and RIXS cross-sections. In section 3, we will show the numerical results of RIXS and its comparison with experimental data for Li_2CO_3 at the oxygen K-edge. In section 4, we will conclude this paper with a summary and discussion.

2. Methods

2.1. Hamiltonians

The tight-binding Hamiltonian of the multi-orbital system can be expressed as:

$$H_0 = \sum_{i,j,\alpha_i,\alpha_j,\sigma} t_{i,j,\alpha_i,\alpha_j} c^\dagger_{i,\alpha_i,\sigma} c_{j,\alpha_j,\sigma} \tag{1}$$

where index i/j is the atom index, α_i represents the orbital index at atom i, σ represents spin. $t_{i,j,\alpha_i,\alpha_j}$ represents the hopping between orbital α_i at atom i and orbital α_j at atom j. When $i = j$ and $\alpha_i = \alpha_j$, $t_{i,i,\alpha_i,\alpha_i}$ represents the site energy. This Hamiltonian applies to both the RIXS/XAS initial state and RIXS final state without core-hole.

For electronic states with a core-hole, the Hamiltonian with one core-hole can be written as:

$$H_i = \sum_{i,j,\alpha_i,\alpha_j,\sigma} t_{i,j,\alpha_i,\alpha_j} c^\dagger_{i,\alpha_i,\sigma} c_{j,\alpha_j,\sigma} - \sum_{\alpha_{i_0},\sigma} U_c n_{i_0,\alpha_{i_0},\sigma} n^h_{i_0,\sigma'} + \epsilon(O_h) n^h_{i_0,\sigma'} \tag{2}$$

where $n^h_{i_0,\sigma'}$ is the number operator for the core-hole with spin σ' at site i_0, $\epsilon(O_h)$ is the core-hole site energy, U_c is the core-hole potential which only applies to the orbitals on the same site i_0, $n_{i_0,\alpha_{i_0},\sigma}$ is the electron number operator for orbital α_{i_0} at site i_0. This Hamiltonian applies to the RIXS intermediate state or XAS final state, under the assumption that the only interaction term brought in by the core-hole is the monopole term of the Coulomb interaction. Please note that systems with higher order interactions such as multiplet interactions can not be described by the above Hamiltonian. For this reason, the Wannier orbital based RIXS algorithm, which we propose here, is suitable for simulating XAS and RIXS at the oxygen K-edge (sulfur K-edge etc), but not transition-metal L-edges where the $3d-2p$ multiplet interaction on the transition metal atom typically plays an important role.

There are a number of ways to obtain the tight-binding model Hamiltonian for a material specific problem. What we use here is the maximally localized Wannier functions method implemented in Wannier90 [13], by projecting *ab initio* density functional theory (DFT) results onto the maximally localized Wannier basis set to obtain the Hamiltonian of interest. Compared with directly fitting the DFT bandstructure, each basis element obtained in the Wannier method has a physical meaning, thus it naturally preserves the symmetries of the problem.

2.2. XAS

The general XAS formalism in the single Slater determinant construction has been discussed previously by Liang *et al* [14]. Here, we rederive this for XAS, and extend the derivation to RIXS.

The initial state of the $N+1$ electron wavefunction for XAS is the ground state of the electronic system of N valence electrons plus one core-electron. Without loss of generality, we only consider one core-electron in the initial state or one core-hole in the final state of XAS. This $N+1$ electron wavefunction can be expressed as $|\Phi_0\rangle = \prod_{v=1}^N b_v h|0\rangle$, where $|0\rangle$ is the empty state, h is the core-electron creation operator (or core-hole annihilation operator), b_v annihilates a valence hole (or creates a valence electron) with single particle index v. $|\Phi_0\rangle$ should be understood as the single Slater determinant of the $N+1$ one-electron wavefunction.

The final $N+1$ electron wavefunction for XAS can be presented in the following:

$$|\Phi_i\rangle = \prod_{\mu=1}^{N+1}\tilde{\phi}^\dagger_{i_\mu}\prod_{v=1}^{N} b^\dagger_v h^\dagger|\Phi_0\rangle = \prod_{\mu=1}^{N+1}(\sum_c \xi_{i_\mu,c}a^\dagger_c + \sum_v \xi_{i_\mu,v}b_v)\prod_{v=1}^{N} b^\dagger_v h^\dagger|\Phi_0\rangle \tag{3}$$

where $|\Phi_0\rangle$ is the initial state $N+1$ electron wavefunction, $a^\dagger_c$ creates an electron with index c in the conduction band, and $\prod_{v=1}^N b^\dagger_v h^\dagger|\Phi_0\rangle$ represents the state with empty electrons in the valence and one core-hole. $\tilde{\phi}^\dagger_{i_\mu}$ creates an electron with state index i_μ for the XAS final state Hamiltonian with core-hole, and $\tilde{\phi}^\dagger_{i_\mu}$ is connected with the initial single particle eigenstates without core-hole by the relation: $\tilde{\phi}^\dagger_{i_\mu} = \sum_c \xi_{i_\mu,c}a^\dagger_c + \sum_v \xi_{i_\mu,v}b_v$.

The XAS cross-section can be expressed using Fermi's Golden rule:

$$I(\omega) = \sum_i |\langle\Phi_i|\epsilon\cdot\mathbf{R}|\Phi_0\rangle|^2\delta(E_i - E_0 - \omega) \tag{4}$$

Here $|\Phi_i\rangle$ is the $N+1$ electron final state wavefunction and $|\Phi_0\rangle$ is the $N+1$ electron initial state wavefunction. E_0 is the sum of the N lowest single-particle energies in the valence $E_0 = \sum_{v=1}^N e_v$, and E_i is the sum of the single-particle energies for final state i: $E_i = \sum_{\mu=1}^{N+1}\tilde{e}_{i_\mu} + e_h$. As stated in the introduction, we will treat them in the Slater determinant of the single particle basis, and consider many-body effects by treating the matrix elements explicitly. The matrix elements can be expressed in the single particle basis without core-hole:

$$\langle\Phi_i|\epsilon\cdot\mathbf{R}|\Phi_0\rangle = \sum_c (A^i_c)^*\langle\phi_c|\epsilon\cdot r|\phi_h\rangle \tag{5}$$

where $|\phi_h\rangle$ is the single-particle wavefunction of the core-electron and $|\phi_c\rangle$ is the single-particle wavefunction of a conduction electron with index c. $\langle\phi_c|\epsilon\cdot r|\phi_h\rangle$ shows the dipole of one electron excited from the core up to a conduction state with index c; the coefficient A_c^i shows the overlap of the $(N+1)$-electron wavefunction between the initial state and the final state, and can be expressed as:

$$A_c^i = det\begin{pmatrix} \xi_{i1,v=1} & \xi_{i1,v=2} & \cdots & \xi_{i1,v=N} & \xi_{i1,c} \\ \xi_{i2,v=1} & \xi_{i2,v=2} & \cdots & \xi_{i2,v=N} & \xi_{i2,c} \\ \vdots & & \ddots & & \vdots \\ \xi_{iN+1,v=1} & \xi_{iN+1,v=2} & \cdots & \xi_{iN+1,v=N} & \xi_{iN+1,c} \end{pmatrix} \tag{6}$$

2.3. RIXS

The RIXS cross-section, as expressed in the Kramer's Heissenberg formula [1]:

$$I(\omega_{in},\omega_{loss}) = \sum_f |\sum_i \langle\Phi_f|\epsilon\cdot\mathbf{R}|\Phi_i\rangle \frac{1}{E_i - E_0 - \omega_{in} - i\Gamma}\langle\Phi_i|\epsilon\cdot\mathbf{R}|\Phi_0\rangle|^2\delta(E_f - E_0 - \omega_{loss}) \tag{7}$$

The RIXS intermediate state wavefunction $|\Phi_i\rangle$ has the same form as the XAS final state wavefunction. The RIXS final state $N+1$ electron wavefunction can be expressed as:

$$|\Phi_f\rangle = a_c^\dagger b_v^\dagger|\Phi_0\rangle + a_{c1}^\dagger a_{c2}^\dagger b_{v1}^\dagger b_{v2}^\dagger|\Phi_0\rangle + \ldots \approx a_c^\dagger b_v^\dagger|\Phi_0\rangle \tag{8}$$

This approximation only keeps the first term (or a single electron-hole pair) in the final state. For intermediate states $|\Phi_i\rangle = \prod_{\mu=1}^{N+1}(\sum_c \xi_{\mu,c}a_c^\dagger + \sum_v \xi_{\mu,v}b_v)\prod_{v=1}^{N} b_v^\dagger h^\dagger|\Phi_0\rangle$ and final states $|\Phi_{f=(v,c)}\rangle = a_c^\dagger b_v^\dagger|\Phi_0\rangle$, the matrix element can be expressed as:

$$\langle\Phi_{f=(v,c)}|\epsilon\cdot\mathbf{R}|\Phi_i\rangle = A_c^i\langle\phi_h|\epsilon\cdot r|\phi_v\rangle \tag{9}$$

By expanding the matrix elements with single-particle wavefunction overlaps, the above RIXS cross-section can be expressed as:

$$I(\omega_{in},\omega_{loss}) = \sum_{f=(v,c)} |\sum_i (A_c^i)\langle\phi_h|\epsilon_{out}\cdot r|\phi_v\rangle \frac{1}{E_i - E_0 - \omega_{in} - i\Gamma}(A_c^i)^*\langle\phi_c|\epsilon_{in}\cdot r|\phi_h\rangle|^2\delta(E_f - E_0 - \omega_{loss}) \tag{10}$$

where the final state energy $E_{f=(v,c)} = \sum_{v'=1}^{N} e_{v'} - e_v + e_c$.

The preceding formulation is a general expression for the RIXS cross-section with single electron-hole pair excitations in the final state. In the following, we will make an additional approximation: for the XAS final state or RIXS intermediate state $|\Phi_i\rangle$, we only will consider those many-body states with the N lowest single-particle levels occupied (the remaining electron can occupy any unfilled single-particle state at arbitrary energy):

$$|\Phi_i\rangle = \tilde{\phi}_\mu^\dagger \prod_{l=1}^{N} \tilde{\phi}_l^\dagger \prod_{v=1}^{N} b_v^\dagger h^\dagger|\Phi_0\rangle. \tag{11}$$

For large band gap insulators, intermediate states with one or more electron-hole pairs corresponds to a much larger excitation energy. [14] As we will show in the next section, these approximations give reasonable results for the incident energy range near the excitation edge.

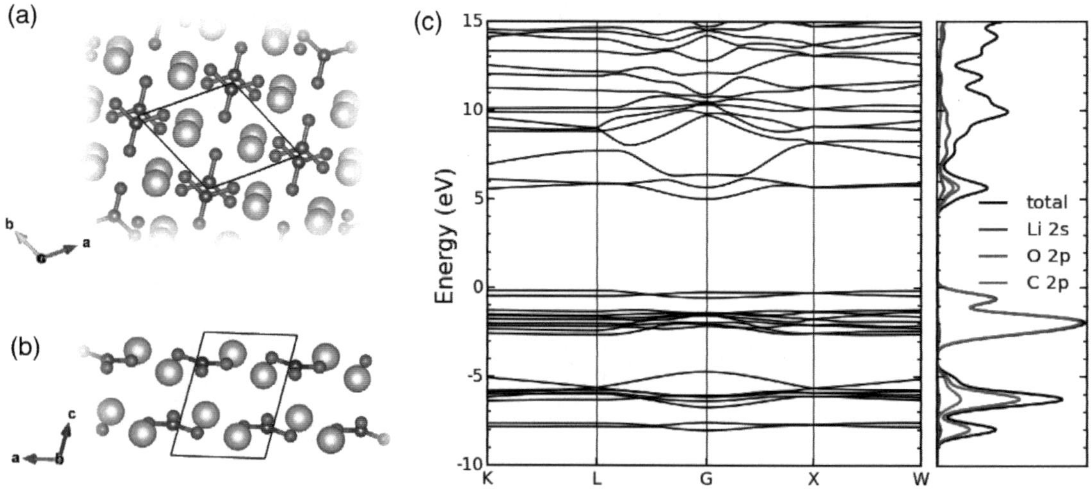

Figure 1. (a) Topview and (b) sideview of Li_2CO_3 crystal structure. Green, brown and red spheres represent Li, C and O atoms respectively. Black frame shows the single primitive cell boundary. (c) DFT calculated band structure and partial density of states of Li_2CO_3. Energy 0 represents fermi level E_f.

3. Results

In this section, we show the numerical simulation of O K-edge RIXS spectra for the example material Li_2CO_3. Experimental RIXS spectra of Li_2CO_3 at the oxygen K-edge have been measured by Zhuo and his colleagues [16]. This material is interesting because of its connection with Li-ion battery materials and possible explanations of anionic redox. The crystal structure of Li_2CO_3 has been shown in Figure 1(a) and (b). Each primitive cell has four Li^+ and two CO_3^{2-}. Solid state Li_2CO_3 is an insulator with band gap $\sim$ 5eV, with its band structure and partial density of states shown in Figure 1(c). The bandstructure and density of states were calculated with the density functional theory package *Quantum Espresso* [17] using a PBE exchange-correlation functional, norm-conserving pseudopotentials, and a $4 \times 4 \times 4$ momentum sampling.

The Wannier downfolding using Wannier90 [13] was implemented with the initial projection onto carbon pz orbitals, oxygen $px/py/pz$ orbitals, Li s orbitals and ten s-like orbitals at random starting positions (a 34-orbital model). Oxygen K-edge RIXS was calculated using Eq. (10) with a $4 \times 4 \times 4$ supercell. Both the incoming and outgoing x-ray photons were taken to be non-polarized. We used core-hole lifetime broadening $\Gamma = 0.25$eV, core-hole potential $U_c = 3.5$eV and $\epsilon(O_{1s}) = 534$eV (relative to E_f) in the calculation.

In Figure 2, the experimental RIXS data shows three main emission features at x-ray emission energies 519eV, 520.5eV and 526eV respectively. Our theoretical simulation reproduced all the three emission features at 519eV, 520.5eV and 525eV respectively. In simulation we see another emission feature at 526.5eV, which is not seen in the experiment. A probable explanation is that in experiment the 526.5eV emission has merged into the main 526eV emission. Simulation with a larger core-hole lifetime broadening $\Gamma = 0.5$eV also reports merging of these two emissions. To understand the nature of these excitations, we can make connection with the calculated partial density of states, as shown in Figure 1(c). These four calculated emission lines at 519eV,

IOP Conf. Series: Journal of Physics: Conf. Series **1290** (2019) 012014 doi:10.1088/1742-6596/1290/1/012014

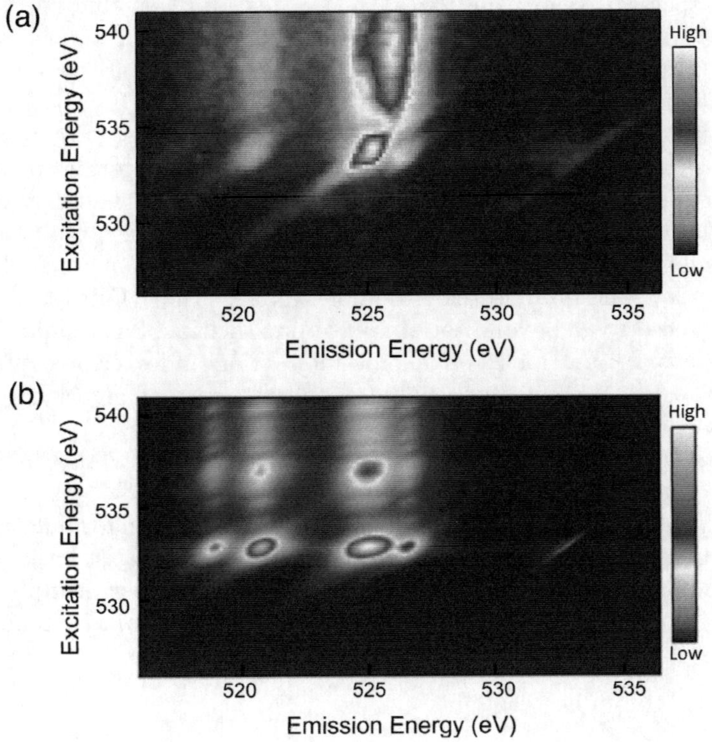

Figure 2. (a) Experimental O *K*-edge RIXS of Li_2CO_3, reprint from Ref.[16]. (b) Theoretical O *K*-edge RIXS spectra of Li_2CO_3 using our method.

520.5eV, 525eV and 526.5eV correspond to the energy for the recombination of valence electrons with oxygen $2p$ orbital content from $\sim$ -8eV, -6eV, -2eV and -0.5eV back to the $1s$ core level.

At incident excitation energy $\sim 533eV$, the experimental RIXS spectra show four Raman-like features (with a finite energy loss) at emission energies at $\sim$ 518eV, 520.5eV, 525eV, and a sharp peak at 526.5eV on the right. Our numerical simulation reports four Raman-like features at $\sim 533eV$ excitation energy, and with emission energies $\sim$ 518.5eV, 520.5eV, 525eV, and 526.5eV. The $\sim 533eV$ incident x-ray energy corresponds to the energy of an electron to be excited from a $1s$ core level to unfilled electronic states with O $2p$ orbital content between 5 - 6.5eV, as shown in Figure 1(c). The energy loss of these four Raman-like features corresponds to the energy of the final state with a single electron-hole pair, where the electron resides in the conduction band with oxygen $2p$ orbital content and the hole resides in one of the four occupied oxygen $2p$ electronic states. For both the emission lines and the Raman features, some of the calculated energies do not line up perfectly with experimental data, mainly due to missing valence Coulomb interactions and excitonic effects in the calculation. However, overall, our numerically simulated Li_2CO_3 RIXS spectra well reproduce the main features of the experimental data.

4. Conclusions and discussions

There are several points to discuss about this algorithm. (1) This algorithm is "almost" *ab initio*, since the only adjustable parameters are U_c and $\epsilon(O_{1s})$, both of which potentially can be evaluated in DFT calculations using an appropriate oxygen pseudopotential with a $1s$ core-hole. In our Li_2CO_3 calculation, we take $\epsilon(O_{1s})$ to be adjustable, fit with the experimental energy scale. The spectral weight only weakly depends on the value of U_c. (2) At higher incoming energies ($>$ 537eV) where the emission lines dominate, our algorithm underestimates the spectral weight, compared with the experiment. The possible explanation is that the approximation in Eq.(11) does not include some of the high energy ($>$ absorption edge + band gap) configurations in the intermediate state. [14] Low energy scales, close to absorption edges, dominate the discussion of most RIXS spectra. They capture rich information about the overlap of the ground, intermediate, and final states of the RIXS process. The current approximation works well for simulating RIXS spectra close to absorption edges for large band gap insulators, while remaining computationally efficient. For semiconductors and metals, the approximation can be changed to allow one or more electron-hole pairs in the RIXS intermediate state. In other words, we may need to go beyond the simple approximations as stated in Eq.(11). (3) The current theory does not account for the Coulomb interaction and excitonic effects. However, in the Wannier orbital basis, one can potentially extend the current algorithm by including the Coulomb interaction (both on-site and at short/intermediate distances). (4) Since the Wannier orbital basis is the most compact, complete basis set for valence electrons, this algorithm is very efficient in terms of both time and space computational complexity.

To summarize, we have developed a Kramer's-Heissenberg-like formulation for the RIXS cross-section by addressing the many-body effects at the matrix element level. Simulated results of oxygen K-edge RIXS for solid state Li_2CO_3 well match the experimental data. Aside from being efficient and reasonably accurate, this algorithm shows potential to extend to more complex RIXS problems.

Acknowledgments

The author thanks Thomas Devereaux and Brian Moritz for enormous support and help. The author thanks Wanli Yang for extensive discussions about the connection between theory and experiment. The author thanks Yuan Chen and Ilkyu Lee for insightful discussions. The author also acknowledges fruitful discussions with Yufeng Liang, whose previous studies and strong passion in physics inspire the author a lot. This work was supported at SLAC and Stanford University by the U.S. Department of Energy, Office of Basic Energy Sciences, Division of Materials Sciences and Engineering, under Contract No. DE-AC02-76SF00515.

References

[1] Ament, L. J. P., van Veenendaal, M., Devereaux, T. P., Hill, J. P. and van den Brink, J. Resonant inelastic x-ray scattering studies of elementary excitations., 2011, *Rev. Mod. Phys.* **83**, 705?767

[2] M. Le Tacon, G. Ghiringhelli, J. Chaloupka, M. Moretti Sala, V. Hinkov, M. W. Haverkort, M. Minola, M. Bakr, K. J. Zhou, S. Blanco-Canosa, C. Monney, Y. T. Song, G. L. Sun, C. T. Lin, G. M. De Luca, M. Salluzzo, G. Khaliullin, T. Schmitt, L. Braicovich and B. Keimer, Intense paramagnon excitations in a large family of high-temperature superconductors, 2011, *Nature Physics*, **7**, 725

[3] M. P. M. Dean, G. Dellea, R. S. Springell, F. Yakhou-Harris, K. Kummer, N. B. Brookes, X. Liu, Y.-J. Sun, J. Strle, T. Schmitt, L. Braicovich, G. Ghiringhelli, I. Bozovic, J. P. Hill, Persistence of magnetic excitations in $La_{2-x}Sr_xCuO_4$ from the undoped insulator to the heavily overdoped non-superconducting metal, 2013, *Nature Materials*, **12**, 1019

[4] HY Huang, CJ Jia, ZY Chen, K Wohlfeld, B Moritz, TP Devereaux, WB Wu, J Okamoto, WS Lee, M Hashimoto, Y He, ZX Shen, Y Yoshida, H Eisaki, CY Mou, CT Chen, DJ Huang, Raman and fluorescence characteristics of resonant inelastic x-ray scattering from doped superconducting cuprates, 2016, *Scientific Reports*, **6**, 19657

[5] Ph Wernet, Kristjan Kunnus, Ida Josefsson, Ivan Rajkovic, Wilson Quevedo, Martin Beye, Simon Schreck, Sebastian Grübel, Mirko Scholz, Dennis Nordlund, Wenkai Zhang, Robert W Hartsock, William F Schlotter, Joshua J Turner, Brian Kennedy, Franz Hennies, Frank MF de Groot, Kelly J Gaffney, Simone Techert, Michael Odelius, Alexander Föhlisch, Orbital-specific mapping of the ligand exchange dynamics of $Fe(CO)_5$ in solution, 2015, *Nature*, **520**, 78

[6] William E. Gent, Kipil Lim, Yufeng Liang, Qinghao Li, Taylor Barnes, Sung-Jin Ahn, Kevin H. Stone, Mitchell McIntire, Jihyun Hong, Jay Hyok Song, Yiyang Li, Apurva Mehta, Stefano Ermon, Tolek Tyliszczak, David

Kilcoyne, David Vine, Jin-Hwan Park, Seok-Kwang Doo, Michael F. Toney, Wanli Yang, David Prendergast and William C. Chueh, 2017, *Nature Communications* **8**, 2091

[7] Wanli Yang and Thomas Devereaux, Anionic and cationic redox and interfaces in batteries: Advances from soft x-ray absorption spectroscopy to resonant inelastic scattering, 2018, *Journal of Power Sources*, **389**, 188

[8] J.J. Rehr, J.J. Kas, F.D. Vila, M.P. Prange, K. Jorissen, Parameter-free calculations of x-ray spectra with FEFF9, *Phys. Chem. Chem. Phys.*, 2010, **12**, 5503

[9] K. Gilmore, J. Vinson, E. L. Shirley, D. Prendergast, C. D. Pemmaraju, J. J. Kas, F. D. Vila, J. J. Rehr, Efficient implementation of core-excitation Bethe-Salpeter equation calculation., 2015, *Comp. Phys. Comm.* **197**, 109

[10] CJ Jia, CC Chen, AP Sorini, B Moritz, TP Devereaux, Uncovering selective excitations using the resonant profile of indirect inelastic x-ray scattering in correlated materials: observing two-magnon scattering and relation to the dynamical structure factor, 2012, *New Journal of Physics*, **14**, 113038

[11] CJ Jia, EA Nowadnick, K Wohlfeld, YF Kung, C-C Chen, S Johnston, T Tohyama, B Moritz, TP Devereaux, Persistent spin excitations in doped antiferromagnets revealed by resonant inelastic light scattering, 2014, *Nature Comm.*, **5**, 3314

[12] Chunjing Jia, Krzysztof Wohlfeld, Yao Wang, Brian Moritz, Thomas P Devereaux, Using RIXS to uncover elementary charge and spin excitations, 2016, *Physical Review X*, **6**, 021020

[13] AA Mostofi, JR Yates, G Pizzi, YS Lee, I Souza, D Vanderbilt, N Marzari, An updated version of wannier90: A tool for obtaining maximally-localised Wannier functions, 2014, *Comput. Phys. Commun.* **185**, 2309

[14] Yufeng Liang and John T Vinson and Sri D Pemmaraju and Walter Drisdell and Eric L Shirley and David Prendergast, x-ray Absorption in Transition Metal Oxides: Self-Consistent Approaches versus a Many-Body Perturbation Theory, 2017, *Physical Review Letters* **118**, 096402

[15] Yufeng Liang and David Prendergast, Quantum many-body effects in x-ray spectra efficiently computed using a basic graph algorithm, 2018, *Physical Review* B **20**, 205127

[16] Zengqing Zhuo, Chaitanya Das Pemmaraju, John Vinson, Chunjing Jia, Brian Moritz, Ilkyu Lee, Shawn Sallies, Qinghao Li, Jinpeng Wu, Kehua Dai, Yi-de Chuang, Zahid Hussain, Feng Pan, Thomas P. Devereaux, Wanli Yang, Spectroscopic Signature of the oxygen States in Peroxides, 2018, *J. Phys. Chem. Lett.*, **9**(21), pp 6378-6384

[17] P. Giannozzi, S. Baroni, N. Bonini, M. Calandra, R. Car, C. Cavazzoni, D. Ceresoli, G. L. Chiarotti, M. Cococcioni, I. Dabo, A. Dal Corso, S. Fabris, G. Fratesi, S. de Gironcoli, R. Gebauer, U. Gerstmann, C. Gougoussis, A. Kokalj, M. Lazzeri, L. Martin-Samos, N. Marzari, F. Mauri, R. Mazzarello, S. Paolini, A. Pasquarello, L. Paulatto, C. Sbraccia, S. Scandolo, G. Sclauzero, A. P. Seitsonen, A. Smogunov, P. Umari, R. M. Wentzcovitch, , QUANTUM ESPRESSO: a modular and open-source software project for quantum simulations of materials, 2009, *Journal of Physics: Condensed Matter*, **21**, 39

IOP Conf. Series: Journal of Physics: Conf. Series **1290** (2019) 012015 doi:10.1088/1742-6596/1290/1/012015

Driving force dependence of the height of a faceted macrostep in non-equilibrium steady-state crystal growth

Noriko Akutsu

Faculty of Engineering, Osaka Electro-Communication University, Hatsu-cho, Neyagawa, Osaka 572-8530, Japan

E-mail: nori3@phys.osakac.ac.jp

Abstract. In order to understand the dynamics of the self-organized macrosteps, the vicinal surface with faceted macrosteps is studied by the Monte Carlo method based on a microscopic lattice model, the restricted solid-on-solid model with point-contact-type step-step attraction (p-RSOS model). We focus on the dynamical effects caused by the change of the surface roughness or the change of the kink density which are masked by the effect of the surface and volume diffusion of the crystal atoms in the ambient phase. Contrast to the step-bunching in the diffusion-limited cases, the height of the faceted macrostep decreases as the driving force for the crystal growth increases.

1. Introduction

Making good quality crystals of SiC is an urgent problem for the production of future power devices with low power consumption. Self-organized macrosteps are known to hinder making good quality of crystal of SiC [1]. We studied the largeness of the macrosteps theoretically based on a lattice model, the restricted solid-on-solid model with point-contact-type step-step attraction (p-RSOS model) [2, 3]. We focus on the surface dynamics with the change of kink density or surface roughness, which has long been overlooked in studies of surface and volume diffusion. In our lattice model, a faceted macrostep exists on a vicinal surface stably, even at equilibrium at low temperatures. The characteristics of the profile of the macrostep are classified by the connectivity of the surface free energy (or the surface tension) which was calculated by the density matrix renormalization group method [4]-[9]. Driving force dependence of the profile of vicinal surface with faceted macrosteps is analyzed by the Monte Carlo method [14, 15]. Contrast to the step-bunching in the diffusion-limited cases[10]-[13], the height of the faceted macrostep decreases as the driving force for the crystal growth increases.

2. The model and statistical mechanical calculations

2.1. The p-RSOS model

The microscopic model considered in this study is the p-RSOS model (Fig. 1) [2, 14]-[20]. The total energy of the (001) surface can be written as

$$\mathcal{H}_{\text{p-RSOS}} = \mathcal{N}\epsilon_{\text{surf}} + \sum_{n,m} \epsilon[|h(n+1,m) - h(n,m)| + |h(n,m+1) - h(n,m)|]$$

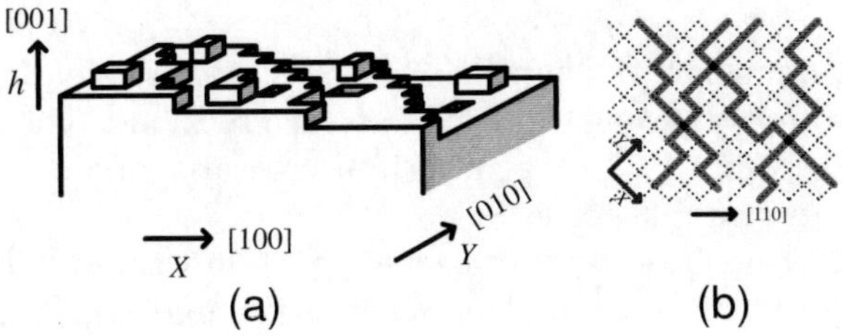

Figure 1. Perspective view of the surface. This figure is taken from [16].

$$+\sum_{n,m} \epsilon_{\rm int}[\delta(|h(n+1,m+1)-h(n,m)|,2)+\delta(|h(n+1,m-1)-h(n,m)|,2)], \quad (1)$$

where $\mathcal{N}$ is the total number of lattice points, $\epsilon_{\rm surf}$ is the surface energy per unit cell on the planar (001) surface, and ϵ is the microscopic ledge energy. The RSOS condition, which is the height difference between the NN sites restricted to $\{0, \pm 1\}$, is required implicitly. Here, $\delta(a, b)$ is the Kronecker delta, and $\epsilon_{\rm int}$ is the microscopic step–step interaction energy. $\epsilon_{\rm int}$ contributes to the surface energy only at the meeting point of neighboring steps because the height difference can be ± 2 only at the meeting points of neighboring steps. The summation with respect to (n, m) is taken over all sites on the square lattice. When $\epsilon_{\rm int}$ is negative, the step–step interaction becomes attractive (sticky steps).

2.2. Calculations of the surface tension: DMRG (PWFRG) calculations

The surface energy of a vicinal surface is calculated by the following equation [21]:

$$\mathcal{H}_{\rm vicinal} = \mathcal{H}_{\rm p-RSOS} - \eta_x \sum_{n,m}[h(n+1,m)-h(n,m)] - \eta_y \sum_{n,m}[h(n,m+1)-h(n,m)], \quad (2)$$

where $\boldsymbol{\eta} = (\eta_x, \eta_y)$ is the the Andreev field [22]. The Andreev field is a chemical potential with respect to a single step.

From statistical mechanics, the grand partition function $\mathcal{Z}$ of the vicinal surface is calculated as follows: $\mathcal{Z} = \sum_{\{h(m,n)\}} \exp[-\beta \mathcal{H}_{\rm vicinal}]$where $\beta = 1/k_{\rm B}T$. The summation with respect to $\{h(m, n)\}$ is taken over all possible values of $h(m, n)$. The thermodynamic grand potential calculated from the grand partition function $\mathcal{Z}$ is the Andreev free energy $\tilde{f}(\boldsymbol{\eta})$ [22], which is obtained as follows:

$$\tilde{f}(\boldsymbol{\eta}) = \tilde{f}(\eta_x, \eta_y) = -\lim_{\mathcal{N}\to\infty} \frac{1}{\mathcal{N}} k_{\rm B}T \ln \mathcal{Z}. \quad (3)$$

In low-dimensional cases, more precision is required than that provided by a mean field calculation of the partition function [23]. Hence, to obtain reliable results, we adopt the DMRG method [4]-[9], which was developed for one-dimensional quantum spin systems. We use the transfer matrix version of the DMRG method, which is known as the product wave-function renormalization group (PWFRG, tensor network) method [7, 8, 9].

The surface free energy per projected x–y area $f(\boldsymbol{p})$ is obtained from $\tilde{f}(\boldsymbol{\eta})$ using the following equation:

$$f(\boldsymbol{p}) = \tilde{f}(\boldsymbol{\eta}) + \boldsymbol{\eta}\cdot\boldsymbol{p}, \quad (4)$$

where $\boldsymbol{p} = (p_x, p_y)$ is the surface gradient of the vicinal surface. The surface tension is the surface free energy per unit normal area. Hence, the surface tension $\gamma_{\rm surf}(\boldsymbol{p})$ was calculated from the surface free energy $f(\boldsymbol{p})$ as [24]

$$\gamma_{\rm surf}(\boldsymbol{p}) = \frac{f(\boldsymbol{p})}{\sqrt{1+p_x^2+p_y^2}}. \quad (5)$$

2.3. Faceting diagram obtained by the anomalous surface tension

In this manner, we obtained the surface tension of the p-RSOS model numerically. The Fig. 2 shows the Wulff figures. The blue lines show the polar graph of surface tensions. The pink lines show the Andreev free energy which is similar to the equilibrium crystal shapes. These light blue lines show the surface tension for metastable states.

As seen from Fig. 2, there are two transition temperatures. At temperatures lower than $T_{f,1}$, the surface tension becomes discontinuous around (111) facet. At temperatures lower than $T_{f,2}$, the surface tension becomes discontinuous around (001) facet. So, we call the zone at the temperatures less than $T_{f,2}$ the step-faceting zone; the zone at the temperatures $T_{f,2} < T < T_{f,1}$, the step droplet zone; and the zone at the temperatures $T_{f,1} < T$, we call the Gruber-Mullins-Pokrovsky-Talapov (GMPT) zone.

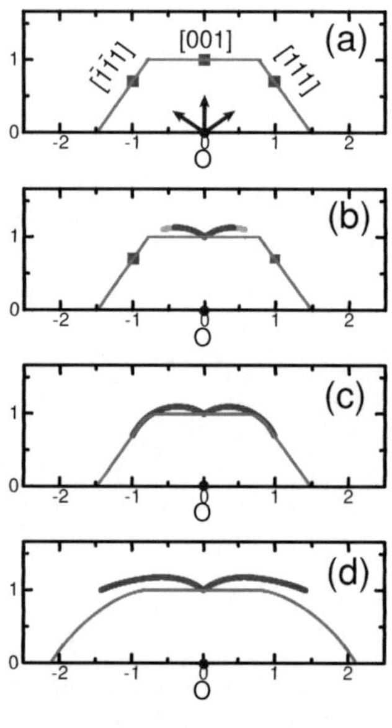

Figure 2. Wulff figures. $\epsilon_{\rm int}/\epsilon = -0.9$. (a) $k_{\rm B}T/\epsilon = 0.6$. (b) $k_{\rm B}T/\epsilon = 0.63$. (c) $k_{\rm B}T/\epsilon = 0.73$. (d) Original RSOS model at $k_{\rm B}T/\epsilon = 0.6$. and $\epsilon_{\rm int} = 0$. This figure is taken from [16].

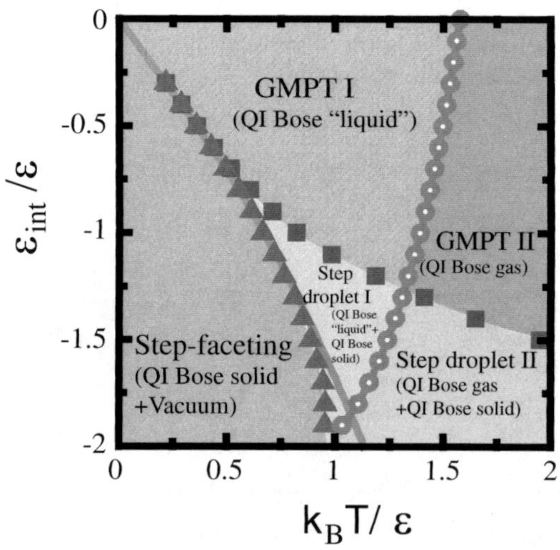

Figure 3. Faceting diagram. The red triangles: $T_{f,2}$. The purple squares: $T_{f,1}$. The pink circles: the roughening transition temperatures of the (001) surface. All marks show the values calculated by the PWFRG method. This figure is taken from [19].

According to the connectivity of the surface tension, we can obtain the faceting diagram (Fig. 3). In the step-faceting zone, only the (001) surface and the (111) surface are thermodynamically stable. Hence, the vicinal surface is formed by the giant steps of the smooth (001) and the (111) surfaces. Whereas, in the step droplet zone, the (111) surface and the vicinal surface with the slope of p_1 are thermodynamically stable. The vicinal surface is formed by macrosteps with the (111) side surface and the stepped surface with the slope of p_1.

The characteristics of the profile of the faceted macrostep is classified according to the connectivity of the surface tension [16].

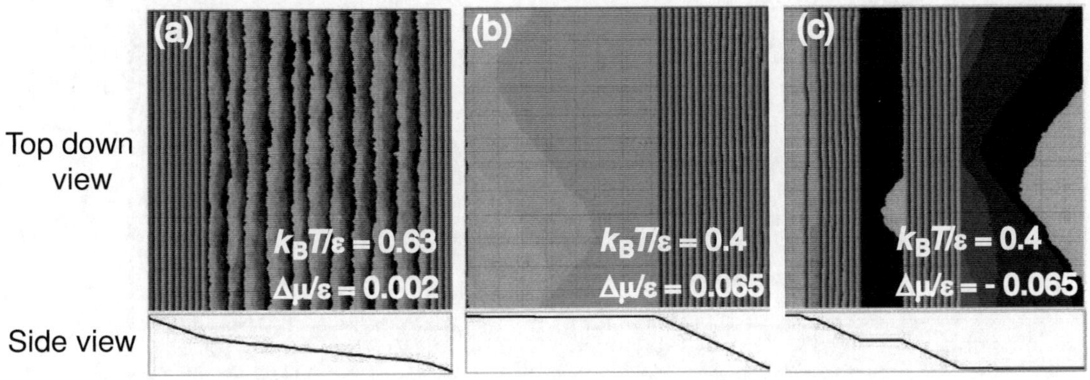

Figure 4. Snapshots of the simulated surfaces at 4×10^8 MCS/site. The height of the surface is shown by the brightness by 10 gradations: thus the brighter the image, the higher the surface. Size: $240\sqrt{2} \times 240\sqrt{2}$. $N_{\rm step} = 240$ for (a), and $N_{\rm step} = 180$ for (b) and (c). $\epsilon_{\rm int}/\epsilon = -0.9$. The faceted side surfaces (the (111) surfaces) remain stably at the equilibrium.

3. Height of the faceted macrostep in the non-equilibrium steady state

3.1. Monte Carlo method

To study the non-equilibrium steady-state with macrosteps, a vicinal surface with the following Hamiltonian and a fixed number $N_{\rm step}$ of steps was investigated using the Monte Carlo method with the Metropolis algorithm:

$$\mathcal{H}_{\rm noneq} = \mathcal{H}_{\rm p-RSOS} - \Delta\mu \sum_{n,m} [h(n,m,t+1) - h(n,m,t)], \tag{6}$$

where t is the time measured by the Monte Carlo steps per site (MCS/site), and Δmu is the driving force of the crystal growth (chemical potential difference between the bulk crystal and the ambient phase). When $\Delta\mu > 0$, the crystal grows, whereas when $\Delta\mu < 0$, the crystal recedes (evaporates, dissociates, or melts).

We required periodic boundary condition in the direction of the mean step-running. In the direction normal to the mean step-running, the surface heights of the lowest side line are connected to the the surface heights at the topmost line by adding the number of elementary steps. As the initial configurations of the surface, the parallel train of steps and the single macrostep are prepared.

In the case of the taking average, the first 2×10^8 Monte Carlo steps per site (MCS/site) were discarded, and then averaged over following 2×10^8 MCS/site. The snapshots of the simulated surfaces are shown in Fig. 4.

3.2. Step droplet zone

In Fig. 5, the driving force dependence of the average height of the macrosteps and the velocity of the surface [14] are shown.

The average height of the macrosteps are obtained as follows:

$$\langle n \rangle = \sum_{\tilde{y}} \sum_{\tilde{x}} |n_{\tilde{x}}(\tilde{y})| / [\sum_{\tilde{y}} n_{\rm step}(\tilde{y})] \approx N_{\rm step}/\langle n_{\rm step} \rangle, \tag{7}$$

where $\tilde{x}$ is selected as the $\langle 110 \rangle$ direction (normal to the mean step-running direction), $\tilde{y}$ is the $\langle \bar{1}10 \rangle$ direction (along the mean step-running direction), $N_{\rm step}$ is the total number of elementary steps, and $n_{\rm step}$ is the number of merged steps.

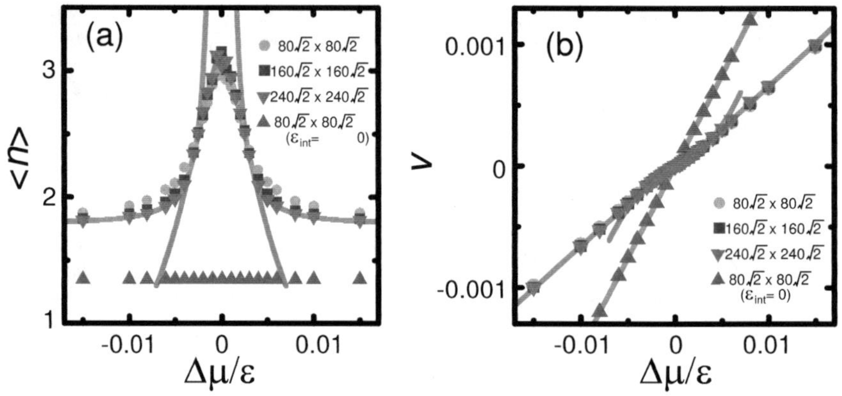

Figure 5. (a) Driving force dependence of the height of the macrostep. (b) Driving force dependence of the growth velocity of the surface. $k_{\rm B}T/\epsilon = 0.63$. $\epsilon_{\rm int}/\epsilon = -0.9$. This figure is taken from ref. [14]

To estimate the growth rate of the surface V, the average surface height $\bar{h}(t)$, was calculated, where

$$\bar{h}(t) = (1/\mathcal{N}) \sum_{n,m} h(n,m). \tag{8}$$

The growth rate of the surface V, is defined as:

$$V = [\bar{h}(t_{\rm max}) - \bar{h}(t_0)]/(t_{\rm max} - t_0), \tag{9}$$

where t_0 and $t_{\rm max}$ are 2×10^8 MCS/site and 4×10^8 MCS/site, respectively.

By analyzing the obtained values, we found a crossover point $\Delta\mu_R/\epsilon = 0.005 \pm 0.002$. As $|\Delta\mu|$ increases, the configuration of the vicinal surface crossover to the two surface co-existent one to the kinetically roughened one. In the two-surface co-existent case, elementary steps attach to or detach from the faceted side surface. To understand the $\Delta\mu$ dependence of the vicinal surface morphology for $|\Delta\mu| < \Delta\mu_R$, we considered a step–attachment–detachment model for the time evolution of $\langle n \rangle$ [14]:

$$\frac{\partial \langle n \rangle}{\partial t} = n_+ - n_-, \tag{10}$$

where n_+ is the rate when the elementary steps catch up to a macrostep, and n_- is the rate when the elementary steps detach from a macrostep. When $n_+ < n_-$, a macrostep dissociates (the case of $\Delta\mu_R < |\Delta\mu|$), whereas when $n_+ > n_-$ (the case of $\Delta\mu_f < |\Delta\mu| < \Delta\mu_{co}$, §3.3), $\langle n \rangle$ increases up to $N_{\rm step}$, where $N_{\rm step}$ is the total number of elementary steps on the surface. In this case, n_- limits the growth/recession rate of the surface. At steady-state, $n_+ = n_- = V/a$, where a is the height of the elementary step.

Then, in the case of $|\Delta\mu| < \Delta\mu_R$, after some calculations [14], V is expressed as follows:

$$\begin{aligned} V &= p_1(\Delta\mu) v_1(\Delta\mu), & (11)\\ p_1(\Delta\mu) &= 0.332 + 15.6|\Delta\mu|/\epsilon + 4.43 \times 10^3 (|\Delta\mu|/\epsilon)^2, & (12)\\ v_1(\Delta\mu) &= 0.132\Delta\mu, & (13) \end{aligned}$$

where $p_1(\Delta\mu)$ is the slope of the "terrace surface", $v_1(\Delta\mu)$ is the mean step velocity of the elementary steps obtained by this Monte Carlo calculations. We show the calculated lines in the Fig. 5 (b) by using Eq. (11) together with Eqs. (12) and (13) by light blue lines. These lines reproduce the Monte Carlo data well.

On the other hand, $\langle n \rangle$ is expressed by z, N_m and N_{step} as follows:

$$\langle n \rangle = \left(z + \frac{N_m}{N_{\text{step}}} \right)^{-1}, \quad z^{-1} = \frac{\bar{p}}{(\sqrt{2} - \bar{p})} \left(\frac{\sqrt{2}}{p_1} - 1 \right) \tag{14}$$

where N_m is the number of the macrosteps, and $\bar{p} = aN_{\text{step}}/L$ is $1/\sqrt{2}$.

The lines of $\langle n \rangle$ are shown by the light blue lines in Fig. 5 (a). These lines also reproduce the Monte Carlo results well.

3.3. Step-faceting zone

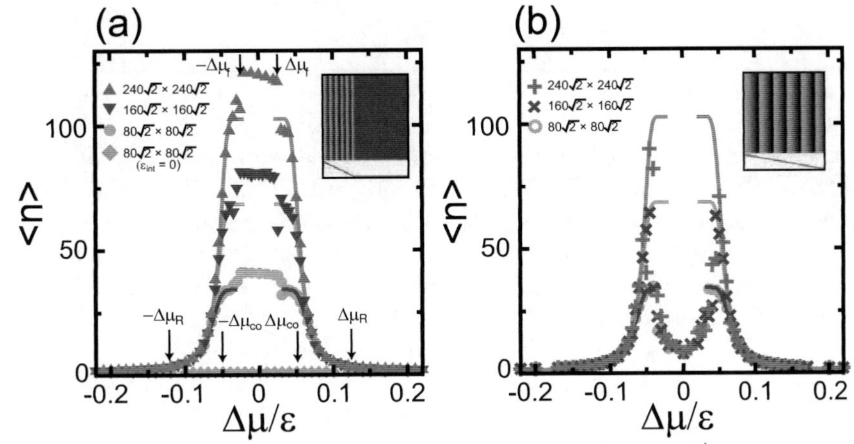

Figure 6. Driving force dependence of the height of the macrostep. Insets show the initial configurations. $k_BT/\epsilon = 0.4$. $\epsilon_{\text{int}}/\epsilon = -0.9$. This figure is taken from ref. [15].

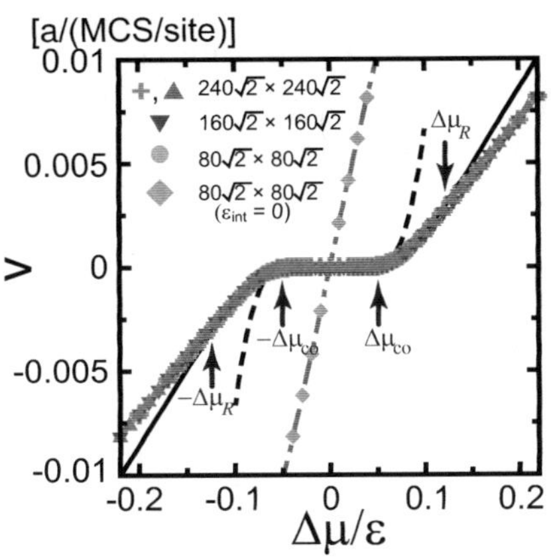

Figure 7. Driving force dependence of the growth velocity of the surface. $k_BT/\epsilon = 0.4$. $\epsilon_{\text{int}}/\epsilon = -0.9$. This figure is taken from ref. [15].

When the temperature is low enough in the step-faceting zone, the driving force dependences of the height of the macrostep and the growth velocity of the surface change from the case in the previous subsection. The vicinal surface at equilibrium consists of the (001) surfaces and the (111) surfaces because only these surfaces are thermodynamically stable.

IOP Conf. Series: Journal of Physics: Conf. Series **1290** (2019) 012015 doi:10.1088/1742-6596/1290/1/012015

Table 1. Characteristic driving forces.($L/(\sqrt{2}a) = 240$, $a = 1$.)

Symbol	value/ϵ	$k_{\rm B}T/\epsilon$	description
$\Delta\mu_f(MCS_{\rm max}, L)$	0.023 ± 0.007	0.4	Freezing point (§3.3)
$\Delta\mu_{co}(L)$	0.050 ± 0.007	0.4	Crossover point from 2D nucleation mode to successive step-detachment mode (§3.3).
$\Delta\mu_R(L)$	0.121 ± 0.005	0.4	Crossover point between the step-detachment mode and kinetically roughened mode.
	0.005 ± 0.002	0.63	§3.2

The most notable characteristic in the dynamics in the step-faceting zone is the freezing of the motion of the surface in the area near equilibrium. Since the both of the (001) and the (111) surfaces are smooth, there are very few kinks on the surface. Hence, two-dimensional nucleation process is needed to grow the surfaces. However, the mean waiting time of the formation of the critical nucleus becomes exponentially longer as the driving force $|\Delta\mu| \to 0$. The waiting time easily exceed the simulation time near equilibrium. In this manner, the dynamics near equilibrium strongly affected by the finiteness of the system size and time.

Then, by analyzing the obtained values, we found several characteristic values for the driving force in addition to $\Delta\mu_R$. We show them in Table 1. For $|\Delta\mu| < \Delta\mu_f$, no nucleation occurs during the simulation time $MCS_{\rm max}$ on the average. Hence, the surface configuration almost freezes and it depends on the initial configuration (Fig. 6 (a) and (b)).

For $\Delta\mu_f < |\Delta\mu| < \Delta\mu_{co}$, the surface grows intermittently in the manner of the single nucleation on the line of the lowest edge of the macrostep.

For $\Delta\mu_{co} < |\Delta\mu| < \Delta\mu_R$, the surface grows with the successive detachment of elementary steps from the lowest edge of the macrostep.

To understand the $\Delta\mu$ dependence of the vicinal surface morphology for $\Delta\mu_{co} < |\Delta\mu| < \Delta\mu_R$, we again considered a step–attachment–detachment model for the time evolution of $\langle n \rangle$. The Eqs. (11) and (14) are also applicable to this case. However, the equations of $p_1(\Delta\mu)$ and $v_1(\Delta\mu)$ become complex due to the two-dimensional multi-nucleation on the edge of the macrostep. After some calculations [15], $p_1(\Delta\mu)$ and $v_1(\Delta\mu)$ are expressed as follows:

$$p_1(\Delta\mu) = \frac{c_p}{\sqrt{|\Delta\mu/\epsilon|}} \exp\left[\frac{-g_p^*/2}{|\Delta\mu/\epsilon| - \Delta\mu_y(L)/\epsilon}\right], \quad (15)$$

$$v_1(\Delta\mu) = \{0.094 + 3.2 \times 10^{-3} \exp[0.18/|\Delta\mu/\epsilon|]\}\Delta\mu/\epsilon,$$

$$g_p^* = 0.423\epsilon, \quad c_p = 0.604, \quad \Delta\mu_y(L)/\epsilon = 0.018 \pm 0.006. \quad (16)$$

Here, $\Delta\mu_y(L)$ is introduced [15] so that the growth velocity of the surface is well explained by the classical nucleation theory qualitatively and quantitatively. Physically, $\Delta\mu_y(L)$ indecates a yield point of the self-detachment of elementary steps from a macrostep. The surface velocity (Eq.(11)) calculated by the Eqs. (15) and (16) is shown in Fig. 7 with black solid lines. The mean height of the macrostep (Eq. 14) are also shown in Fig. 6 with colored solid lines. In both cases, the lines reproduce the Monte Carlo results well.

3.4. Kinetically roughened surface
For $\Delta\mu_R < |\Delta\mu|$, the surface roughens kinetically. Both of the surface velocity and the mean height of the macrostep obey power law behaviors.

In the step droplet zone ($k_B T/\epsilon = 0.63$), we obtained from the Monte Carlo values as follows:

$$\begin{aligned} \langle n \rangle &= 1.79 + 9.16 \times 10^{-6}(|\Delta\mu|/\epsilon)^{-1.89}, \\ v &= 0.0855 \ \mathrm{sign}(\Delta\mu)(|\Delta\mu|/\epsilon)^{1.06}, \end{aligned} \quad (17)$$

where $\mathrm{sign}(x) = 1$, 0, and -1 for $x > 0$, $x = 0$, and $x < 0$, respectively. The pink solid lines in Fig. 5 shows the lines calculated by Eq. (17).

In the step-faceting zone ($k_B T/\epsilon = 0.4$), we obtained

$$\begin{aligned} \langle n \rangle &= 0.0310(|\Delta\mu|/\epsilon - \Delta\mu_{co}(L)/\epsilon)^{-1.57} + 1.33, \\ |V| &= 0.0677(|\Delta\mu|/\epsilon - \Delta\mu_{co}(L)/\epsilon)^{1.19}. \end{aligned} \quad (18)$$

Therefore, the height of the faceted macrostep decreases as $|\Delta\mu|$ increases. This is because, intuitively, the frequent nucleations on the edge of the faceted macrostep makes detach elementary steps in the high rate.

4. Discussions
It is interesting that the figure for V (Fig. 7) is analogous to Fig. 4(b) in Ref. [25], which shows the velocity of the particles where plastic depinning occurs. The system has a depinning threshold, $V \propto (F_D - F_c)^\beta$ with $\beta = 1.5$, where V is the average velocity of the particles, F_D is the driving force, and F_c is the depinning threshold. The step attachment-detachment motion on the terrace is analogous to the motion of particles with plastic depinning. However, a common mathematical framework has not yet been obtained.

5. Conclusion
The height of the faceted macrostep decreases as the absolute value of the driving force of the crystal growth $|\Delta\mu|$ increases.

Acknowledgments
This work is supported by Kakenhi Grants-in-Aid (Nos. JP25400413 and JP17K05503) from the Japan Society for the Promotion of Science (JSPS).

References
[1] Takeshi Mitani, Naoyoshi Komatsu, Tetsuo Takahashi, Tomohisa Kato, Shunta Harada, Toru Ujihara, Yuji Matsumoto, Kazuhisa Kurashige, and Hajime Okumura, "Effect of aluminum addition on the surface step morphology of 4H–SiC grown from Si–Cr–C solution", J. Cryst. Growth, **423**, 45 –49 (2015).
[2] N. Akutsu, "Thermal step bunching on the restricted solid-on-solid model with point contact inter-step attractions", Applied Surface Science, **256**, 1205–1209 (2009). doi:10.1016/j.apsusc.2009.05.080.
[3] N. Akutsu, "Zipping process on the step bunching in the vicinal surface of the restricted solid-on-solid model with the step attraction of the point contact type", J. Cryst. Growth, **318**, 10–13 (2011). doi: 10.1016/j.jcrysgro.2010.10.088.
[4] S. R. White, "Density matrix formulation for quantum renormalization groups", Phys. Rev. Lett., **69**, 2863–2866 (1992).
[5] T. Nishino, "Density matrix renormalization group method for 2D classical models", J. Phys. Soc. Jpn., **64**, 3598–3601 (1995).
[6] U. Schollwöck, "The density-matrix renormalization group, Rev. Mod. Phys., **77**, 259–315 (2005).
[7] T. Nishino, K. Okunishi, "Product wave function renormalization group", J. Phys. Soc. Jpn., **64**, 4084–4087 (1995).

[8] Y. Hieida, K. Okunishi, and Y. Akutsu, "Magnetization process of a one-dimensional quantum antiferromagnet: The product-wave-function renormalization group approach", Phys. Lett. A, **233**, 464–470 (1997).
[9] Y. Hieida, K. Okunishi, and Y. Akutsu, "Numerical renormalization approach to two-dimensional quantum antiferromagnets with valence-bond-solid type ground state", New J. Phys., **1**, 7 (1999).
[10] A. A. Chernov, "The spiral growth of crystals", Sov. Phys. USP, 116–148 (1961).
[11] T. Nishinaga, C. Sasaoka, and A. A. Chernov, " A numerical analysis for the supersaturation distribution around LPE macrostep", Ed. I. Sunagawa, *Morphology and Growth Unit of Crystals* (Terra Scientific Publishing Company, Tokyo, 1989).
[12] T. Nishinaga and T. Suzuki, "Towards understanding the growth mechanism of III-V semiconductors on an atomic scale", J. Cryst. Growth, **128** 37-43 (1993).
[13] M. Sato, M. Uwaha and Y. Saito, "Instabilities of steps induced by the drift of adatoms and effect of the step permeability", Phys. Rev. B **62**, 8452–8472 (2000).
[14] N. Akutsu, "Disassembly of Faceted Macrosteps in the Step Droplet Zone in Non-Equilibrium Steady State", Crystals, **7**, Article ID cryst7020042, (2017). doi:10.3390/cryst7020042.
[15] N. Akutsu, "Height of a faceted macrostep for sticky steps in a step-faceting zone", Phys. Rev. Materials **2**, 023603 (2018), 1-15. DOI: 10.1103/PhysRevMaterials.2.023603.
[16] N. Akutsu, "Profile of a Faceted Macrostep Caused by Anomalous Surface Tension", Adv. Condens. Matter Phys. **2017**, Article ID 2021510 (2017). doi:10.1155/2017/2021510.
[17] N. Akutsu, " Non-universal equilibrium crystal shape results from sticky steps", J. Phys. Condens. Matter, **23**, 485004 (2011). doi:10.1088/0953-8984/23/48/485004.
[18] N. Akutsu, "Sticky steps inhibit step motions near equilibrium", Phys. Rev. E **86**, 061604 (2012). DOI: 10.1103/PhysRevE.86.061604.
[19] N. Akutsu, "Faceting diagram for sticky steps", AIP Adv., **6**, 035301 (2016). dx.doi.org/10.1063/1.4943400.
[20] N. Akutsu, "Effect of the roughening transition on the vicinal surface in the step droplet zone", J. Cryst. Growth **468**, 57–62 (2017). doi:10.1016/j.jcrysgro.2016.10.014.
[21] N. Akutsu,Y. Akutsu, "Thermal evolution of step stiffness on the Si (001) surface: Temperature-rescaled Ising-model approach", Phys. Rev. B, **57**, R4233–R4236 (1998).
[22] A. F. Andreev, "Faceting phase transitions of crystals", Sov. Phys. JETP, **53**, 1063–1069 (1981).
[23] N. D. Mermin, H. Wagner, " Absence of ferromagnetism or antiferromagnetism in one-or two-dimensional isotropic Heisenberg models", Phys. Rev. Lett. **1966**, **17**, 1133–1136 (1966). "Erratum: Absence of ferromagnetism or antiferromagnetism in one-or two-dimensional isotropic Heisenberg models. N. D. Mermin and H. Wagner [Phys. Rev. Letters 17, 1133 (1966)]." Phys. Rev. Lett., **17**, 1307 (1966).
[24] N. Akutsu, Y. Akutsu, "Roughening, faceting and equilibrium shape of two-dimensional anisotropic interface. I. Thermodynamics of interface fluctuations and geometry of equilibrium crystal shape", J. Phys. Soc. Jpn.,**56**, 1443–1453 (1987).
[25] C. Reichhardt, C. J. Olson Reichhardt, "Depinning and nonequilibrium dynamic phases of particle assemblies driven over random and ordered substrates: A review", Reports on Progress in Physics **80**, 026501 (2017).

IOP Conf. Series: Journal of Physics: Conf. Series **1290** (2019) 012016 doi:10.1088/1742-6596/1290/1/012016

A Brief History of *Strukturbericht* Symbols and Other Crystallographic Classification Schemes

Michael J. Mehl
Physics Department, United States Naval Academy, Annapolis MD 21402, USA
E-mail: mmehl@usna.edu

Abstract. In 1913, x-ray crystallography was first used to determine the crystal structure of diamond. Since that time hundreds of thousands, if not millions, of crystal structures have been determined. Published structures require critical review and indexing to make them easily available to other researchers. One of the first systems for doing this was *Strukturbericht* (Structure Reports), published in Germany from 1931-1943 and covering research from 1913-1939, and now best known for the *Strukturbericht* symbols which label common crystal structures. A comprehensive history of *Strukturbericht* has not been written. This brief report sketches the early history of *Strukturbericht*, as well as its post-World War II successor publications, handbooks, and online resources.

1. Introduction

In 1912, Max van Laue suggested that the diffraction of x-rays through crystals would produce a pattern which could be used to determine the crystal structure.[1] Within the year, this technique was used by Bragg and Bragg[2] to determine the structure of diamond. Later Hull[3] determined the structure of graphite, and the race was on to determine the structure of any compound that could be studied.[1]

It quickly became obvious that nature frequently repeats itself. The metallic elements mostly take close-packed structures, either face-centered cubic or hexagonal close-packed, or the nearly close-packed body-centered cubic lattice. Of the remaining elements, the early row 14 elements all take on the diamond structure. We only see other patterns when we go to elements on the far right-hand side of the periodic table or in the actinides.[2]

A similar trend shows in compounds. Many AB type binaries take either the sodium chloride, cesium chloride or zincblende structures, and A_2B compounds can frequently have the positions found in fluorite (CaF_2).

All of this means that we can think (hope) that elements and most compounds exist in a limited number of structures, and that compounds having similar structures are somehow related to one another. This suggests that we can categorize them according to their atomic configuration. One possibility for doing this is to note that all periodic crystals must belong to one of the 230 three-dimensional space groups, and only occupy allowed positions within those space groups.[5] Of course this allows an infinite number of possible structures, and so is not

[1] Of interest is the work of Davey[4], who found that tungsten was too opaque for standard X-ray diffraction techniques, and so diluted his sample with ten parts wheat flour in order to get a diffraction pattern.
[2] The one exception being manganese, which takes on two unique atomic arrangements.

IOP Conf. Series: Journal of Physics: Conf. Series **1290** (2019) 012016 doi:10.1088/1742-6596/1290/1/012016

an optimal characterization scheme, although it can make a useful indexing scheme. A better scheme that this needs to be devised.

Scientists being human, and therefore fallible, it is also necessary to have the published structures reviewed for accuracy. Such a review is also useful for those who might have missed the original research in the literature.

In 1931 Ewald and Hermann[6] published the first such review, "*Strukturbericht* (Structure Reports) 1913-1928" which was introduced

> ... to provide a compilation and critical review of all previously published articles on X-ray structure determination. The arrangement was made according to the chemical formula, along with *a summary by type* [emphasis mine] which allows a survey of the known structural types and their representatives. Unlike other published structural compositions, the present one is characterized by the fact that the individual works are discussed. It is expected that this will improve the accuracy of the determinations, the exact nature of the material used, the corrections to be made to older reports based on newer work, and other similar points.[3]

From the date it can be seen that this was intended to cover the field of structure determination from the beginning.

As indicated, this volume was a critical review of x-ray crystallographic literature, but it also attempted to group crystal structures into types. This was the introduction of the *Strukturbericht* symbol: $A1$ was given to close-packed face-centered cubic crystals, $A2$ for body-centered cubic, $A3$ for hexagonal close-packed, *etc.* For binary materials, $B1$ represented the sodium chloride structure, $B2$ cesium chloride, and $B3$ zincblende. More complicated structures were given other alphabetic indices with other letters.

The publication of *Strukturbericht* continued through the middle of World War II, and included research through 1939. Its successor, "Structure Reports," published by the International Union of Crystallography (IUCr), began immediately after the War and continues. Post-War it was realized that the number of *Strukturbericht* symbols was growing uncontrollably, and so the practice of giving new structures symbols was dropped, although Smithells[7] and Pearson[8] did add new symbols to cover structures of some of the transuranic elements and intermetallic compounds, as we will discuss below.

Even though the generation of new symbols ceased, the ones found in the *Strukturbericht* volumes, Smithells, and Pearson are still in use. Everyone interested in superconductivity knows what an $A15$ (Cr_3Si or β-W) structure is, as compounds with this structure became known as the first "high temperature" (20K) superconductors.[9] However, to our knowledge no complete history of *Strukturbericht* has been written. There is not even a complete index of all the symbols, the last one being published with volume II in 1937.[10] Smithells, Pearson, and Villers,[11] among others, have published partial lists, but these are weighted toward the intermetallics and essentially ignore silicates (*Strukturbericht* S) and other systems.

This brief paper is by no means a complete history of *Strukturbericht* and the associated symbols, but it is meant to give a description of the symbols, some of the historical development and changes through the years, and, finally, note that we are in the process of actually producing a definitive index of *Strukturbericht* symbols as an online resource.[12]

2. *Strukturbericht Band* (Volume) *I* 1913-1928

The first volume of *Strukturbericht*[6], edited by Peter Paul Ewald and Carl Hermann, was published as a supplement to *Zeitschrift für Kristallographie* in 1931 and covered research from 1913 (the crystallographic dawn of time) through 1928. This introduced the original form for *Strukturbericht* symbols: a letter, representing the compound type (element, binary, ternary,

[3] All translations from the original German were done by the author with the assistance of Google Translate.

IOP Conf. Series: Journal of Physics: Conf. Series **1290** (2019) 012016 doi:10.1088/1742-6596/1290/1/012016

etc.), followed by a number, representing the order of compounds in the list. The original list of *Strukturbericht* letters were

- Type A, elements. As noted above, the first elements were
  - $A1$, copper, representing close-packed face-centered cubic structures
  - $A2$, tungsten, representing body-centered cubic structures
  - $A3$, magnesium, representing hexagonal close-packed structures
  - $A4$, diamond
- Type B, binary compounds with stoichiometry AB, including
  - $B1$, NaCl (rock salt)
  - $B2$, CsCl
  - $B3$, Zincblende, the cubic ZnS phase analogous to diamond
- Type C, binary compounds with stoichiometry A_2B, including
  - $C1$, fluorite, CaF_2
  - $C2$, pyrite, FeS_2
  - $C3$, cuprite, Cu_2O
- Type D, binary compounds with stoichiometry A_mB_n. These were divided into several ranges:
  - $D1 - D10$: AB_3 compounds, including $D1$, ammonia (NH_3)
  - $D11 - D20$: AB_n with $n > 3$, starting with $D11$, SnI_4
  - $D31 - D50$: $(AB_n)_2$, including $D31$, $(HgCl)_2$ and D41, $(BH_3)_2$
  - $D51$ and up: more general A_mB_n compounds, including $D51$, corundum (Al_2O_3) and $D61$ La_2O_3

  this scheme proved to be limiting and was modified in volume II.
- Type E, compounds with more than two atom types but without radicals. No E type symbols were defined in the first volume.
- Type F, compounds with two- and three-atom radicals. This was again divided into two parts,
  - $F1 - F50$, compounds of the form $A_m(BC)_n$, such as $F1$, CoAsS, and
  - $F50 - F100$, compounds with form $A_m(BC_2)_n$ and $A_m(BCD)_n$. Examples included $F51$, $NaHF_2$, and $F61$, chalcopyrite, $CuFeS_2$.

  This division would also prove to be limiting.
- Type G, compounds with four-atom radicals. This was divided into several categories, ending with $G31 - G40$, "complex formula." Examples included $G1$, calcite $CaCO_2$, and $G31$, beryl, $Be_3Al_2(SiO_3)_6$.
- Type H, which was divided in to $H1 - H50$, compounds with five-atom radicals, and $H51 - H100$, containing all of the higher-order radicals. (One might imagine Ewald and Hermann throwing up their hands at this point, as they realized what they had gotten into.) The range $H41 - H50$ was set aside for hydrated materials, *e.g.* $H41$, $K_2CuCl_4{\cdot}2H_2O$.
- Type L, "alloys," (*Legierungen* in German) which included
  - $L1 - L19$, ordered alloys based on the fcc lattice, including $L10$, CuAu, and $L12$, Cu_3Au, and
  - $L20 - L39$, ordered alloys based on the bcc lattice, ($L21$, Heusler alloy, prototype Cu_2AlMn).
- Type M, "miscellaneous," a category so nebulous that it was never used, and
- Type O, organic systems, a subject all its own which is beyond the scope of this article.

Each section of *Strukturbericht* was divided into a listing of the elements and prototype compounds indexed by *Strukturbericht* symbols, including the space group, occupied Wyckoff positions, lattice constants, and atomic positions, as well as drawings of the crystal.

At the end of each alphabetic *Strukturbericht* section the editors placed a long review of the literature for each type.

3. *Strukturbericht* Band II, 1928-1932

The second volume of *Strukturbericht*[10] was published in 1937 and covered the years 1928 through 1932. In 1937 Ewald emigrated to England, having resigned his post at the University of Munich in protest over the firing of all Jewish professors,[13] and so the volume was edited by Hermann, Otto Lohrmann, and Hans Philipp.

The new publication included the only index of *Strukturbericht* symbols ever published in the original volumes. It also made some changes in the format of the labels: as an example, by this time the editors had identified thirteen compounds with composition AB_3. These were impossible to fit in the limited range $D1 - D10$ of volume I, so the format "letter-number-subscript" was introduced. Ammonia, NH_3, formerly $D1$, became $D0_1$, and skutterudite, formerly $D2$, became $D0_2$. (Eventually the $D0$ range would extend to $D0_{24}$, Ni_3Ti.) This practice was extended to all the other types.

The second volume also introduced the first fourteen entries in the rather general E category, including $E0_1$, PbFCl. Other E entries included chalcopyrite ($CuFeS_2$), cubic perovskite ($CaTiO_3$), and ilmenite ($FeTiO_3$), which were reclassified from the original $F6_1$, $G0_5$, and $G0_4$ designations, respectively, to $E1_1$, $E2_1$, and $E2_2$.

Other new labels were added:

- Type I, compounds with BX_6 radicals. These were mostly renamed from type H, *e.g.* K_2PtCl_6, which was originally $H6_1$ and now became $I1_1$.
- Type K, "complex radicals," including $K1_1$, $K_2S_2O_5$, $K1_1$, $K_2S_2O_6$, and $K1_2$, $Cs_2S_2O_6$.
- Type S, the silicates. Several of these had been previously listed in other categories, such as kyanite, Al_2SO_5, which originally was labeled $H5_1$ and now became $S0_1$. In all, thirty two compounds were identified, all but six of these new to this volume.

4. *Strukturbericht* Band III, 1933-1935

Volume III was the last multi-year volume of the series, published in 1937.[14] Edited by Carl Gottfried and F. Schlossberger, it covered papers from 1933 through 1935.

This volume did not include any new *Strukturbericht* types, but, perhaps reflecting the confusion between the letters in all Latin alphabets, it renamed class I as class J.

5. The War Years: *Strukturbericht* Band IV-VII

The remaining volumes each covered a single year. Remarkably, most were published while World War II was in progress, and despite the fact that many German scientists, following Ewald, had fled the country, or like Hermann, had been imprisoned by the Nazi regime.[15]

- Band IV (covering 1936),[16] edited by Gottfried, published in 1938.
- Band V (1937),[17] again edited by Gottfried, was published in 1940.
- Band VI (1938),[18] was edited by K. Herrmann, was published in 1941, and the last volume,
- Band VII (1939),[19] was again edited by Herrmann and published in 1943, while the Eastern Front with the Soviet Union was collapsing, the allies were invading Italy, and preparations for D-Day were underway.

  Much of the final volume concerns updating the list of compounds belonging to a previously defined *Strukturbericht* symbol. It opens by noting that scandium takes on the face-centered

IOP Conf. Series: Journal of Physics: Conf. Series **1290** (2019) 012016 doi:10.1088/1742-6596/1290/1/012016

cubic $A1$ structure, and ends with an entry for cronstedtite, $Fe_4^{II}Fe_2^{III}[(OH)_8—Fe_2^{III}Si_2O_{10}]$, assigning it to the $S5_1$ type along with the prototype, muskovite ($KH_2Al_3Si_3O_{12}$).

In spite of the War, these later volumes were distributed throughout the world. In the United States, Edwards Brothers (Ann Arbor) published all seven volumes in 1943, "in the Public Interest by Authority of the Alien Property Custodian," an office created by Presidential order to manage enemy alien property.[20]

6. Post-War: Continuation and Extension of *Strukturbericht*

With the end of the War, the task of collecting structure reports was transferred to the newly-created International Union of Crystallography (IUCr),[15] which began publishing yearly *Structure Reports* in English. These volumes emphasize the connection with *Strukturbericht* by starting with Volume 8, published in 1956 but covering research from 1940-1941,[21] and Volume 9 (1942-1944, published in 1955.[22] Volumes 8 & 9 were actually preceded by Volume 10, published in 1953 and covering the years 1945-1946.[23] Since then the *Reports* have been published more-or-less regularly.[24]

Structure Reports did not continue the use of *Strukturbericht* symbols to designate structure types, but this did not end the generation of new labels. In 1949 Colin J. Smithells published his first *Metals Reference Book*,[25] which included a partial index of *Strukturbericht* symbols and compounds assigned to each symbol. This list concentrated on metallic systems, and so left out some elements and compounds. In addition, Smithells listed two new compounds, $ThSi_2$, and $TeCu_2$, provisionally giving them the *Strukturbericht* designation C_x. In the second edition of *Metals Reference*,[7] Smithells extended this, adding many new *Strukturbericht.* While still following the letter-number-subscript scheme started in *Strukturbericht* II, the subscripts for the new entries were alphabetic rather then numeric. Thus $ThSi_2$ became *Strukturbericht* type C_c and $TeCu_2$ became C_g. Reflecting the advances in physics in the war and postwar era, the list included the unique structures for uranium and the transuranic elements: A_a (Pa), A_b (β-U), A_c (α-Np), A_d (β-Np), *etc.* Altogether, Smithells added seventy-five new structures to the list.

In 1958 W. B. Pearson published the first edition of his influential *Handbook of Lattice Spacings and Structures of Metals and Alloys.*[8] The *Handbook* included most of Smithells' *Strukturbericht* index and added new structures, using the letter-number-alphabetic subscript format, with subscripts chosen to avoid overlap with Smithells. Pearson soon realized the further extensions would become unwieldy, and in the next edition of his Handbook[26] he dropped the *Strukturbericht* index and listed structures by their Pearson symbols, which have the advantage of making the crystal type and number of atoms readily apparent, at the cost of having numerous structures of wildly different types using the same symbol.

7. Modern indices of compounds

Despite the demise of the *Strukturbericht* labeling system, it is still useful to have lists of possible crystal structures.

- Trotter and Bree[27] published an index to the original seven volumes of *Strukturbericht*, but regrettably did not include an index by *Strukturbericht* type.
- Modern versions of *Smithells Metals Reference Book*[28] and *Pearson's Handbook*[11] include indices of metallic structures by *Strukturbericht*, Pearson symbol, and space group.
- The *Gmelin Handbook of Inorganic and Organometallic Chemistry*[29] also includes a *Strukturbericht* symbols, including the Pearson symbol, space group, and occupied Wyckoff positions for each structure. Unlike the Pearson and Smithells Handbooks, this list includes many non-metallic systems, and shows duplicate structures, which frequently show up in the original *Strukturbericht.*

- Electronic databases are also available:
  - The *The American Mineralogist Crystal Structure Database*[4][30] gives structural information for a large number of minerals, including mineral names, but does not include the *Strukturbericht* designation. When possible, it also links to PDF versions of the original publications.
  - *Springer Materials*[5][31] maintains a constantly updated list of crystal structures. Edited by P. Villars, this is essentially an electronic form of *Pearson's Handbook*, but it includes non-metallic systems as well.

None of these lists gives a comprehensive list of *Strukturbericht* symbols, which one would think should at least have been included in the 1976 index. In 1995, work was begun to collect the common *Strukturbericht* structures into a web page, eventually called *Crystal Lattice Structures*. The web page was eventually expanded to nearly three hundred structures, both in and out of *Strukturbericht*, but eventually went off line. Recently, the database has been revised and updated as part of a collaboration between the United States Naval Academy and the Curtarolo Research Group at *Duke University*. The resulting page, now known as the Library of Crystallographic Prototypes,[12] is integrated with the AFLOW system for materials discovery.[32]

The site[6] currently includes 590 unique crystal structures, and we are constantly adding to it. We index structures by *Strukturbericht*, Pearson symbol, space group and prototype material. Each entry includes

- The *Strukturbericht* designation of the structure, if one has been given,
- the Pearson symbol
- the space group,
- occupied Wyckoff positions,
- primitive vectors for the lattice,
- basis vectors for each atom in the primitive unit cell,
- a rotatable view of the primitive or conventional unit cell, which can be expanded to show multiple unit cells,
- a standardized Crystallographic Information File (CIF), and
- a structure file which can be used to as input for a variety of electronic structure codes.

We will continue to update the database. Our next goal is to include all of the inorganic structures found in the original *Strukturbericht*, Smithells, and Pearson, though we will as always include other systems of interest. We invite suggestions for structures that should be listed.

8. Summary

This has been a very brief review of the history of *Strukturbericht* symbols. This will be expanded in the future, including more historical data. In particular, research is needed to determine how publication of *Strukturbericht* was continued during the Second World War, and to find out why Colin Smithells decided to expand the system when the crystallographic community, as represented by the IUCr, decided not to continue it.

[4] http://rruff.geo.arizona.edu/AMS/amcsd.php
[5] https://materials.springer.com/
[6] http://www.aflowlib.org/CrystalDatabase/

Acknowledgments
I would like to thank Barry Klein for inviting me to present this history at the XXX IUPAP Conference on Computational Physics (CCP 2018, Davis CA). R. Benjamin Young, a summer intern at the Naval Research Laboratory in 1995, did the initial work constructing the *Crystal Lattice Structure Web Page*. Support was provided by the Kinnear Foundation and Duke University. Special thanks are due to L. Mehl, S. Mehl, and G. Hart for the gift of all seven *Strukturbericht* volumes as well as the 1976 index.

References

[1] Bragg L 1968 *Scientific American* **219** 58–74
[2] Bragg W H and Bragg W L 1913 *Proceedings of the Royal Society of London, Series A* **89** 277–291
[3] Hull A W 1917 *Physical Review* **10** 661–696
[4] Davey W P 1925 *Physical Review* **26** 736–738
[5] Wyckoff R W G 1922 *The Analytical Expression of the Results of the Theory of Space-Groups* vol 318 (Washington DC: Carnegie Institution of Washington) URL `https://books.google.com/books?id=HTRUAAAAMAAJ`
[6] Ewald P P and Hermann C (eds) 1931 *Strukturbericht 1913-1928* (Leipzig: Akademische Verlagsgesellschaft M. B. H.)
[7] Smithells C J 1955 *Metals Reference Book* 2nd ed (London: Butterworths Scientific)
[8] Pearson W B 1958 *A Handbook of Lattice Spacings and Structures of Metals and Alloys* (*International Series of Monographs on Metal Physics and Physical Metallurgy* no N.R.C. No. 4303) (Oxford, London, Edinburgh, New York, Paris, Frankfort: Pergamon Press) 1964 reprint with corrections
[9] Stewart G R 2015 *Physica C: Superconductivity and its Applications* **514** 28–35
[10] Hermann C, Lohrmann O and Philipp H (eds) 1937 *Strukturbericht Band II 1928-1932* (Leipzig: Akademische Verlagsgesellschaft M. B. H.)
[11] Villars P and Calvert L 1991 *Pearson's Handbook of Crystallographic Data for Intermetallic Phases* 2nd ed (Materials Park, OH: ASM International)
[12] Mehl M J, Hicks D, Toher C, Levy O, ert M Hanson R, Hart G and Curtarolo S 2017 *Computational Materials Science* **136** S1–S828
[13] Editorial U 1986 *Acta Crystallographica A* **42** 409–410
[14] Gottfried C and Schossberger F (eds) 1937 *Strukturbericht Band III 1933-1935* (Leipzig: Akademische Verlagsgesellschaft M. B. H.)
[15] Kamminga H 1989 *Acta Crystallographica A* **45** 581–601
[16] Gottfried C (ed) 1938 *Strukturbericht Band IV 1936* (Leipzig: Akademische Verlagsgesellschaft M. B. H.)
[17] Gottfried C (ed) 1940 *Strukturbericht Band V 1937* (Leipzig: Akademische Verlagsgesellschaft M. B. H.)
[18] Herrmann K (ed) 1941 *Strukturbericht Band VI 1938* (Leipzig: Akademische Verlagsgesellschaft M. B. H.)
[19] Herrmann K (ed) 1943 *Strukturbericht Band VII 1939* (Leipzig: Akademische Verlagsgesellschaft M. B. H.)
[20] Roosevelt F D 1942 Executive order 9095 establishing the office of alien property custodian placed online by Gerhard Peters and John T. Woolley, *The American Presidency Project* URL `http://www.presidency.ucsb.edu/ws/?pid=16232`
[21] Wilson A J C, Baenziger N C, Bivoet J M and Roberson J M (eds) 1956 *Structure Reports for 1940-1941* vol 8 (Utrecht: Oosthoek)
[22] Wilson A J C, Baenziger N C, Bivoet J M and Roberson J M (eds) 1955 *Structure Reports for 1942-1944* vol 9 (Utrecht: Oosthoek)
[23] Wilson A J C, Baenziger N C, Bivoet J M and Roberson J M (eds) 1953 *Structure Reports for 1945-1946* vol 10 (Utrecht: Oosthoek)
[24] Donohue J 1979 *Acta Crystallographica A* **35** 346–347
[25] Smithells C J 1949 *Metals Reference Book* (London: Butterworths Scientific)
[26] Pearson W B 1967 *A Handbook of Lattice Spacings and Structures of Metals and Alloys, Volume 2* (*International Series of Monographs on Metal Physics and Physical Metallurgy* no N.R.C. No. 8752) (Oxford, London, Edinburgh, New York, Paris, Frankfort: Pergamon Press)
[27] Trotter J and Bree J M (eds) 1976 Strukturbericht *Cumulative Index for Volumes 1-7 (1913-1939)* (Utrecht: Bohn, Scheltema & Holkema)
[28] Brandes E A and Brook G B (eds) 1992 *Smithells Metals Reference Book* seventh ed (Oxford, Auckland, Boston, Johannesburg, Melbourne, New Delhi: Butterworth Heinemann)
[29] Parthé E, Gelato L, Chabot B, Penso M, Cenzula K and Gladyshevskii R 1993 *Standardized Data and Crystal Chemical Characterization of Inorganic Structure Types* 8th ed (*Gmelin Handbook of Inorganic and Organometallic Chemistry* vol 2) (Berlin, Heidelberg: Springer-Verlag)

[30] Downs R T and Hall-Wallace M 2003 *American Mineralogist* **88** 247–250
[31] Materials S The research solution for identifying material properties https://materials.springer.com/
[32] Center for Material Genomics, Material Science, Duke University, *AFLOW: Automatic-flow for materials discovery, including references therein* URL `http://aflowlib.org/`

IOP Conf. Series: Journal of Physics: Conf. Series **1290** (2019) 012017 doi:10.1088/1742-6596/1290/1/012017

DCA++ project: Sustainable and scalable development of a high-performance research code

U R Hähner[1], G Balduzzi[1], P W Doak[2], T A Maier[2], R Solcà[3] and T C Schulthess[1,3]

[1] Institute for Theoretical Physics, ETH Zurich, 8093 Zurich, Switzerland

[2] Computational Science and Engineering Division, Center for Nanophase Materials Sciences, Oak Ridge National Laboratory, Oak Ridge, Tennessee 37831, USA

[3] Swiss National Supercomputing Center, ETH Zurich, 6900 Lugano, Switzerland

E-mail: haehneru@itp.phys.ethz.ch

Abstract. Scientific discoveries across all fields, from physics to biology, are increasingly driven by computer simulations. At the same time, the computational demand of many problems necessitates large-scale calculations on high-performance supercomputers. Developing and maintaining the underlying codes, however, has become a challenging task due to a combination of factors. Leadership computer systems require massive parallelism, while their architectures are diversifying. New sophisticated algorithms are continuously developed and have to be implemented efficiently for such complex systems. Finally, the multidisciplinary nature of modern science involves large, changing teams to work on a given codebase. Using the example of the DCA++ project, a highly scalable and efficient research code to solve quantum many-body problems, we explore how computational science can overcome these challenges by adopting modern software engineering approaches. We present our principles for scientific software development and describe concrete practices to meet them, adapted from agile software development frameworks.

1. Introduction

High-performance computing (HPC) is approaching the exascale era, which will open up great opportunities in various research areas of computational science. Exascale supercomputers could provide the performance boost required for more precise weather forecasts [1], accelerate the computational design and discovery of functional materials [2], and make simulations of the full human brain possible [3]. The transition to exascale computing will involve a diversification of computer architectures and a vastly increasing demand on parallelism [4]. Thus, to live up to the exascale promises, computational science is confronted with the tasks of porting existing application codes, developing new scalable algorithms and addressing issues such as fault tolerance [5].

The evolution of leadership computing systems is only one facet of the complex, dynamic environment computational science is embedded in. The exploratory and iterative nature of science requires the continued development of domain specific algorithms, which tend to grow in complexity. For scientific success and impact, these algorithms have to be turned rapidly into production code. Academic integrity requires verification and validation of the codes, and reproducibility of simulation results they produce. The high personal turnover in academia and

IOP Conf. Series: Journal of Physics: Conf. Series **1290** (2019) 012017 doi:10.1088/1742-6596/1290/1/012017

the volatility of funding sources, in turn, cause research teams to constantly change in size and composition. This implies that new team members need to be trained and the contributions of leaving team members preserved.

To maintain scientific productivity, software development in computational science needs to sustain the changes in computer architectures, algorithms and teams. To further foster scientific productivity, software development in research groups should likewise feature scalability in the number of developers such that a growing team indeed accelerates software development and thereby the scientific progress. Sustainable and scalable software development accompanied by sufficient quality assurance can be supported by appropriate practices. Traditional software development processes that require up-front specification of requirements, such as the waterfall model [6], are in contrast to the uncertainty of the scientific discovery process [7]. Scientific software development, however, can benefit from modern software engineering methods, such as the agile methodology, which promotes an iterative, adaptive and collaborative approach [8].

The term *agile software development* describes practices and frameworks that embrace the values and follow the principles formulated by the *Manifesto for Agile Software Development* [9]. Besides the anticipation of change, the agile methodology values the people involved in the development process and emphasizes continuous improvement. While there are various successful agile frameworks such as *Scrum* [10] and *Lean software development* [11], we found the concept of *Extreme Programming (XP)* [12] most useful to our project. XP defines a set of practices that improve software development with respect to the agile values. These practices can be applied consecutively to the development process according to the project's goals and needs.

In the development of the DCA++ project [13], a high-performance research software framework in the field of condensed matter physics, we are faced with the challenges of scientific software development as described above. The codebase is developed and maintained by a multi-site team located at ETH Zurich, the Oak Ridge National Laboratory (ORNL) and the affiliated HPC centers, the Swiss National Supercomputing Center and the Oak Ridge Leadership Computing Facility. Accordingly, the team's competences range from condensed matter and computational physics to applied mathematics and computer science. The realm of the DCA++ project is the development of modern quantum cluster methods, such as the dynamical cluster approximation (DCA) [14, 15, 16] and its DCA$^+$ extension [17, 18], and their efficient implementation for leadership computing systems. These methods can be applied to models of high-temperature superconductivity to gain understanding of superconducting materials such as the copper-oxide based compounds, usually referred to as cuprates [19, 20]. At the heart of the DCA and the DCA$^+$ algorithms is a continuous-time quantum Monte Carlo (QMC) solver [21], which consumes most of the compute time of a typical DCA++ production run. For this reason, continuous effort is put into the optimization of these QMC algorithms to increase performance and to reduce time-to-solution. Feature extensions to the DCA++ code are usually driven by new research questions and novel theoretical models, and agility is required to develop code that satisfies the needs of our domain scientists and collaborators. To achieve sustainability and scalability, and to incorporate systematic testing in the DCA++ project, we have explored, integrated and adapted agile software development practices.

In the remainder of this paper, we first introduce four principles - inspired by the Manifesto for Agile Software Development - that we believe are necessary for successful and sustainable development of a high-performance research code. Some of the principles, such as the design principles for sustainable software, translate into immediate actions, while others, such as transparency and reproducibility in the context of open science, require suitable practices. In section 3, we describe concrete practices that we apply in the DCA++ project to comply with our principles. We conclude by describing challenges when adopting new practices in general and agile practices in a research group in particular.

IOP Conf. Series: Journal of Physics: Conf. Series **1290** (2019) 012017 doi:10.1088/1742-6596/1290/1/012017

2. Principles

2.1. Communication and feedback

To benefit from the multidisciplinarity of the DCA++ team, active and frequent communication between team members is at the heart of the development process. The spatial distance between the development sites makes messaging and collaboration tools like Slack [22] as well as video conferences important everyday channels of communication. To overcome language barriers that naturally exist between different disciplines, face-to-face meetings are still preferred and in many cases absolutely necessary to make efficient progress.

Developmental work on the DCA++ code can generally be divided into two categories: (1) implementation of new science capabilities, and (2) optimization of existing code for either new computing architectures or for allowing access to larger problem sizes. An initial implementation of a new science capability is usually performed by the team members closest to the science. In short feedback loops, students and postdocs develop the new feature in collaboration with their domain science advisors. To optimize existing code for execution on leadership computing hardware, expertise from all disciplines is required: domain scientists provide insight into domain specific algorithms, applied mathematicians translate the performance critical parts into the language of linear algebra, and computer scientists estimate computational work and memory requirements, and evaluate parallel programming models and design choices. The collective effort results in initial ideas for algorithmic improvements and code optimizations. As for the implementation of new capabilities, these initial ideas are followed up by an iterative and collaborative development process.

To gain confidence in suitability and correctness of developed code, it is essential to supplement feedback through discussions by verification and validation (V&V). The characteristics of scientific software are captured by the computational science specific definition of V&V given by Babuska and Oden [23]:

- Verification: "The process of determining if a computational model obtained by discretizing a mathematical model of a physical event and the code implementing the computational model can be used to represent the mathematical model of the event with sufficient accuracy."
- Validation: "The process of determining if a mathematical model of a physical event represents the actual physical event with sufficient accuracy."

2.2. Technical excellence

The fast evolution of leadership computer systems is both a challenge and a chance for computational science. Iterative development of scalable and efficient algorithms is required to use the growing capacity of these machines and to progress in science. DCA++ was able to exploit the first petascale supercomputer available to open science, ORNL's Jaguar, by featuring multi-level parallelism, cache-friendly delayed Monte Carlo updates [24] and mixed precision support. This made DCA++ the first scientific application to break the sustained petaflop barrier on a general-purpose supercomputer [25]. Later, DCA++ successfully responded to the emergence of hybrid CPU-GPU systems and sustained 15 petaflop/s on ORNL's Titan supercomputer, the successor to Jaguar [26]. The increasing diversity of architectures, from CPU-only to GPU-dominated node configurations, demands that we continue to develop the DCA++ code with regard to evolving supercomputing capabilities.

At the same time, we seek for algorithmic improvements that overcome current limitations and make parameter ranges accessible that have been out of reach before. The DCA^+ algorithm, for example, introduces a continuous lattice self-energy in the DCA formalism, which further cures the fermionic sign problem of the embedded quantum Monte Carlo method [17]. This advancement enables large cluster calculations of the 2D Hubbard model with values

IOP Conf. Series: Journal of Physics: Conf. Series **1290** (2019) 012017 doi:10.1088/1742-6596/1290/1/012017

of the Coulomb interaction that are relevant for the study of models of high-temperature superconductors [26].

Technical excellence is not limited to algorithm development, but also applies to software design. We strive for designs that permit clear expression of algorithms, and feature flexibility to support various types of theoretical models and alternative algorithms to solve them. Designs also need to ensure extensibility to new models and algorithms, and allow easy adaptation to changing requirements (see section 2.3). The solution is an object-oriented and generic programming model. The C++ programming language provides us *inheritance* and *templates* to implement it.

2.3. *Sustainability*

We pursue sustainability as the ability to continuously develop software that meets our needs of today and that is prepared to adapt to our needs of tomorrow. The anticipated change of needs is driven by both the progress in science (e.g. development of new algorithms) and the evolution of leadership computing platforms (e.g. emergence of new architectures). Consequently, we characterize sustainable software as being *maintainable* and *extensible*, in line with current suggestions of the software engineering community [27].

Development of sustainable software can be enabled through appropriate (design) principles and practices. In object-oriented programming, the *SOLID* design principles provide a guideline for writing maintainable and extensible code. The acronym SOLID encapsulates five individual principles [28]:

- ***S**ingle responsibility principle*:
  "A class should have only one reason to change."
- ***O**pen/closed principle*:
  "Software entities (classes, modules, functions, etc.) should be open for extension but closed for modification."
- ***L**iskov substitution principle*:
  "Subtypes must be substitutable for their base types."
- ***I**nterface segregation principle*:
  "Clients should not be forced to depend on methods they do not use."
- ***D**ependency inversion principle*:
  A. "High-level modules should not depend on low-level modules. Both should depend on abstractions."
  B. "Abstractions should not depend upon details. Details should depend upon abstractions."

We refer to the literature for the details of each principle, but we want to emphasize that *SOLID code* features *high cohesion* and *low coupling*, attributes of a maintainable and extensible design.

In addition to these concrete principles for object-oriented design, sustainability can be fostered by the following general design principles:

- *KISS* ("Keep it simple, stupid"): Simple designs are easier to understand and maintain.
- *YAGNI* ("You aren't gonna need it"): Incremental design [12] combined with refactoring (see section 3.4), instead of up-front design, is in accord with the exploratory nature of research.
- *DRY* ("Don't repeat yourself") [29]: Duplication multiplies the cost of change.

A further key principle to sustainable software development is valuing *defect prevention* over *defect detection* [30]. The cost of a defect ranges from fixing it, to repeating expensive production runs, to possibly retracting publications, and increases significantly with the time between deployment and discovery. Thus, catching defects early and preventing them from entering production code minimizes maintenance cost and ensures sustainability. In practice, defect prevention can be accomplished by creating comprehensive automated tests for any functional code (see 3.3. Test-driven development) and having code pass two or more pairs of eyes (see 3.4. Code reviews).

2.4. Open science

Transparency and reproducibility of scientific research is of paramount importance for scientific integrity. Routines that aid in transparency and reproducibility are adopted by an increasing number of journals. Many publications now contain raw data associated with published results in additional online material, sometimes even integrating code, as well as details of the analysis that was used in additional supplemental material.

Scientific work produced with the DCA++ application follows most of the "ten simple rules for reproducible computational research" set up by Sandve et al. [31]. A typical workflow consists of running the DCA++ code on an HPC system, generating data, which then in most cases is transferred to a local workstation for analysis. Since this workflow does not employ an integrated workflow environment, most of the the rules are followed explicitly and sometimes manually. As a minimal requirement, we track sufficient information during the workflow so that we can reproduce any results presented in published work at a later date. The DCA++ output files include full build details: source code version information (see section 3.6), loaded modules and configuration options (see section 3.2). These output files, the raw data as well as the input files used for generating them are stored and archived together with the manuscript files, as are the scripts that create published figures.

As part of open science, any new developments in the DCA++ code are publicly available as open source, or are made available with a certain delay. By open sourcing the code, we want to further increase transparency and reproducibility, drive transfer of knowledge and promote software reuse in our domain. Open source scientific software likewise facilitates the important process of verifying and validating codes. An independent implementation of the same algorithm provides a useful baseline for verification, while validation can be performed against predictions of alternative methods.

3. Practices

3.1. Version control system

The benefits of using a *version control system (VCS)* are substantial to the success of a long-lived software project. The full long-term history of every file in terms of documented *commits* allows tracking changes and rolling back to a specific version if necessary. This helps to understand how the design of the software has evolved in general and to analyze and resolve bugs in particular. Storing versions properly likewise contributes to reproducibility of simulation results.

A VCS also provides the means for scalable collaborative software development. When team members work concurrently on the same part of the code, the VCS can often automatically merge these changes into a common version. If conflicting changes arise, the VCS assists to detect and resolve them. Most VCS allow tracking an arbitrary number of *branches* of a codebase, where a branch is a particular history of changes made to the code. The process of bringing two branches into the same state, where they contain identical code, is referred to as a *merge*. By branching and merging, independent features can be implemented in isolation in dedicated branches and integrated back into the main trunk, usually called *master* branch, when they are ready. That way, the master branch, holding the production code, is operational at all times.

For the DCA++ project, we chose the widely-used setup of a Git repository [32] hosted on the GitHub platform [33]. Git is a *distributed* VCS, where every developer's local working copy of the codebase is a full-fledged clone of the repository. Git excels at performance and makes branching and merging fast and easy. Besides hosting Git repositories, GitHub offers a full stack of features to maintain, manage and share a software project through an accessible web interface.

The full benefits of a version control system can only be experienced when the entire codebase is maintained in a single repository. Within this repository, branching capabilities should not be abused to create de facto divergent production codebases on long-lived branches. This one codebase per researcher issue is a familiar and avoidable waste in computational research that prevents progressive improvements of scientific codes as well as scalability and sustainability of scientific software projects. Instead, as outlined in section 3.6, we recommend a single production branch, the master branch, and few short-lived feature extension branches that will be rapidly integrated into master.

3.2. Automated build and test

Scientific software is usually built from source by the user. Diverse computing environments, long lists of dependencies and various possible configurations, however, can cause the build process to be time-consuming and error-prone, likely overburdening the user.

CMake [34] is a tool to automate configuring and building software, while being cross-platform and flexible to handle the variety of systems. CMake provides mechanisms to automatically detect compilers and external libraries, but also offers the possibility to manually specify these and other configuration options.

On HPC systems, we complete the build automation with the EasyBuild framework [35]. EasyBuild allows us to pack all dependencies of DCA++, i.e. CMake, compilers and all external libraries, into a single module. This module can be loaded like any other system module to obtain the full software stack required to build DCA++, as shown in the example steps of Listing 1.

```
$ module load DCA++/1.0-CrayGNU-17.08
$ git clone https://github.com/CompFUSE/DCA.git dca_source
$ mkdir build && cd build
$ cmake ../dca_source/ \
        -DCMAKE_BUILD_TYPE=Release \
        -DDCA_HAVE_LAPACK=TRUE
$ make
```

Listing 1. CMake based building procedure using Make as the native build system. We start by loading the DCA++ module, which provides all dependencies and has been installed with EasyBuild. Next, we clone the DCA++ repository to obtain the source code. In the third line, we create an empty build directory and change to it. Then, we execute the `cmake` command with the path to the source and appropriate options to configure the build, which generates a hierarchy of Unix Makefiles. Finally, we call `make` to build the applications.

Automated tests that run fast are a great way for developers to receive quick feedback about the changes they have made. Test-driven development (see section 3.3) applies this to the extreme. In the DCA++ project, all tests are written with Google Test [36], a C++ xUnit testing framework. Building and running these tests is easily automated by CMake and its test driver program CTest.

Automated continuous testing, i.e. executing the full test suite automatically within the development process, is a cornerstone in our workflow (see section 3.6) and performed by a Jenkins automation server [37]. Jenkins is connected to the GitHub repository that hosts the source code and triggered by pull requests and whenever commits are pushed to the master branch. Jenkins then builds all applications and tests on a production system, runs the tests and reports the results. Once per week, Jenkins does a full build, which includes all dependencies.

IOP Conf. Series: Journal of Physics: Conf. Series **1290** (2019) 012017 doi:10.1088/1742-6596/1290/1/012017

3.3. Test-driven development

Test-driven development (TDD) is a software development practice that produces self-testing code and fosters simple designs by following two basic rules [38]:

- Write new code only in response to a failing automated test.
- Eliminate duplication.

TDD implements these two rules with the *Red-Green-Refactor* development cycle:

1. Red: Create a small, automated test that fails.
2. Green: Write just enough code to make the new test and all other tests pass.
3. Refactor: Clean up and remove duplication.

The three steps are repeated continuously. The amount of functionality covered by a cycle and the number of intermediate steps in the refactoring stage depend on the complexity of the problem and the experience of the developer. Smaller steps, which can be taken with higher confidence, are generally favorable, though [38].

Software written with TDD features automated unit tests with 100% code coverage. Besides delivering assurance on correctness of new code and preventing regression of existing code, these tests are prerequisite for safe and fast refactoring (see section 3.4). Unit tests also provide an extremely valuable form of documentation. They describe specifications and usage by example and, as long as they pass, can never get out of date.

TDD often has a positive effect on the design of the code, as well. The design principles KISS and YAGNI (see section 2.3) are built into TDD: Code driven by tests does not do more than the tests require. "Don't repeat yourself" (DRY) is the main goal of the *Refactor* step. TDD does not enforce the SOLID design principles (see again section 2.3), yet testable code is often SOLID. The SOLID principles, in turn, promote testability. The Liskov substitution principle (the L of SOLID), for example, enable us to substitute a dependency with a mock or stub without altering the behavior of the object to test.

The TDD cycle not only produces simple and testable code, its three steps also focus software development on one task at a time. In the *Red* step, we are concerned with the public interface of the new feature, which we describe with an automated test. The failing of the new test confirms the test's validity for the feature. During the *Green* step, what only matters is getting the new test (and all other tests) to pass. The *Refactor* step is the moment to get the design right. The tests ensure that we do not break functionality in the process.

3.4. Refactoring

It is common that a software project accumulates *technical debt* over time, knowingly and unknowingly. We introduce technical debt when we explicitly or implicitly duplicate code, use an incomprehensible naming convention or add an unnecessary dependency, that is whenever we make a poor design choice. Paying off this technical debt and improving readability, maintainability and extensibility of the code is the goal of refactoring. By refactoring, we improve the internal design of existing code without changing the external behavior.

Well-practiced refactoring is a process of many tiny, behavior-preserving code changes. In isolation, these changes are almost unnoticeable, but their cumulative effect leads to a significant design improvement. Small, focused changes, compared to large, sweeping transformations, are simpler to implement, thus minimizing the risk of introducing bugs. Small changes are also easier to revert in case they do not provide the intended enhancement or do alter the external behavior.

There is a large catalog of standard refactoring techniques [39], often simply referred to as *refactorings*. Each of these techniques, such as *Move Method* or *Remove Parameter*, performs a small, concrete code transformation. Many integrated development environments (IDEs) automate several of the refactorings, making them faster and safer to apply.

To ensure that external behavior is truly preserved, fast, automated tests are required, that sufficiently cover the part of the code to be refactored. In a short feedback loop, we first make a small change, then run the tests and only repeat if they pass. If a test fails, we need to revisit the change we have just made, and fix or undo it.

Refactoring code without adequate tests, *legacy code* according to the definition of Feathers [40], often puts us in a chicken-and-egg dilemma: We need to refactor to get tests in place to refactor. These situations require extra careful techniques and practices to safely break out of this cycle.

In the DCA++ project, we frequently refactor code before we add new functionality. It is a way to get a better understanding of the existing code that interfaces the new feature and to restructure it such that adding the new code is easy. Refactoring can also be an instrument for tracking down a bug in the code. Refactoring techniques like *Decompose Conditional* reduce complexity, while others like *Extract Method* isolate the defect [39].

3.5. Mutual code reviews

While code reviews have become standard practice in the software industry, it is our experience that scientific software development often lacks this activity. However, particularly when non-software engineers write code, reviews as part of the development cycle provide a way to guarantee code quality.

Code quality evidently includes correctness and sufficient test coverage to avoid the worst case scenario of having to retract one or more publications due to erroneous numerical results. Code reviews should likewise cover design, readability and conformity with an agreed coding style to ensure long-term maintainability of the software project, especially when the original author some day leaves the team.

Benefits of code reviews exceed quality assurance, though. The code review process offers a bidirectional learning opportunity. By receiving feedback on the submitted code, the author can improve their software development skills. The reviewer, on the other end, might see an as yet unknown programming language feature or is able to transfer an elegant solution to their own problem.

Code reviews also propagate knowledge of and behind the codebase across the team. That way, duplication of code and work is avoided and knowledge preserved in changing teams. Having multiple informed developers gives more flexibility in assigning tasks and allows to discuss important design decisions.

In the DCA++ project, we follow a strict policy and require everyone's code to go through review. Code reviews are integrated in our workflow as part of the pull request (see section 3.6). This final examination before a feature gets merged into the master branch, however, does not replace early and frequent discussions as well as feedback.

Every developer on the team is a potential reviewer and can self-assign themselves to do a review. To get feedback on performance critical parts of the code, we also involve software engineers of the HPC centers we work with.

3.6. Issue branch workflow

A workflow to develop a high-performance research code has to be able to rapidly turn new algorithms and features into quality code that produces reliable results. At the same time, we want to continuously deliver improvements, as result of refactorings (see section 3.4) or in terms of bug fixes, to internal and external users. Like all other practices, the workflow has to be scalable in terms of developers and sustainable in a multidisciplinary, changing team.

We deal with these needs by breaking down all development work into tasks that are tracked as *issues* in the GitHub repository. Issues can be feature requests by group leaders or collaborators, bug reports by users, or a developer's remark about the need to refactor a certain function or class. To ensure quick feedback and to prevent the so-called *merge hell*, issues are sized such that they can be implemented and integrated into the main line within a few weeks. Large or complex features, which we refer to as *epics*, such as the implementation of a new solver, are decomposed into small, independent tasks.

An open issue is assigned to a team member, who works on it in a dedicated branch (see section 3.1). When the work is completed, the code is submitted via a pull request into the master branch. A code review (see section 3.5) and automated testing (see section 3.2) control code quality before the changes become production code. Fig. 1 shows an exemplary network of branches corresponding to the described workflow.

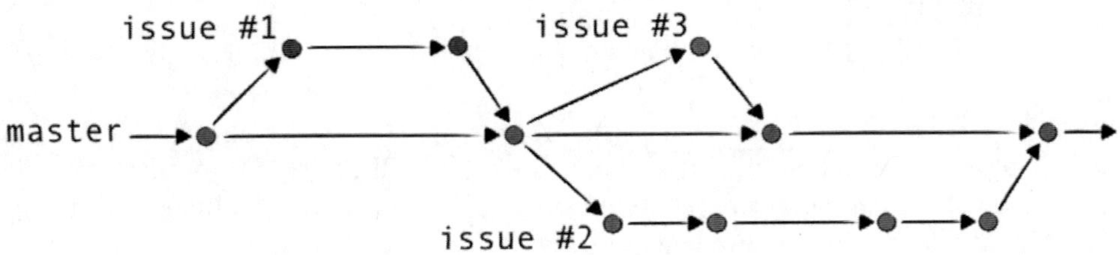

Figure 1. (Color online) Issue branch workflow. To guarantee stability of the master branch and to facilitate concurrent development, each issue is implemented in its own branch.

A large software project involves many independent as well as dependent tasks, either left to do or worked on in parallel. To visually track all tasks and to coordinate the development work in the team, we use a virtual Kanban board. Each task, i.e. open issue in the repository, is represented by a card that is moved across this board through four different stages: *Backlog*, *In progress*, *Review*, and *Done* (see Fig. 2). From our experience, the board is a great help in limiting work in progress, detecting impediments and prioritizing work.

4. Concluding remarks

We have presented four principles and six practices to address the challenges of developing a high-performance research code. The principles, communication and feedback, technical excellence, sustainability, and open science, have been treated separately for purposes of focus and clarity. However, they are not independent, but influence one another. Communication between team members is essential to achieve technical excellence, defect prevention as part of sustainable software development requires feedback through tests and from people, technical excellence includes sustainable software solutions, and open science allows for feedback from the entire community. Likewise, some practices are prerequisites for, or elements of, other practices. Test-driven development requires a framework for automated tests. Refactoring is a step of the test-driven development cycle and an essential practice on its own. Automated testing and code reviews are part of the issue branch workflow, which requires the branching capabilities of a version control system.

To be effective, the practices need to be adopted by the whole team. However, applying a new practice within a team is a difficult task. Practices cannot be imposed. Adopting new practices often implies changing the work routine and therefore requires the team's openness for change and motivation to improve. The motivation to improve can be fostered through leading by example and demonstrating the benefits of a particular practice. Moreover, to preserve

IOP Conf. Series: Journal of Physics: Conf. Series **1290** (2019) 012017 doi:10.1088/1742-6596/1290/1/012017

Figure 2. (Color online) Version of a Kanban board for software development as it is used in the DCA++ project. Each card represents an open issue and can be labeled according to type or priority. All cards start in the *Backlog* column. When a team member begins working on an issue, its card is moved to *In progress*. The card proceeds to the *Review* column as soon as the corresponding pull request is opened. Closed issues are dropped in the *Done* column.

openness for change and to prevent overwhelming people, new practices should be introduced one at a time.

The iterative and collaborative nature of research meets well with the ideas of agile software development. Yet the surrounding conditions in research projects can hinder the successful application of agile practices. Team members and collaborators are often split across different institutions. This prevents regular face-to-face meetings as, for example, required for Scrum's daily stand-up. Researchers usually work on multiple projects or have teaching responsibilities, which leads to scheduling conflicts for practices like XP's pair programming. Adaption of practices is also necessary due to the differences between commercial and scientific software development. Agile software development values customer collaboration. To adopt corresponding practices in a research group, the customer role has to be defined.

Despite the challenges of introducing new practices into a team and applying agile practices in a research environment, the usage of the practices described in this paper contributed substantially to sustainability and scalability in the development of the DCA++ project. Believing in the principles of agile software development, we still want to improve. For example, we want explore more extreme programming practices such as pair programming.

Acknowledgments

U.R.H. and G.B. acknowledge the support by the NCCR MARVEL, funded by the Swiss National Science Foundation. The work of P.W.D. and T.A.M. was supported by the Scientific Discovery through Advanced Computing (SciDAC) program funded by U.S. Department of Energy, Office of Science, Advanced Scientific Computing Research and Basic Energy Sciences, Division of Materials Sciences and Engineering. This research used resources of the Oak Ridge Leadership Computing Facility (OLCF) awarded by the INCITE program, and of the Swiss National Supercomputing Center (CSCS). OLCF is a DOE Office of Science User Facility supported under Contract DE-AC05-00OR22725.

References

[1] Schulthess T C, Bauer P, Fuhrer O, Hoefler T, Schär C and Wedi N 2018 *Comput. Sci. Eng.* 10.1109/MCSE.2018.2888788
[2] MaX Centre of Excellence – Materials Design at the Exascale `http://www.max-centre.eu`
[3] Jordan J, Ippen T, Helias M, Kitayama I, Sato M, Igarashi J, Diesmann M and Kunkel S 2018 *Front. Neuroinform.* **12** 2
[4] Shalf J, Dosanjh S and Morrison J 2010 *Proc. 9th Int. Conf. on High Performance Computing for Computational Science (Berkeley)* (Berlin, Heidelberg: Springer) pp 1–25
[5] Alexandrov V 2016 *J. Comput. Sci.* **14** 1–4
[6] Royce W W 1987 *Proc. 9th Int. Conf. on Software Engineering (Monterey)* (Los Alamitos: IEEE Computer Society Press) pp 328–338
[7] Segal J and Morris C 2008 *IEEE Softw.* **25** 18–20
[8] Sletholt M T, Hannay J, Pfahl D, Benestad H C and Langtangen H P 2011 *Proc. 4th Int. Workshop on Software Engineering for Computational Science and Engineering (Honolulu)* (New York: ACM Press) pp 1–9
[9] Beck K, Beedle M, van Bennekum A, Cockburn A, Cunningham W, Fowler M, Grenning J, Highsmith J, Hunt A, Jeffries R, Kern J, Marick B, Martin R C, Mellor S, Schwaber K, Sutherland J and Thomas D Manifesto for Agile Software Development `http://agilemanifesto.org`
[10] Schwaber K and Beedle M 2001 *Agile Software Development with Scrum* (Upper Saddle River: Prentice Hall PTR)
[11] Poppendieck M, Poppendieck T D and Poppendieck T 2003 *Lean Software Development: An Agile Toolkit* (Boston: Addison-Wesley Professional)
[12] Beck K and Andres C 2004 *Extreme Programming Explained: Embrace Change (2nd Edition)* (Boston: Addison-Wesley Professional)
[13] Hähner U R, Alvarez G, Maier T A, Solcà R, Staar P, Summers M S and Schulthess T C 2019 *Comput. Phys. Commun.* 10.1016/j.cpc.2019.01.006
[14] Hettler M H, Tahvildar-Zadeh A N, Jarrell M, Pruschke T and Krishnamurthy H R 1998 *Phys. Rev.* B **58** R7475–9
[15] Hettler M H, Mukherjee M, Jarrell M and Krishnamurthy H R 2000 *Phys. Rev.* B **61** 12739–56
[16] Maier T, Jarrell M, Pruschke T and Hettler M H 2005 *Rev. Mod. Phys.* **77** 1027–80
[17] Staar P, Maier T and Schulthess T C 2013 *Phys. Rev.* B **88** 115101
[18] Staar P, Maier T and Schulthess T C 2014 *Phys. Rev.* B **89** 195133
[19] Macridin A, Jarrell M, Maier T and Sawatzky G A 2005 *Phys. Rev.* B **71** 134527
[20] Staar P, Mishra V, Chatterjee U, Campuzano J C, Scalapino D J and Maier T A 2016 *Nat. Commun.* **7** 11875
[21] Gull E, Millis A J, Lichtenstein A I, Rubtsov A N, Troyer M and Werner P 2011 *Rev. Mod. Phys.* **83** 349–404
[22] Slack `https://slack.com`
[23] Babuska I and Oden J T 2004 *Comput. Methods Appl. Mech. Eng.* **193** 4057–66
[24] Alvarez G, Summers M S, Maxwell D E, Eisenbach M, Meredith J S, Larkin J M, Levesque J, Maier T A, Kent P R C, D'Azevedo E F and Schulthess T C 2008 *Proc. 2008 ACM/IEEE Conf. on Supercomputing (Austin)* (Piscataway: IEEE Press) pp 61:1–10
[25] Summers M S, Alvarez G, Meredith J, Maier T A and Schulthess T C 2009 *J. Phys. Conf. Ser.* **180** 012077
[26] Staar P, Maier T A, Summers M S, Fourestey G, Solca R and Schulthess T C 2013 *Proc. Int. Conf. on High Performance Computing, Networking, Storage and Analysis (Denver)* (New York: ACM) pp 1:1–11
[27] Venters C C, Capilla R, Betz S, Penzenstadler B, Crick T, Crouch S, Nakagawa E Y, Becker C and Carrillo C 2018 *J. Syst. Softw.* **138** 174–188
[28] Martin R C 2002 *Agile Software Development: Principles, Patterns, and Practices* (Upper Saddle River: Prentice Hall PTR)
[29] Hunt A and Thomas D 1999 *The Pragmatic Programmer: From Journeyman to Master* (Boston: Addison-Wesley Professional)
[30] Tate K 2005 *Sustainable Software Development: An Agile Perspective* (Boston: Addison-Wesley Professional)
[31] Sandve G K, Nekrutenko A, Taylor J and Hovig E 2013 *PLoS Comput. Biol.* **9**(10) e1003285
[32] Git `https://git-scm.com`
[33] GitHub `https://github.com`
[34] Martin K and Hoffman B 2015 *Mastering CMake: A Cross-Platform Build System (CMake 3.1)* (New York: Kitware, Inc.)
[35] Hoste K, Timmerman J, Georges A and Weirdt S D 2012 *Proc. 2012 SC Companion: High Performance Computing, Networking Storage and Analysis (Salt Lake City)* (Washington: IEEE Computer Society) pp 572–582

IOP Conf. Series: Journal of Physics: Conf. Series **1290** (2019) 012017 doi:10.1088/1742-6596/1290/1/012017

[36] Google Test https://github.com/google/googletest
[37] Jenkins https://jenkins.io
[38] Beck K 2002 *Test Driven Development: By Example* (Boston: Addison-Wesley Professional)
[39] Fowler M 1999 *Refactoring: Improving the Design of Existing Code* (Boston: Addison-Wesley)
[40] Feathers M 2004 *Working Effectively with Legacy Code* (Upper Saddle River: Prentice Hall PTR)

IOP Conf. Series: Journal of Physics: Conf. Series **1290** (2019) 012018 doi:10.1088/1742-6596/1290/1/012018

Crambin Homologues in the H0P Lattice Model

Zewen Zhang,[1*] Alfred C.K. Farris,[1] Guangjie Shi,[1] Thomas Wüst,[2] David P. Landau[1]

[1]Center for Simulational Physics, University of Georgia, Athens, GA 30602, USA
[2]Scientific IT Services, ETH Zürich, 8092 Zürich, Switzerland

E-mail: *zewenzhang@physast.uga.edu

Abstract. To compare folding behavior among lattice proteins which have similar corresponding structures in nature, Crambin homologues are tested in the semi-flexible H0P lattice model using replica-exchange Wang-Landau sampling. Our simulation shows that, at low temperature, these lattice homologues have two common signals in their specific heat curves, implying similarity in the thermodynamic behaviors; while the structural behaviors are more diverse, showing the different stability of their ground state structures at very low temperature. The ground state structures of different homologues can also vary dramatically.

1. Introduction

The problem of protein folding has been studies for over 80 years. Progress in simulations, as well as experimental and theoretical works, have deepened our understanding of the folding pathway[1], structural prediction[2] and molecular biological function[3]. However, it is still not clear why the folding process is so cooperative[4, 5, 6]. To solve this puzzle, it is usually more important to make use of some coarse-grained models than to attempt to use complex atomic details[7, 8, 9]. A representative model is the original hydrophobic-polar (HP) model[10], in which 20 amino acids are condensed into 2 different categories: hydrophobic (H) and polar (P). In this model, hydrophobic interactions among amino acids are considered as the key driving force during the folding process. The further simplification of the HP model is that it places monomers on lattice sites, and only accounts for interactions between nearest neighbor non-bonded HH contacts. This model turns out to be very helpful to our understanding of the protein folding problem, giving descriptions about collapse transition[11], structural stability[12, 13], cooperativity[14], etc.

Experimental evidence indicates that proteins with over 40% sequence identity often have the same native structures. Correspondingly, some previous research on single-site mutation in HP model[13] are partially consistent with these experimental results. However, even in the H and P "degeneracy" (which have 10 kinds of amino acids for either H or P category), the differences between two sequences can still be much larger than a single site. Do homologues with a difference of more than two monomers still have stable native states in the lattice model? In this paper we test five homologues of Crambin to explore this question.

Meanwhile, to overcome high degeneracies of ground states in the original HP model[15, 16], we introduced a third kind of monomer, "neutral" (0) and a stiffness energy into our lattice protein model. In the resulting H0P model, energy ranges of lattice proteins are usually wider

IOP Conf. Series: Journal of Physics: Conf. Series **1290** (2019) 012018 doi:10.1088/1742-6596/1290/1/012018

than those in the HP model. To keep the sampling efficient, here we apply replica-exchange Wang-Landau (REWL)[17] sampling to the simulation. We also extract some structural behaviors from replica-exchange multicanonical (REMUCA) sampling[18].

2. Models and Simulation Methods

2.1. Semi-flexible H0P Lattice Protein Model

In consideration of the important role that hydrophobic driving force plays in folding of globular proteins, the original HP model is designed to reduce the complexity of atomic details in a protein sequence, allowing for exploring the full conformational space extensively and efficiently[19]. Previous research indicates that HP model leads to high degeneracy in native states[13, 15, 20]. Our semi-flexible H0P model shows great improvement on reducing such degeneracy, by adding more detail on the monomeric scale. The model classifies 20 amino acids into three different types of "beads", based on their hydrophobicity levels, and all "beads", or monomers, in a chain are put on different sites of a simple cubic lattice. In this H0P polymer system, only contacts between two non-bonded nearest neighbors are taken into consideration. While in principle there could be interactions between all types of monomers, here we only choose HH contacts and H0 contacts, which are supposed to be the major contributions to free energy in globular polymers, for calculation of system energy. We also add an energetic term for any "bend" in the protein structure to show the internal rigidity of polypeptide chain[22]. Therefore, the Hamiltonian of our semi-flexible H0P lattice model is given by[15]:

$$\mathcal{H} = -\epsilon_{HH} n_{HH} - \epsilon_{H0} n_{H0} - \epsilon_{\theta} n_{\theta}, \tag{1}$$

where n_{HH} is the number of HH contacts, n_{H0} is the number of H0 contacts and n_θ is the number of bends; ϵ_{HH} is the HH contact energy, ϵ_{H0} is the H0 contact energy and ϵ_θ is the bending energy.

2.2. Crambin Homologues

Crambin is a short protein which is often used to test simulation algorithms[21]. It has also been mapped and studied in the lattice protein model[15, 16, 23]. Meanwhile, sequences and structures of Crambin and its homologues (small plant toxins, such as purothionin and viscotoxin) have long been known by experiment[24, 25]. Therefore, these small proteins are an ideal starting point for us to look into similarities among homologues in the H0P model.

In this paper, one purothionin (PDB entry *1bhp*), one hordothionin (*1wuw*) and three viscotoxins (*1orl*, *1jmn* and *3c8p*) are chosen to be tested as Crambin homologues. Their sequences are mapped into an H0P form, as shown in **Table 1**. Although there are only three kinds of monomers in the H0P representation, these homologues still have significant differences in their sequences. However, comparing to their differences in real sequences, which are around 50%, they are more "alike" in the H0P model.

2.3. Replica-exchange Wang-Landau Sampling

Traditional Monte Carlo methods such as Metropolis sampling are sometimes inefficient at low temperature; for systems with complex free energy landscapes like lattice proteins, these sampling methods are likely to be trapped at some local minima. Instead, Wang-Landau sampling[26] is used to get around these difficulties.

The partition function (Z) at a certain temperature T can be derived from a summation over the energy space:

$$Z(T) = \sum_{\{E_i\}} g(E_i) \exp(-E_i/k_B T), \tag{2}$$

Table 1. Sequences of Crambin homologues in the H0P Model. In the sequences of homologues, monomers different from the corresponding sites in Crambin sequence are marked in red. The dash (-) here represents for one missing monomer, which helps to align the sequences.

Protein	Sequence in H0P Model								
Crambin	PPHH0	0HHHP	0PHPH	HPH00	00P0H	H0000	0HHHH	0000H	00P00	P
1bhp	P0HHP	00H0P	PH0PH	HP0P0	0PP-H	H0PHH	PHPH0	00H0H	0PPH0	P
1wuw	P0HHP	00H0P	PH0PH	HPHP0	0PP-H	H0P0H	PHPH0	00HPH	000H0	P
1orl	P0HH0	P000P	PH0P0	HPH00	00PPP	H0PH0	0HPHH	0000H	00P00	P
1jmn	P0HH0	P000P	PH0P0	HPH00	00PPH	H00H0	0HPHH	0000H	00P00	P
3c8p	P0HH0	0000P	PH0P0	HPH00	00PP0	H0PH0	0HPHH	0000H	00P00	P

where g(E) is the temperature-independent density of states (DoS) at a certain energy (E), and k_B is the Boltzmann constant. Wang-Landau sampling provides an unbiased estimation of g(E) by iteratively performing a random walk in energy space. During simulation, the acceptance probability of one MC trial move relating the current configuration A and proposed configuration B is given by:

$$P(A \to B) = min\{1, \frac{g'(E_A)}{g'(E_B)}\}, \tag{3}$$

where $g'(E_A)$ and $g'(E_B)$ are current estimators of the DoS for energy E_A and E_B, respectively. After a trial move, the DoS for the accepted energy level E_n will be updated according to $g'(E_n) \to f \times g'(E_n)$, where f is a modification factor. A histogram H(E) is also kept during the simulation to count all the energy levels visited. After every trial move, it is updated according to $H(E_n) \to H(E_n)+1$. For a preset number of trial moves, the H(E) is checked for its flatness. If it is flat enough, which indicates all energy levels are almost evenly visited, the modification factor is then adjusted to a lower value such as $\sqrt{f}$, and the histogram is reset to zero.

The replica-exchange Wang-Landau algorithm improves the efficiency of WL sampling by taking advantage of parallel computing[17]. It divides the whole energy range into multiple overlapping energy windows; in each window, there are one or more walkers sampling independently. Replica-exchanges between two walkers in the overlapping area of two neighboring windows i and j are proposed regularly at fixed intervals, with the acceptance probability given as:

$$P_{acc} = min\{1, \frac{g_i'(E_A)}{g_i'(E_B)} \frac{g_j'(E_B)}{g_j'(E_A)}\}, \tag{4}$$

where $g_k'(E_X)$ donates the estimator for the density of states from walker k in configuration X. In this way, REWL allows us to efficiently search the whole complex energy landscapes with consistency among all overlapping energy windows. To further improve the efficiency of sampling, special trial moves are applied in this research[27, 28].

3. Results

3.1. Degeneracy

In this research, the values of contact and bending energy ($\epsilon_\theta = -0.1\epsilon_{H0} = -0.05\epsilon_{HH}$) are chosen specifically to keep the degeneracy of Crambin native states low[16, 22]. With this set of parameters, degeneracies of native states and some low lying excited states of all homologues are searched for with multicanonical sampling. As listed in **Table 2**, the degeneracies of all native states are low, which supports that the parameters chosen here are suitable for simulating this family of lattice proteins. However, some homologues still show high degeneracy in their low lying excited states.

Table 2. Degeneracy of ground states and low excited states. g(E_i) represents the DoS of the i-th excited state from the ground state. E_0 is the ground state. 0 in g(E_2) of Crambin indicates no state has been found at this energy level, which might be a gap in the energy spectrum.

Energy Level	Crambin	*1bhp*	*1wuw*	*1orl*	*1jmn*	*3c8p*
g(E_0)	1	3	2	3	2	5
g(E_1)	6	26	10	24	24	47
g(E_2)	0	84	17	133	118	362
g(E_3)	6	119	24	633	358	1810

3.2. Specific Heat

Once the DoS is well estimated from REWL, we can directly extract the specific heat of the tested lattice protein systems. Specific heat curves of three homologues are plotted in **Fig.1** as examples. For all tested homologues as well as Crambin itself, their specific heat curves show two signals at high temperature (peaks and shoulders), corresponding to the coil-globule transition and folding transition. Meanwhile, at low temperature, there are also two signals at fixed temperature (T = 0.02 and T = 0.05), which indicate that all these homologues start rearranging their structures at the same temperature. Therefore, in the H0P lattice model, these homologues show similar features in their thermodynamic behaviors.

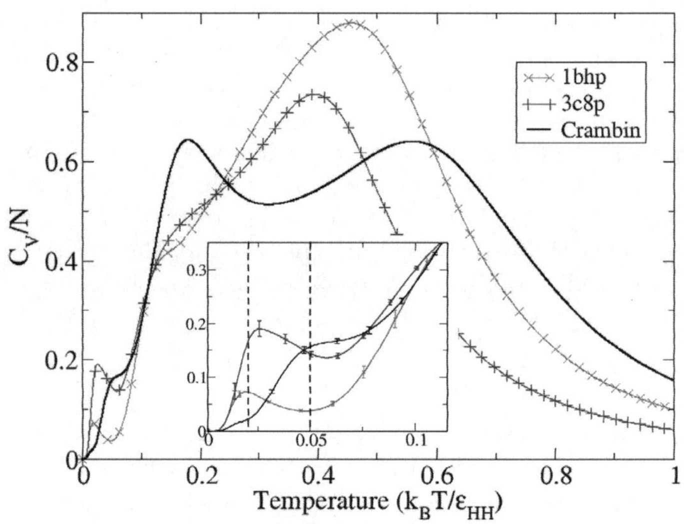

Figure 1: Specific Heat (C_V) curves for Crambin Homologues

Low temperature region is magnified in the bottom. For clarity, only three of the C_V curves for six tested proteins are shown in the graph. Two dashed lines here help to point out signals at T=0.02 and T=0.05. Error bars smaller than the size of the points are not shown.

3.3. Radius of Gyration

The radius of gyration for different homologues is also measured. Their first derivatives are plotted in **Fig.2**. At low temperature, some of the homologues, such as *1wuw*, still keep two signals at T = 0.02 and T = 0.05; however, other homologues, such as *1orl*, do not show signals at these temperatures. Therefore, the structural behaviors of these homologues at low temperature region are dissimilar to each other in the H0P model.

IOP Conf. Series: Journal of Physics: Conf. Series **1290** (2019) 012018 doi:10.1088/1742-6596/1290/1/012018

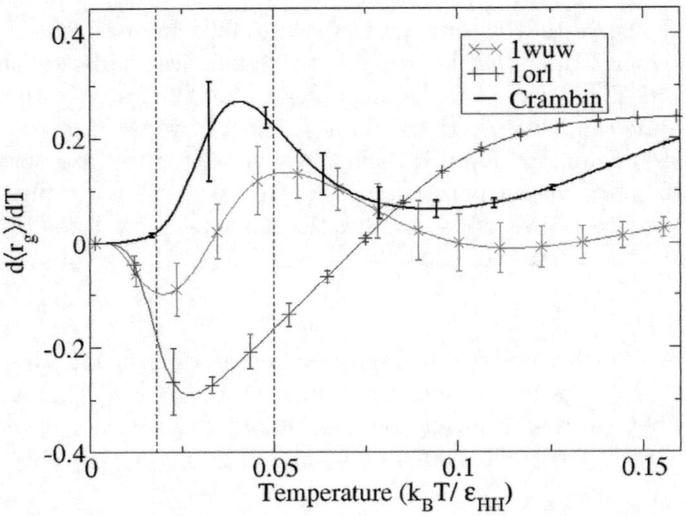

Figure 2: First detivative of the average radius of gyration $\langle r_g \rangle$ for Crambin Homologues

For clarity, only three of the $\langle r_g \rangle$ curves for six tested proteins are shown in the graph. Two dashed lines here help to point out signals at T=0.02 and T=0.05. Error bars smaller than the size of the points are not shown.

3.4. Native Structures

All of the tested lattice proteins in this paper have a small number of ground states, as shown in 3.1. Therefore, it is possible to compare the contact maps of their ground state structures. The results show that, even though their structures in nature are similar[25], in the H0P model the similarity between their native structures are mostly low. The identity between the native structures of different homologues are mostly less than 50%. There is only one case where two homologues have same native structures, as shown in **Fig.3**. In this example, the monomers that are different in these two sequences are all on the surface of the globular structures, which agrees with predictions[7].

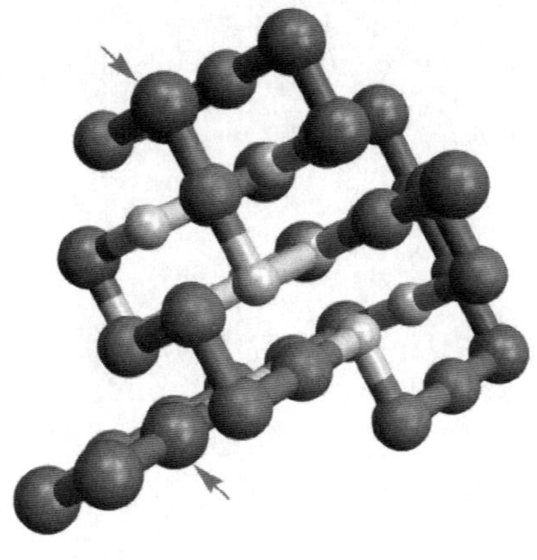

Figure 3.1 Native structure of *1orl*

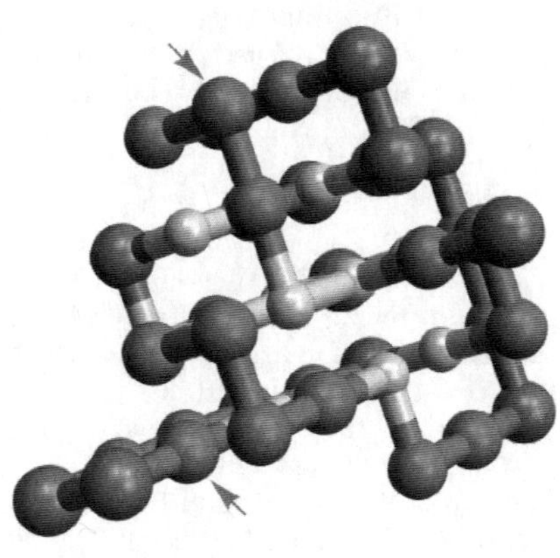

Figure 3.2 Native structure of *3c8p*

The native states for *1orl* and *3c8p* have same ground state structures. Different monomers in sequences are indicated by red arrows. White represents H monomers; red represents 0 monomers; grey represents P monomers.

IOP Conf. Series: Journal of Physics: Conf. Series **1290** (2019) 012018 doi:10.1088/1742-6596/1290/1/012018

4. Conclusion

With the REWL sampling method, we tested six similar lattice proteins in the semi-flexible H0P lattice model. The results show that their thermodynamic behaviors are similar, with two signals in the low temperature region. These common signals indicate that the native structures of these lattice homologues start "melting down" at the same temperature. However, their structural behaviors have very little in common. Even though the same structures are observed for different sequences in one case, most of the native structures of these short homologues have very limited similarities in the H0P lattice model, which implies the impact of mutations is much stronger in the H0P model than it in the 20 amino acid model.

Acknowledgements

The computing sources of this work is provided by Georgia Advanced Computing Resource Center, a partnership between the University of Georgias Office of the Vice President for Research and Office of the Vice President for Information Technology; and Texas Advanced Computing Center under XSEDE Grant No. PHY130014.

References

[1] Wallin S and Shakhnovich E I 2008 *J. Phys.: Condens. Matter.* **20** 283101
[2] Hoque T, Chetty M and Sattar A 2009 *J. Comput. Biol.* **16** 85-103
[3] Koga N, Tatsumi-Koga R, Liu G, Xiao R Acton T B, Montelione G T and Baker D 2012 *Nature.* **491** 222
[4] Davidson A R 2008 *Proc. Natl. Acad. Sci.* **105** 2759-2760
[5] Lammert H, Wolynes P G and Onuchic J N 2012 *Proteins* **80** 362-373
[6] Muñoz V, Campos L A and Sadqi M 2016 *Curr. Opin. Struct. Biol.* **36** 58-66
[7] Dill K A 1999 *Protein Sci.* **8** 1166-1180
[8] Kolinski A and Skolnick J 2004 *Polymer* **45** 511-524
[9] Clementi C 2008 *Curr. Opin. Struct. Biol.* **18** 10-15
[10] Dill K A 1985 *BioChemistry* **24** 1501-1509
[11] Shakhnovich E I 1997 *Curr Opin Struct Biol.* **7** 29-40
[12] Shortle D, Chan H S and Dill K A 1992 *Protein Sci.* **1** 201-215
[13] Shi G, Vogel V, Wüst T, Li Y W and Landau D P 2014 *Phys. Rev.* E **90** 033307
[14] Chan H S, Zhang Z, Wallin S and Liu Z 2011 *Annu. Rev. Phys. Chem.* **62** 301-326
[15] Shi G, Farris A CK, Wüst T and Landau D P 2016 *J. Phys. Conf. Ser.* **686** 012001
[16] Farris A CK, Shi G, Wüst T and Landau D P 2018 *J. Phys. Conf. Ser.* **1012** 012008
[17] Li Y W, Vogel T, Wüst T and Landau D P 2014 *J. Phys. Conf. Ser.* **510** 012012
[18] Sugita Y and Okamoto Y 2000 *Chem. Phys. Lett.* **329** 261-270
[19] Dubey S P, Kini N G, Balaji S and Kumar M S 2018 *Crit. Rev. Biomed. Eng.* **46**
[20] Yue K, Fiebig K M, Thomas P D, Chan H S, Shakhnovich E I and Dill K A 1995 *Proc. Natl. Acad. Sci.* **92** 325-329
[21] Shi G, Wüst T, Li Y W and Landau D P 2015 *J. Phys. Conf. Ser.* **640** 012017
[22] Farris A CK, Shi G, Wüst T and Landau D P 2018 *J. Chem. Phys.* **149** 125101
[23] Dill K A, Fiebig K M and Chan H S 1993 *Proc. Natl. Acad. Sci.* **90** 1942-1946
[24] Debreczeni J É, Girmann B, Zeeck A, Krätzner R and Sheldrick G M 2003 *Acta. Crystallogr. D Biol. Crystallogr.* **59** 2125-2132
[25] Schmidt A, Teeter M, Weckert E and Lamzin V S 2011 *Acta. Crystallogr. Sect. F Struct. Biol. Cryst. Commun.* **67** 424-428
[26] Landau D P, Tsai S H and Exler M 2004 *Am. J. Phys.* **72** 1294-1302
[27] Wüst T and Landau D P 2009 *Phys. Rev. Lett.* **102** 178101
[28] Wüst T and Landau D P 2012 *J. Chem. Phys.* **137** 064903

IOP Conf. Series: Journal of Physics: Conf. Series **1290** (2019) 012019 doi:10.1088/1742-6596/1290/1/012019

A First Look at Lattice Effects in Coarse-Grained Protein Models via Wang-Landau Simulations

Alfred C.K. Farris[1], Daniel T. Seaton[2], and David P. Landau[1]

[1] Center for Simulational Physics, Department of Physics and Astronomy, The University of Georgia, Athens, GA 30602, USA

[2] Office for the Vice Provost for Advances in Learning, Harvard University, Cambridge, MA 02138, USA

E-mail: alfred.farris@uga.edu

Abstract.
In order to study the effects of lattice constraints on coarse-grained protein models, we apply Wang-Landau sampling to the continuum analogue of the hydrophobic-polar (HP) lattice protein model. The continuum version is inspired by the AB polymer model but incorporates potentials chosen specifically to mimic those of the lattice case. Because of their relative simplicity, both the lattice and continuum models offer significant computational advantage over all-atom simulations, but the impact of the additional lattice constraint on generic folding behavior is unknown. In this preliminary study, we compare and contrast thermodynamics during the folding process of the continuum model to the original HP lattice protein model for sequences mapped from Crambin, a 46 amino acid plant protein. We find that the folding process for both of these coarse-grained models is quite similar, with major structural transitions occurring at almost exactly the same temperatures.

1. Introduction

Protein folding has been studied for over 50 years, but we still cannot explain how such complex biomolecules fold so efficiently [1]. Coarse-grained models analyzed using statistical mechanics have greatly increased our understanding of protein folding [2], but it is not clear how much detail to include [3, 4]. Thus, one persistent issue is deciding where the ideal balance between model simplicity and physical realism lies. Among coarse-grained models, one substantial difference is that of a lattice vs. continuum representation of the spatial degrees of freedom. The simplification of lattice models allows for efficient computation and more precise interactions, but clearly cannot represent all spatial configurations. Continuum models have the perceived advantage of full spatial resolution, but suffer from the complications associated with empirical potentials and higher computational cost.

The simplest protein models incorporate a reduction of the 20 amino acids into a limited number of types, such as in the hydrophobic-polar (HP) lattice model [5, 6], and the continuum AB model [7]. Both of these models use monomers to represent only two distinct amino acids types (H,P or A,B), allowing for a reduction in the different kinds of interactions, but a seemingly substantial difference remains with regard to spatial representation (lattice vs. continuum). It has been difficult to quantify the effects associated with lattice constraints on such models because the potentials used in the HP and AB models are often different enough

IOP Conf. Series: Journal of Physics: Conf. Series **1290** (2019) 012019 doi:10.1088/1742-6596/1290/1/012019

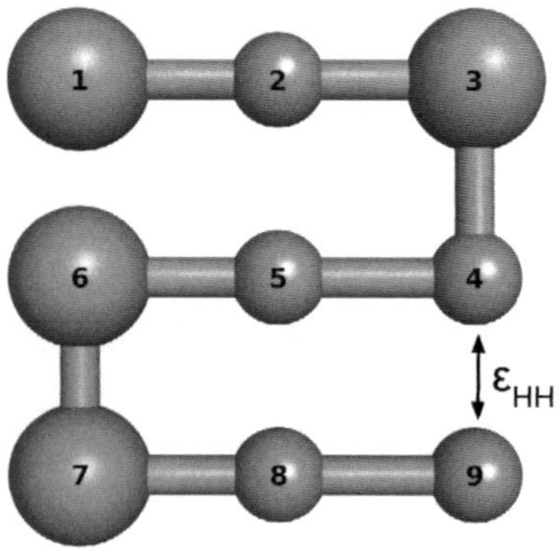

Figure 1. Sample HP lattice model configuration. Hydrophobic and polar monomers are colored in gray and orange, respectively. The interaction between non-bonded hydrophobic monomers, such as monomers 4 and 9, contributes energy ϵ_{HH}. In total, there are 3 HH contacts (n_{HH}).

that discrepancies in the results can be attributed to the potentials themselves.

In order to efficiently sample the complex free-energy landscape associated with coarse-grained protein models, here we use Wang-Landau sampling [8, 9] to offer a first look into the effects of lattice constraints on coarse-grained models by comparing and contrasting the original HP lattice model to an off-lattice analogue, inspired by the AB model, but for which the potentials are chosen to mimic the lattice case. We compare these models for sequences mapped from Crambin – a small (46 amino acid) plant protein [10, 11]. One note regarding the HP model is that it is an NP-complete optimization problem and has been used as an algorithmic test-bed over a wide range of disciplines; see [12] for an overview of these studies.

2. Models and Simulation Methods

2.1. HP Lattice Model

The HP model simplifies the complexity of protein folding by classifying each of the twenty amino acids as either hydrophobic (H) or polar (P), based on their desire to interact with water [5, 6]. The protein is then modeled as a polymer performing a self-avoiding random walk with H and P monomers placed at sites of a simple cubic lattice, automatically accounting for excluded volume effects found in real proteins. To model the hydrophobic "driving force", each non-bonded, nearest neighbor HH contact contributes energy ϵ_{HH} (= 1 in this study), resulting in the following Hamiltonian:

$$\mathcal{H} = -\epsilon_{HH} n_{HH}, \tag{1}$$

where n_{HH} is the number of non-bonded HH contacts. Thus, each such contact lowers the internal energy of the system, and leads to the formation of a hydrophobic core in the native state, as in real proteins. The major advantages of this model are that it allows for efficient sampling of conformation space and avoids the uncertainty associated with empirical potentials. Fig. 1 shows a sample HP model configuration with energy $E = -3$ based on three HH interactions.

2.2. HP Model Continuum Analogue

The AB continuum protein model shares many similarities with the HP model, such as the same reduced amino acids and simple interactions. However, the potentials in the AB model are not constructed to mimic the HP model case, and thus side-by-side comparisons are never made. In this study we introduce a continuum analogue to the HP model, where non-bonded hydrophobic-hydrophobic interactions are modeled via a Lennard-Jones potential [13], bonds are connected via a finitely-extensible-nonlinear-elastic (FENE) potential [14], and all other monomer interactions are purely repulsive – using the repulsive component of the Lennard-

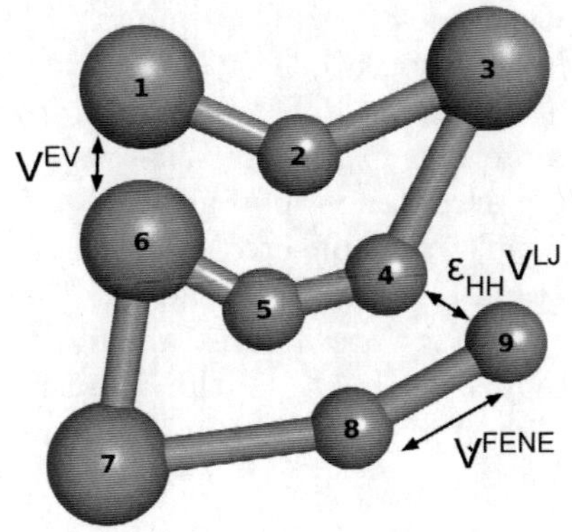

Figure 2. Sample continuum HP model configuration. Hydrophobic and polar monomers are colored in gray and orange, respectively. The interaction between non-bonded hydrophobic monomers, such as monomers 4 and 9, contributes energy $\epsilon_{HH} V^{LJ}$. The interaction between non-bonded, non-hydrophobic monomers, such as monomers 1 and 6, contributes energy V^{EV}. The interaction between bonded monomers, such as monomers 8 and 9, contributes energy V^{FENE}.

Jones potential, to mimic the excluded volume effects inherent to the lattice HP model. Thus, our continuum Hamiltonian is given as follows:

$$\mathcal{H} = \epsilon_{HH} \sum_{\langle i_H, j_H \rangle} V^{LJ}(r_{ij}) + \sum_{\langle bonded \rangle} V^{FENE}(r_{bond}) + \sum_{\langle other \rangle} V^{EV}(r_{other}), \tag{2}$$

where

$$V^{LJ}(r_{ij}) = \left(\frac{r_0}{r_{ij}}\right)^{12} - 2\left(\frac{r_0}{r_{ij}}\right)^{6}, \tag{3}$$

$$V^{FENE}(r_{bond}) = -0.5k\,(r_{max} - r_0)^2\, ln\left(1 - \left(\frac{r_{bond} - r_0}{r_{max} - r_0}\right)^2\right), \tag{4}$$

and

$$V^{EV}(r_{other}) = \left(\frac{r_0}{r_{other}}\right)^{12}, \tag{5}$$

where the values ϵ_{HH} and r_0 set the energy and length scales of our system, respectively, and in this study are both set to 1 using dimensionless units, r_{ij} is the distance between non-bonded monomers i and j, r_{bond} is the distance between bonded monomers, and r_{other} is the distance between non-bonded non-hydrophobic pairs. We choose for the stiffness $k = 120\frac{\epsilon}{r_0^2}$ (to achieve similar stiffness as in [15]), and for the the maximum displacement from equilibrium $r_{max} = 1.2r_0$. Moreover, we use a truncated, shifted Lennard-Jones potential such that $V^{LJ}(r \geq r_c) = 0$, where the cutoff distance $r_c = 3.0r_0$. One important note is that the reference energy value, ϵ_{HH}, is set to 1 in both the lattice and continuum case. Fig. 2 shows a sample continuum HP model configuration and Fig. 3 shows a plot of the continuum potentials used in this study. In contrast to a study using a similar model investigating homopolymer systems, where bonds were connected via a combination of FENE and Lennard-Jones potentials [15], we have found the quantitative differences to be quite small when using only the FENE potential between bonds and thus follow the latter approach here for simplicity.

In this study we investigate Crambin, a slightly hydrophobic protein found in cabbage [10,16]. Due to its relatively small size ($N = 46$ amino acids), Crambin is one of the more computationally tractable proteins to model. Here we use the same HP model mapping as in [17].

2.3. Wang-Landau Sampling

Traditional Monte Carlo methods like Metropolis sampling perform poorly at low temperatures and fail for systems with complex free energy landscapes. Wang-Landau sampling

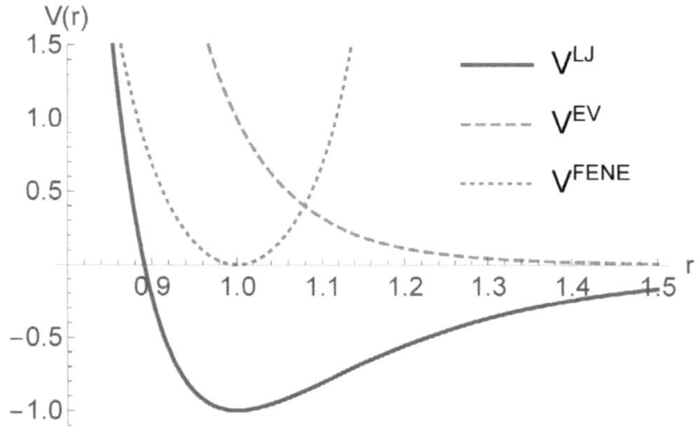

Figure 3. Continuum HP model potentials. Non-bonded hydrophobic-hydrophobic interactions are given via a Lennard-Jones potential (V^{LJ}), bonds are connected via a FENE potential (V^{FENE}), and all other monomer interactions are purely repulsive (V^{other} – the repulsive r^{-12} term from the Lennard-Jones potential).

[8,9] is an iterative Monte Carlo method that obtains a system's density of states (DOS), $g(E)$, by performing a random walk in energy space. During the simulation, a trial move to change the system from state A to B is proposed, and accepted with probability

$$P(A \to B) = min\left\{1, \frac{g(E_A)}{g(E_B)}\right\}, \tag{6}$$

where $g(E_A)$ and $g(E_B)$ are the current estimates of the DOS for states A and B, respectively. After a trial move where state n is accepted, $g(E_n)$ is updated via $g(E_n) \to f \cdot g(E_n)$, where f is a modification factor. At this point, a histogram $H(E)$ of visited system states updated via $H(E_n) \to H(E_n) + 1$ after state n is accepted. The histogram is checked at regular intervals, and reset to zero when all system states have been visited such that all histogram entries are at least 80% of the average histogram value. The modification factor is then updated via $f \to \sqrt{f}$, and this process is repeated until $f < f_{min}$, some predefined minimum value. Since the Wang-Landau acceptance probability is independent of temperature, this sampling scheme is much more efficient at sampling systems with complex free energy landscapes.

In this study, f was initially set to e^1, and $f_{min} = exp(10^{-6})$. In the lattice model, our trial moves were chosen randomly such that the percentage of each move type were as follows: 75% pull moves [18], 23% bond-rebridging moves [19], and 2% pivot moves [20]. This combination of Monte Carlo moves has proven very efficient in studying lattice proteins together with Wang-Landau sampling [12,21]. In the continuum model, we use N diffusion moves for one of each of the following: reptation, bond-rebridging, crankshaft, and random pivot moves.

2.4. Thermal Properties and Protein Structures

With $g(E)$ obtained from Wang-Landau sampling, the partition function (Z) can be computed at any temperature (T) simply as

$$Z(T) = \sum_E g(E) e^{-E/k_B T}, \tag{7}$$

where E is the system energy, and k_B is the Boltzmann constant ($= 1$ in this study). We then calculate the ensemble average of any quantity Q as follows:

$$\langle Q \rangle = \frac{\sum_E Q \cdot g(E) e^{-E/k_B T}}{Z}. \tag{8}$$

The heat capacity (C_V) can be calculated from fluctuations in the energy via

$$C_V(T) = \frac{\langle E^2 \rangle - \langle E \rangle^2}{k_B T^2}. \tag{9}$$

IOP Conf. Series: Journal of Physics: Conf. Series **1290** (2019) 012019 doi:10.1088/1742-6596/1290/1/012019

3. Results

3.1. Specific Heat

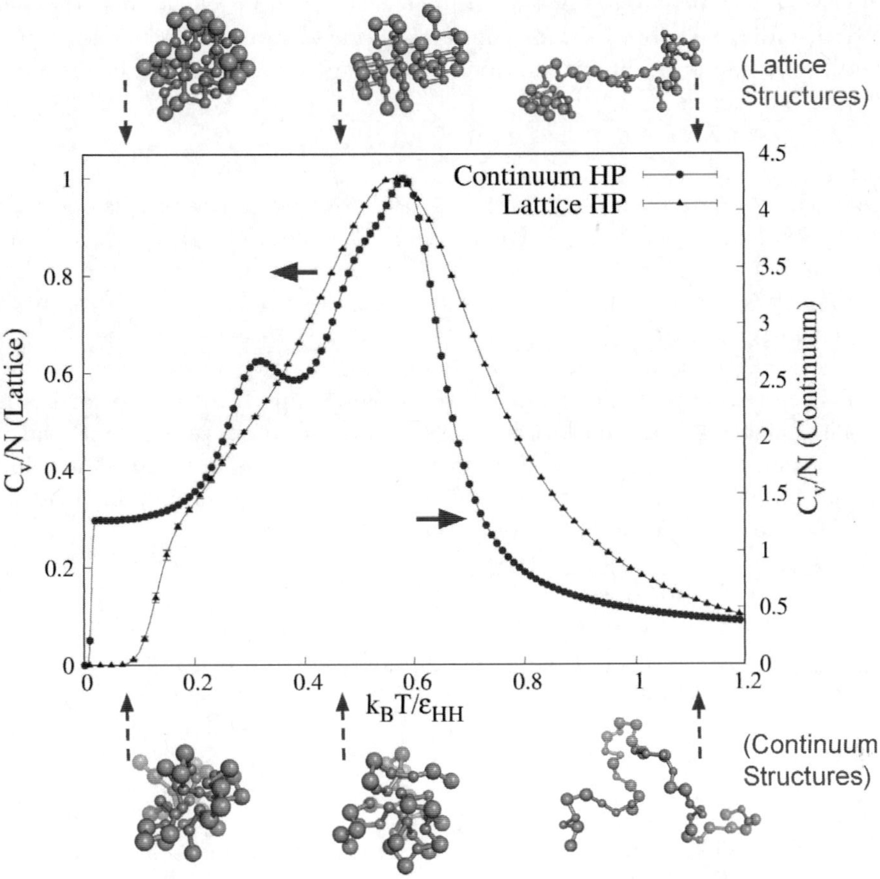

Figure 4. Specific heat curves for the lattice and continuum HP models for Crambin, with characteristic structures shown relative to peak locations. Error bars smaller than the points are not shown.

The specific heat (C_V/N) offers insight into the folding process, with peaks indicating structural transitions. In smaller proteins, the collapse process occurs in two steps [22]: Firstly a coil-globule transition, in which the protein folds from a random coil to a globule. Secondly, the "folding transition" consists of the protein becoming even more compact, and ultimately relaxing to the native state. Fig. 4 shows the specific heat curve for the both the lattice and continuum HP models of Crambin, along with characteristic structures relative to the peak locations. Both models exhibit these two structural transitions (consistent with all-atom simulations of Crambin [23]): the lattice HP model displays a peak and broad shoulder, but in the continuum case they both appear as peaks. While the magnitude of these curves differ between models, the temperature scale is common, and thus the "transition" temperatures (and overall results) are strikingly similar. Note that although the specific heat curve for the continuum model appears to go to zero at low temperature, this is a non-physical result of cutting the sampled energy range off at a certain minimum. The true specific heat will level off to a value given by the equipartition theorem at zero temperature. In the lattice case, the drop-off to zero is because there is a gap between the ground-state and first-excited state energy.

The main difference between the two models is the emergence of a faint transition signal in the form of a small shoulder between the two peaks in the continuum case. We have not found substantial numerical evidence in any order parameter to explain this shoulder (no signal in the radius of gyration, radial distribution function, end-to-end distance, or core density), and thus this shoulder still remains somewhat mysterious. Its origin is currently under investigation and will be included in a further study.

4. Conclusion

We have used Wang-Landau sampling to offer a first look into lattice effects on coarse-grained protein models by examining the original HP model and its continuum analogue for a sequence mapped from Crambin. Analysis of the specific heat curves reveals that the folding process shares many similarities between the two models. The coil-globule transition occurs at almost exactly the same temperature and is marked by a peak in both cases. While the specific heat displays quantitative differences at the folding transition (peak vs. shoulder), the qualitative folding behavior is the same and the temperature at which this transition occurs is similar for both models. This suggests that the lattice constraint does not have a substantial impact on the generic folding behavior in this case. A more detailed study is in progress in which we aim to uncover the quantitative meaning of the small shoulder that appears in the continuum case, as well as investigate lattice effects on semi-flexible protein models.

5. Acknowledgements

This study was supported in part by resources and technical expertise from the Georgia Advanced Computing Resource Center, a partnership between the University of Georgia's Office of the Vice President for Research and Office of the Vice President for Information Technology.

References

[1] Dill K A and MacCallum J L 2012 *Science* **338** 1042–6
[2] Dill K A and Chan H S 1997 *Nat. Struct. Mol. Biol.* **4** 10–19
[3] Dill K A, Bromberg S, Yue K, Chan H S, Fiebig K M, Yee D P and Thomas P D 1994 *Protein Sci.* **4** 561–602
[4] Kolinski A and Skolnick J 2004 *Polymer* **45** 511–524
[5] Dill K A 1985 *Biochemistry* **24** 1501–9
[6] Lau K F and Dill K A 1989 *Macromolecules* **22** 3986–3997
[7] Stillinger F H, Head-Gordon T and Hirshfeld C L 1993 *Phys. Rev. E* **48**(2) 1469–1477
[8] Wang F and Landau D P 2001 *Phys. Rev. Lett.* **86** 2050–2053
[9] Wang F and Landau D P 2001 *Phys. Rev. E* **64** 056101
[10] Hendrickson W A and Teeter M M 1981 *Nature* **290** 107 – 113 ISSN 0031-9422
[11] Schmidt A, Teeter M, Weckert E and Lamzin V S 2011 *Acta Crystallographica Section F* **67** 424–428
[12] Wüst T and Landau D P 2009 *Phys. Rev. Lett.* **102**(17) 178101
[13] Lennard-Jones J E 1924 *Proc. Royal Soc. Lond. A* **106** 463–477
[14] Kremer K and Grest G S 1990 *J. Chem. Phys.* **92** 5057–5086
[15] Seaton D T, Wüst T and Landau D P 2010 *Phys. Rev. E* **81**(1) 011802
[16] Vanetten C, Nielsen H and Peters J 1965 *Phytochemistry* **4** 467 – 473 ISSN 0031-9422
[17] Lattman E E, Fiebig K M and Dill K A 1994 *Biochemistry* **33** 6158–6166
[18] Lesh N, Mitzenmacher M and Whitesides S 2003 *RECOMB* 188–195
[19] Deutsch J M 1997 *J. Chem. Phys.* **106** 8849
[20] Sokal A D 1995 *Monte Carlo and Molecular Dynamics Simulations in Polymer Science* (Oxford University Press) edited by K. Binder
[21] Wüst T and Landau D P 2012 *J. Chem. Phys.* **137** 064903
[22] Jackson S E 1998 *Folding and Design* **3** R81 – R91 ISSN 1359-0278
[23] Shimada J, Kussell E L and Shakhnovich E I 2001 *J. Mol. Biol.* **308** 79 – 95

IOP Conf. Series: Journal of Physics: Conf. Series **1290** (2019) 012020 doi:10.1088/1742-6596/1290/1/012020

Free energy profiles of lipid translocation across pure POPC and POPC/CHOL bilayer: all-atom molecular dynamics study

Hiroaki Saito[1,*], Tetsuya Morishita[2], Taku Mizukami[3], Ken-ichi Nishiyama[4], Kazutomo Kawaguchi[5], and Hidemi Nagao[5]

[1]RIKEN Center for Biosystems Dynamics Research (BDR), Suita, Japan,
[2]CD-FMat, and MathAM-OIL, National Institute of Advanced Industrial Science and Technology (AIST), Tsukuba, Japan,
[3]Japan Advanced Institute of Science and Technology (JAIST), Nomi, Japan,
[4]Cryobiofrontier Research Center, Faculty of Agriculture, Iwate University, Morioka, Japan,
[5]Institute of Science and Engineering Kanazawa University, Kanazawa, Japan.

[*]hiroaki.saito@riken.jp

Abstract. All-atom molecular dynamics (MD) simulations of pure 1-palmitoyl-2-oleoyl-*sn*-glycero-3-phosphocholine (POPC) bilayer and POPC bilayer containing cholesterol (POPC/CHOL) with CHARMM36 force field were carried out to investigate the effects of CHOL on structure and lipid translocation across the membranes. We calculated the potential mean force (PMF) profiles for translocation of POPC and CHOL along the pure POPC and the POPC/CHOL bilayer normal by umbrella sampling method. The obtained PMF profile of the POPC translocation was in good agreement with that of the previous MD study, showing that the estimated PMF profiles with the CHARMM36 force field should be reasonable. We found that the PMF peak appears slightly beyond the bilayer center. The CHOL effects on the PMF profile of POPC translocation were clearly observed; the free energy barrier for the flip-flop of lipid increased and, the energy for the desorption of lipid decreased. These changes in the PMF profile should be responsible for the tight packing of the POPC/CHOL bilayer. The PMF profiles for the CHOL translocation showed that the free energy barriers at the bilayer center were sufficiently smaller than those of POPC translocation, indicating that the CHOL can easily flip over compared to the POPC in the membrane. In the case of PMF of CHOL translocation in the POPC/CHOL system, while the PMF energy in the hydrophobic core region was higher than that of the pure POPC bilayer, the free energy difference at bilayer center was small.

IOP Conf. Series: Journal of Physics: Conf. Series **1290** (2019) 012020 doi:10.1088/1742-6596/1290/1/012020

1. Introduction

Biomembrane is a kind of membrane to separate boundary in organelle or cell and has important functions such as selective permeation of ions/molecules and activation of the membrane proteins. The biomembrane mainly consists of amphiphilic molecules such as lipids and forms bilayer structure, and the structure, dynamics and molecular permeability of the membrane depend on the kinds of lipids and lipid composition in the membrane. The cholesterol (CHOL) is an example molecule which changes the structure and dynamics of the membrane. It has been known that the orientation of lipid chain increases by adding CHOL to the lipid bilayer and changes the membrane state to the liquid-ordered phase above 20-30 mol% of cholesterol [1, 2]. The dynamics of lipid transports in the membrane such as the flip-flop changes in the lipid bilayer containing CHOL [3]. Also, cholesterol blocks spontaneous insertion of membrane proteins into liposomes [4]. Since these dynamical properties of lipids is related to important properties such a membrane fusion, the effect of CHOL on the dynamical structure and transport properties of lipids or CHOL along the membrane normal should be investigated in detail. However, due to the difficulty of experimental observation, the details of these properties are not well known yet.

In the computational studies, potential mean force (PMF) profiles of the lipid translocation across the membrane has been investigated by molecular dynamics (MD) simulations [5-8]. The calculated PMF profiles showed the large energy barriers for the flip-flop and desorption of lipids from the membrane, and these results were qualitatively consistent with the experimental observation of lipid transfer from solution to micelle [5]. In their MD studies, they have adopted the united atom model, which is a sort of coarse-grained particle model, with GROMACS force field [5-8]. On the other hand, the CMARMM force field has been standardly used for the all-atom MD simulations for lipid bilayer systems [9-11]. The MD simulations for the several kind of lipid bilayers including CHOL with the CHARMM36 force filed have been verified so far [9-11], and the estimated membrane structures such as area per lipids and order parameters have been consistent to those of experimental observations. However, in our knowledge, the free energy changes related to the lipid translocation across the membrane have not been investigated yet.

In this study, we thus carried out MD simulations of the pure 1-palmitoyl-2-oleoyl-sn-glycero-3-phosphocholine (POPC) bilayer and the POPC bilayer containing CHOL (POPC/CHOL) with the CHARMM36 force field and calculated the PMF profiles for the POPC translocation along the pure POPC and POPC/CHOL bilayer normal. The effects of CHOL on structure and lipid translocation across the membranes were investigated in detail. We carefully verified the POPC flip-flop level position around the bilayer center by long-term position constrained MD simulations. The PMF profiles of CHOL across the pure POPC bilayer and the POPC/CHOL bilayer were also estimated to investigate the differences with those of POPC translocation and the effects of CHOL on the PMF profile.

2. Material and method

2.1. Molecules and membrane systems

The molecular structures of the POPC and CHOL are shown in Figure 1. POPC has a 16:0 palmitic acid (PA) and an 18:1 oleic acid (OA) linking to the glycerol backbone and has a phosphorylcholine (PC) head group. CHOL has a rigid sterol ring and a short hydrocarbon chain. A hydroxyl group is linked to the sterol ring of CHOL, and this hydroxyl group is oriented to the water bulk region of the membrane. Initial bilayer structures of the pure POPC and the POPC/CHOL mixtures for MD simulations consist of 128 lipids and 8,192 water molecules. In the POPC/CHOL bilayer, thirteen POPC were replaced to CHOL in each leaflet of the bilayer, corresponding to about 20 mol % of CHOL addition. We also prepared the POPC bilayer containing only one CHOL to investigate the PMF profile of CHOL across the POPC bilayer. In this system, one CHOL was replaced to POPC in the upper leaflet of the POPC bilayer. The initial coordinates of each system were prepared by using CAHRMM-GUI [11].

IOP Conf. Series: Journal of Physics: Conf. Series **1290** (2019) 012020 doi:10.1088/1742-6596/1290/1/012020

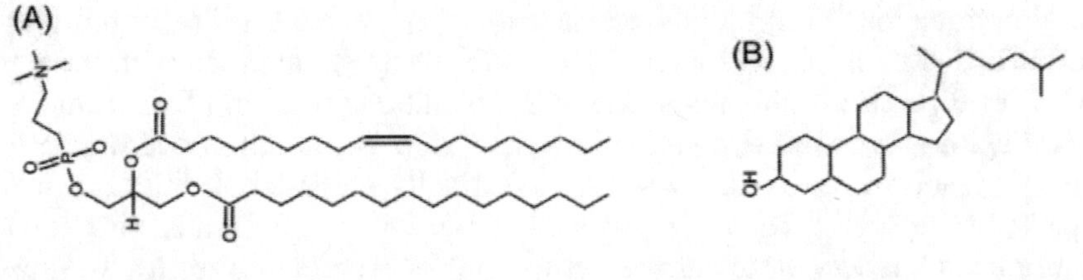

Figure 1. Molecular structures of (A) 1-palmitoyl-2-oleoyl-*sn*-glycero-3-phosphocholine (POPC), and (B) cholesterol (CHOL).

2.2. *Molecular dynamics simulations*

MD simulations of the pure POPC bilayer and POPC/CHOL bilayers were carried out for 200 ns under the constant temperature (T = 303.15 K) and pressure (P = 1 atm) condition. Nose-Hoover thermostat [12, 13] and Parinnelo-Rahman type borostat [14] were used to control the system temperature and the pressure. The periodic lengths of thermostat and barostat were both 1 ps. CHARMM36 force field [9, 10] and TIP3P water model [15] were adopted for lipids and water molecule, respectively. Cutoff for the van der Waals (vdW) interaction was applied using a switching scheme, within a range of radius from 10 to 12 Å. Particle mesh Ewald (PME) method [16] was adopted for the calculation of the coulomb electrostatic interactions. The time step for numerical integration of equations of motion was 2 fs, and the coordinates of the systems were stored every 1 ps. Figure 2 shows the area per molecule A_m as a function of MD time for each system. A_m at each MD step was calculated by dividing the xy area of MD box by the lipid number in the leaflet of bilayer. The results showed that all molecular areas were sufficiently equilibrated after 100 ns. Thus, the last 100 ns of the MD runs were used for the structure analysis of the membranes. All MD simulations were done by GROMACS 5.1.2 [17].

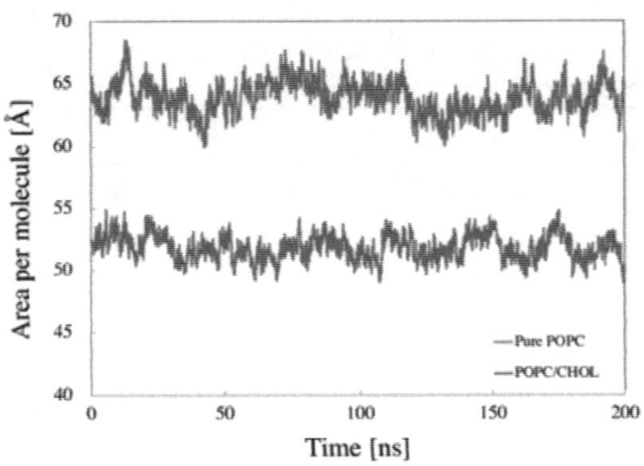

Figure 2. Area per molecule as a function of MD time for the pure POPC and the POPC/CHOL bilayers.

2.3. *PMF calculations*

PMF profiles for the translocation of POPC and CHOL along the membrane normal were evaluated by the umbrella sampling method [18]. In this study, the reaction coordinate for the PMF calculations was defined as the difference between the z elements of picked atom and the center of mass of the bilayer. The region for the umbrella sampling calculations ranges from the water bulk region (z = 45 Å) to the region beyond the bilayer center (z = –5 Å). The spring constant of harmonic potential for the position restrain was 3000 $\mathrm{kJ/mol/nm^2}$ and the window width was 1 Å.

First, one POPC was selected in the upper leaflet of the bilayer and the carried out the MD simulations to sample the POPC coordinates along the reaction coordinate. The harmonic potential was applied to the phosphor (P) atom of the POPC, and the picked P atom was transferred with the

IOP Conf. Series: Journal of Physics: Conf. Series **1290** (2019) 012020 doi:10.1088/1742-6596/1290/1/012020

velocity of 1 Å/ns from the equilibrium position of P atoms in the upper leaflet to the bilayer center and to the water bulk region direction. In this study, to carefully verify the z level position where the transferred POPC actually flips over in the membrane, 100ns term MD simulations with the position restrain were run at each z level position. This long-term MD simulations were run first from $z = 0$ Å, and the last snapshot coordinate of the MD simulation was then used as the initial coordinate for the MD simulation at the next z level position (ex. $z = -1$ Å). We finally prepared 50 initial coordinates that the P atom of the picked POPC is on each z level. Each window was equilibrated first for 10 ns, and the production runs for each window were carried out for 100 ns. The PMF profiles as a function of the reaction coordinate z were calculated by the weighted histogram analysis method (WHAM) [18]. The energy peak of the obtained PMF around the bilayer center corresponds to the free energy barrier for flip-flop, and the PMF energy in the water bulk region can be seen as the desorption energy of the lipid from the membrane.

The PMF profiles of CHOL translocation across the pure POPC and the POPC/CHOL bilayer were also evaluated to investigate the translocation properties of CHOL. We applied the same harmonic potential to the oxygen atom of the hydroxyl group of the picked CHOL in this case. The calculations of PMF profiles were done by the *g_wham* program [18] in the GROMACS 5.2.1 package.

3. Results and discussions

The average area per molecule A_{mol}, membrane thickness $d_{P\text{-}P}$, and order parameter $-S_{CD}$ for each system were listed in Table 1. The estimated A_{mol} of pure POPC bilayer was 63.5 Å^2, which is comparable with experimental value, 68.3Å^2 [19]. The $d_{P\text{-}P}$ were calculated as the time averaged distances between z elements of the P atoms in each leaflet of the bilayer. The order parameters $-S_{CD}$ shown in Table 1 were estimated by averaging order parameters of the two acyl chains (*sn*-1 and *sn*-2) of POPC in each system. The results showed that the order parameter $-S_{CD}$ and the membrane thickness $d_{P\text{-}P}$ increased by adding CHOL. These structural changes were also observed in the previous experiment and the MD studies [20], showing that our MD simulations with the CHARMM36 force field reproduced the relevant membrane structures for each system.

Table 1. The average area per molecule A_{mol}, membrane thickness $d_{P\text{-}P}$, and order parameter $-S_{CD}$ for pure POPC and POPC/CHOL bilayer. The parenthesis shows the standard deviations of the values. The listed $-S_{CD}$ in the table were average values of $-S_{CD}$ at each carbon of *sn*-1 and *sn*-2 chains.

	A_{mol} [Å^2]	$d_{P\text{-}P}$ [Å]	$-S_{CD}$
Pure POPC	63.5 (1.3)	39.1 (0.7)	0.15
POPC/CHOL	51.8 (1.1)	43.3 (0.7)	0.21

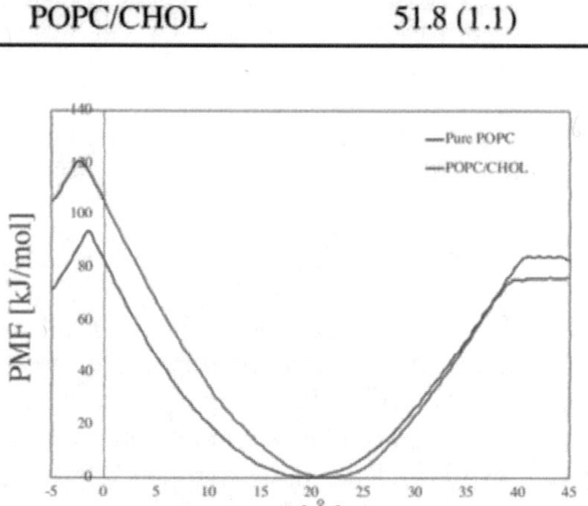

Figure 3. PMF profile of POPC across the pure POPC and POPC/CHOL bilayer.

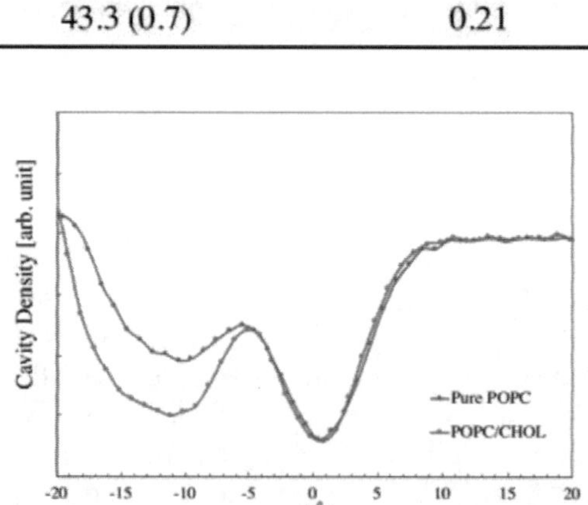

Figure 4. Cavity distribution profiles as a function of reaction coordinate z.

IOP Conf. Series: Journal of Physics: Conf. Series **1290** (2019) 012020 doi:10.1088/1742-6596/1290/1/012020

Figure 3 shows the PMF profiles for POPC translocation as a function of reaction coordinate z. The minimum of the PMF profile at z = 19.4 Å corresponds to the equilibrium level position of P atoms in the upper leaflet of the bilayer, and the PMF energy at this z was defined as zero in this study. The PMF energy increases as the POPC transfers from the equilibrium position and reaches the plateau region in the water bulk region (z > 40 Å).

In this study, the PMF energy at the center of bilayer is given as ΔG_{center}. In the previous MD studies, the free energy at the bilayer center was assumed as the primary free energy barrier for flip-flop of lipids in the membrane [5-8]. However, the POPC does not always flip over at the z = 0 Å where the P atom of picked POPC is on the bilayer center. In fact, our long-term MD simulation shows that the picked POPC did not flip over at z = 0 Å; it flipped over slightly beyond the bilayer center in this study. The peak of PMF was found at z = –1.4 Å and the corresponding free energy barrier from the equilibrium level position, ΔG_{flip}, was 94 kJ/mol. The free energy difference at the bilayer center ΔG_{center} was 83 kJ/mol, and this result was close to that of the previous study, 89 kJ/mol [8]. On the other hand, the desorption energy of POPC ΔG_{desorp} was 84 kJ/mol, being in good agreement with the previous study [8]. We thus conclude that the estimated PMF profiles with the CHARMM36 force field should be reasonable results.

Next, we discuss the effect of CHOL on the POPC translocation in the membrane. The PMF profile of POPC across the POPC/CHOL bilayer showed that the free energy barrier for flop-flop, ΔG_{flip}, increased and, the energy for desorption of lipid, ΔG_{desorp}, decreased. These PMF changes were also shown in the previous MD study for the DPPC/CHOL mixture bilayer [6]. The peak of PMF profile was found at z = –2.4 Å, and the energy barrier for flip-flop, ΔG_{flip}, was 120.7 kJ/mol. The desorption energy ΔG_{desorp} was 76 kJ/mol. As shown in the structural analysis of the membranes, since the membrane structures such as the order parameter, area per molecules and membrane thickness changed by adding CHOL to the POPC bilayer, the changes of the PMF in the POPC/CHOL bilayer should also be responsible for the cholesterol condensing effect in the membrane. We thus analyzed cavity distribution along the bilayer normal to investigate the molecular packing in each system. For this calculation, the systems were divided into grid with size of 1 Å at each MD step, and those grids were checked if atom exits or not in the bins. Figure 4 shows the cavity distribution profile in the pure POPC and the POPC/CHOL bilayer along z coordinate of the system. In this figure, the equilibrium position of P atoms in the upper leaflet of the bilayer is defined as z = 0 Å. The bilayer center and the water bulk region correspond to around z = –20 Å and 20 Å, receptivity. We found that the cavity distribution in the POPC/CHOL bilayer largely decreased around the hydrophobic core region of the membrane. The molecular packing of the POPC/CHOL bilayer thus increase, resulting in the increase of the PMF energy in the corresponding region.

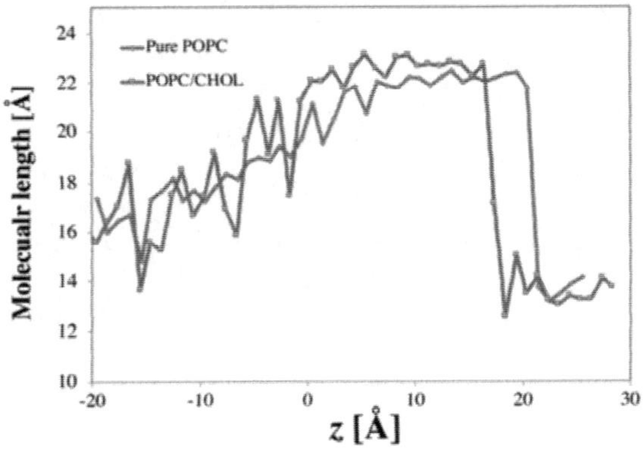

Figure 5. Molecular length of the picked POPC as a function of reaction coordinate z.

Next, we discuss the free energy differences of ΔG_{desorp} between the pure POPC and the POPC/CHOL bilayer systems. We considered that the POPC structures transferred to the water bulk region changes, and these changes of lipid conformations should contribute to the changes of PMF profile. We thus plotted the molecular length of the picked POPC as a function of reaction coordinate z in Figure 5. In this figure, the molecular length of POPC is defined as the distance between the P atom and the sn-2 chain tail, and the reaction coordinate z is defined as the difference between z elements of the P atom of the picked POPC and the equilibrium level position of P atoms in the upper leaflet of bilayer. The molecular length of the POPC slowly increased until the membrane surface region, and it then suddenly decreases when the picked POPC came out from the membrane surface. If once the POPC completely came out from the membrane, the interaction between the picked POPC and the bilayer becomes less, resulting in the plateau shape of the PMF profile in the water bulk region. On the other hand, we found that the molecular length of POPC in the POPC/CHOL bilayer decreased at lower z level. In the case of POPC/CHOL bilayer, since the highly order lipid chains are tightly packed in the hydrophobic, the structural entropy should be smaller than that of the normal POPC bilayer. The disorder chain tail of POPC is thus thought to come out at smaller z level to avoid the entropic cost [6].

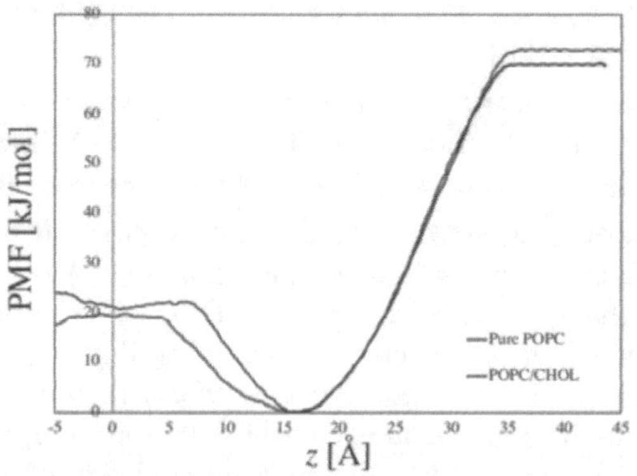

Figure 6. PMF profile of CHOL across the pure POPC and POPC/CHOL bilayer.

Figure 6 shows the PMF profiles of CHOL across the pure POPC and the POPC/CHOL bilayer. The result shows that the free energy barrier at the bilayer center, ΔG_{center}, for the pure POPC bilayer was 19.4 kJ/mol, which is sufficiently smaller than that of POPC, 83 kJ/mol, indicating that CHOL easily can flip over than POPC in the membrane. These results are corresponding the experiments [21] and previous MD study for the DPPC/CHOL bilayer [6]. The CHOL has a smaller head polar group and molecular size compared to the POPC, and these differences should be factors for the smaller ΔG_{center}. On the other hand, the evaluated free energy barrier for desorption of CHOL from the membrane, ΔG_{desorp}, was 70 kJ/mol, showing comparable free energy barrier to that of POPC, 84 kJ/mol. In the POPC/CHOL bilayer system, the PMF profile in the membrane inertia region was steeper than that for the pure POPC bilayer, and a small PMF peak was observed around z = 6 Å. Since the z region where the PMF profile increases corresponds to that the cavity distribution largely decreases in the hydrophobic core of the bilayer (cf. Figure 4), this energy increase should also be responsible for the tight packing of POPC/CHOL bilayer. The free energy then gently decreased as the CHOL approached to the bilayer center. We found that the free energy difference at bilayer center ΔG_{center} was small, indicating that effect of CHOL on the transferring of CHOL across the bilayer center should also be small.

4. Conclusion

We carried out all-atom molecular dynamics (MD) simulations of the pure POPC and the POPC bilayer containing cholesterol (POPC/CHOL) with the CHARMM36 force field and investigated the effects of CHOL on structure and lipid translocation along the membrane normal. The MD simulations showed that the relevant cholesterol condensing effects such as the increase of order parameter of lipid chains and the tight packing in the POPC/CHOL bilayer. We calculated the PMF profiles of POPC and CHOL translocation along the pure POPC and POPC/CHOL bilayer normal by umbrella sampling method. The obtained PMF profile of POPC across the pure POPC bilayer was in good agreement with that of the previous MD study [8], showing that the PMF profiles with the CHARMM36 force field should also be reasonable. We carefully verified the level position of flipping position of POPC around the bilayer center by long-term MD simulations and found that the PMF peak appears slightly beyond the bilayer center. The CHOL effects on the PMF profile of POPC transport across the membrane were clearly observed; the free energy barrier for lipid flipping increased and, the energy for desorption of lipid decreased. The changes of PMF in the POPC/CHOL bilayer should be due to the tight packing of lipid chains in the hydrophobic core of bilayer. The PMF profiles of CHOL across the pure POPC and the POPC/CHOL bilayer were also estimated to investigate the differences with those of POPC translocation and the effects of CHOL on the PMF profile. The obtained PMF profiles of CHOL in each system showed that the free energy barrier at the bilayer center were sufficiently smaller than those of POPC, indicating that CHOL can easily flip over than POPC in the membrane. In the case of PMF profile of CHOL across the POPC/CHOL system, while the PMF energy in the hydrophobic core region is higher than that of the pure POPC bilayer, the free energy difference at bilayer center was small.

5. Acknowledgment

This work was supported by JSPS KAKENHI (Grant Numbers JP15KT0073, JP 17H02209 to K.N., JP17K05620 to T.M. and JP16K05648 and JP18H04636 to H.S.). The computations in this study were performed using the advanced center for computing and communication of RIKEN, the research center for computational science of Institute for Molecular Science (IMS), the research center for advanced computing infrastructure of Japan Advanced Institute of Science and Technology (JAIST), and the center for computational science of Tsukuba university.

References

[1] Ohvo-Rekila H., Ramstedt B., Leppimaki P. and Slotte J. P. 2002 *Prog Lipid Res* **41**, 66-97.
[2] Vist M. R. and Davis J. H. 1990 *Biochemistry* **29**, 451-464.
[3] Kol M. A., van Laak A. N. C., Rijkers D. T. S., Killian J. A., de Kroon A. I. P. M. and de Kruijff B. 2003 *Biochemistry* **42**, 231-237.
[4] Nakamura S., Suzuki S., Saito H. and Nishiyama K. 2018 J Biochem 163, 313-319.
[5] Tieleman D. P. and Marrink S. J. 2006 *J Am Chem Soc* **128**, 12462-12467.
[6] Bennett W. F., MacCallum J. L. and Tieleman D. P. 2009 *J Am Chem Soc* **131**, 1972-1978.
[7] Bennett W. F., MacCallum J. L., Hinner M. J., Marrink S. J.and Tieleman D. P. 2009 *J Am Chem Soc* **131**, 12714-12720.
[8] Sapay N., Bennett W. F. D. and Tieleman D. P. 2009 *Soft Matter* **5**, 3295.
[9] Klauda J. B., Venable R. M., Freites J. A., O'Connor J. W., Tobias D. J., Mondragon-Ramirez C., Vorobyov I., MacKerell A. D. and Pastor R. W. 2010 *J Phys Chem B* **114**, 7830-7843.
[10] Lim J. B., Rogaski B. and Klauda J. B. 2012 *J Phys Chem B* **116**, 203-210.
[11] Lee J., Cheng X., Swails J. M., Yeom M. S., Eastman P. K., Lemkul J. A., Wei S., Buckner J., Jeong J. C., Qi Y., Jo S., Pande V. S., Case D. A., Brooks C. L., MacKerell A. D., Klauda J. B. and Im W. 2016 *J Chem Theory Comput* **12**, 405-413.
[12] Nosé S. 1984 *Molecular Physics* **52**, 255-268.
[13] Hoover W. G. 1985 *Physical Review A* **31**, 1695-1697.
[14] Parrinello M. and Rahman A. 1981 *Journal of Applied Physics* **52**, 7182-7190.
[15] Jorgensen W. L., Chandrasekhar J., Madura J. D., Impey R. W. and Klein M. L. 1983 *J Chem Phys* **79**, 926-935.

IOP Conf. Series: Journal of Physics: Conf. Series **1290** (2019) 012020 doi:10.1088/1742-6596/1290/1/012020

[16] Essmann U., Perera L., Berkowitz M. L., Darden T., Lee H. and Pedersen L. G. 1995 *J Chem Phys* **103**, 8577-8593.
[17] Abraham M. J., Murtola T., Schulz R., Páll S., Smith J. C., Hess B.and Lindahl E. 2015 *SoftwareX* **1-2**, 19-25.
[18] Hub J. S., de Groot B. L. and van der Spoel D. 2010 *J Chem Theory Comput* **6**, 3713-3720.
[19] Kucerka N., Tristram-Nagle S. and Nagle J. F., 2005 *J Membr Biol* **208**, 193-202.
[20] Dai J., Alwarawrah M. and Huang J. Y. 2010 *J Phys Chem B* **114**, 840-848.
[21] Steck T. L., Ye J. and Lange Y., 2002 *Biophys J* **83**, 2118-2125.

IOP Conf. Series: Journal of Physics: Conf. Series **1290** (2019) 012021 doi:10.1088/1742-6596/1290/1/012021

Quantum chemical analysis of reaction indices and reaction path for drug molecules

Takao Otsuka[1], Noriaki Okimoto[1], Hiroaki Saito[1] and Makoto Taiji[1]

[1] RIKEN Center for Biosystems Dynamics Research (BDR), 6-2-4 Furuedai, Suita, Osaka 565-0874, Japan

totsuka@riken.jp

Abstract. We performed a quantum chemical study on evaluating chemical reactivity for drug molecules in Cytochrome P450 (CYP) metabolic reaction. In this study, we focused on two insights for analysing the chemical reactivity: one is the Fukui reaction indices for a molecule, and the other is the minimum energy paths of hydrogen transition reaction between drug molecule and FeO-Porphyrin ring part on the CYP metabolic reaction. The Fukui indices are obtained numerically by using the familiar population analysis method in quantum chemical calculations. We performed the assessments of some population analyses to clarify the numerical behaviour clearly and then evaluated the potential of the reactivity of drug molecule. On the other hand, in the analysis of the minimum energy path for the CYP metabolic reaction, we performed the nudged elastic band (NEB) calculations for the hydrogen transition reaction in the first step of the CYP metabolic reaction. We report the findings for the reactivity from the Fukui indices method and also show the NEB results for the reaction paths.

1. Introduction

In the field of drug discovery, molecular simulation technique become widely used as a means of understanding the mechanism of biological reaction and exploring novel drug molecules. For example, one of the challenging study is the one on Cytochrome P450 (CYP) metabolic reaction. The CYP reaction has two categorized metabolic reaction, Phase I (oxidation, reduction, hydrolytic reaction) and Phase II (conjugation). A lot of drug candidate molecules that cause the metabolic reaction on a specific atom have been synthesized from the detailed analysis for those metabolic reactions mechanism (typically, Phase I reaction) between CYP and a substrate molecule by using quantum mechanics and molecular mechanics (QM/MM) hybrid methods [1]. On the other hand, along with the rapid development of computer and/or computing technology, the knowledge-based studies on the prediction of drug metabolism have increased [2].

Along this line, we have been trying to establish the prediction scheme for the site of metabolism (SOM) of drug molecule by using molecular dynamics (MD) simulation and QM calculation technique [3]. Our strategy for predicting the site of metabolic reaction is to consider "accessibility" and "reactivity" on the CYP metabolic reaction. In accessibility analysis, we evaluate the "accessibility", that is, the distribution probability against the distance between atoms in a drug molecule and FeO-Porphyrin ring part of CYP by the combination of molecular docking and MD calculations. In reactivity analysis, we evaluate the "reactivity" by using the empirical QM-based transition energies of atoms in a drug molecule, which has already been tabulated in previous study [4].

IOP Conf. Series: Journal of Physics: Conf. Series **1290** (2019) 012021 doi:10.1088/1742-6596/1290/1/012021

We found that these combination methodology works very well for predicting the SOM of a drug molecule [3].

As one of our successive studies, we performed a quantum chemical study on evaluating chemical reactivity for drug molecules in CYP metabolic reaction in order to explore further non-empirical parameters. In this study, we focused on two insights for analyzing the chemical reactivity for a drug molecule: one is the Fukui reaction indices for a molecule, and another is the minimum energy paths of hydrogen transition reaction between drug molecule and FeO-Porphyrin ring on the CYP metabolic reaction. Traditionally, the Fukui indices have been used as one of the potential index of chemical reaction for molecule only. The Fukui indices are obtained numerically by using the familiar population analysis method such as Mulliken, Löwdin, Hirschfeld, Iterative-Hirschfeld, and Natural population analysis in quantum chemical calculations. We performed the assessment of these population analyses to clarify the numerical behavior clearly and then evaluated the potential of the reactivity of drug molecule. On the other hand, in order to understand the first step mechanism of the metabolic reaction, we performed the nudged elastic band (NEB) calculations for the hydrogen transition reaction and analyzed the minimum energy path. Because the hydrogen transition reaction from a drug molecule to FeO-Porphyrin is the first step of the metabolic reaction and is also the rate-limiting step. In the NEB calculations, we used the cluster models for some structure poses of the drug molecule and FeO-Porphyrin ring part in CYP. We report some findings for the metabolic reactivity. Those were somewhat contrary results to the understanding from the Fukui indices method. We also show the NEB results for the reaction path for the hydrogen transition reaction.

2. Methods

In this study, in order to analyze the chemical reactivity of drug molecules on the CYP metabolic reaction, we focused on two quantum chemical methods. Here, we briefly describe some key points of our computational approaches, that is, the Fukui reaction indices (or Fukui function) method for the potential index for chemical reaction, and the NEB method for the evaluation of hydrogen transition. Although these methods are independent each other, we assessed and evaluated the possibility for applying the understanding the metabolic reaction of drug molecules.

2.1. Fukui reaction indices

The Fukui reaction indices (or Fukui function) is well known as one of the analysis method for a molecular reactivity [5] and can evaluate the site of high or low reactivity in a molecule. Theoretically, the Fukui function is derived from the framework of density functional theory (DFT) exactly,

$$f(r) = \left(\frac{\delta\mu}{\delta v(r)}\right)_N = \left(\frac{\delta\rho(r)}{\delta N}\right)_v,$$

where the Fukui function f is evaluated by the electron density ρ and the number of electrons N in a molecule. If some appropriate approximations are applied, the Fukui function can evaluate the nucleophilic or the electrophilic reactions, respectively. In practical quantum chemical calculations, the Fukui function can be obtained from the electron density of the Fukui's frontier orbitals [6,7] by using ϕ_{HOMO} and ϕ_{LUMO}, or the population analysis on *i*-th atoms in a molecule. The former is expressed as the electron density map. The latter is expressed by the population analysis numerically,

$$f^{+}(r) \cong q_i\ (N+1) - q_i\ (N), (Nucleophilic\ attack)$$
$$f^{-}(r) \cong q_i\ (N) - q_i\ (N-1), (Electrophilic\ attack)$$

IOP Conf. Series: Journal of Physics: Conf. Series **1290** (2019) 012021 doi:10.1088/1742-6596/1290/1/012021

where the Fukui function f^+ (or f^-) represents the case of the nucleophilic attack (or the case of the electrophilic attack), and the charge q_i of i-th atom is evaluated by the quantum chemical calculations of a molecule for N, $N+1$ and $N-1$ electrons systems [8].

From the definition of the Fukui function, the numerical value by the Fukui function has positive value conventionally, and therefore, it has been considered traditionally that the negative value by the Fukui function is meaningless. However, some recent studies discuss the negative values of Fukui function [9].

2.2. *Nudged elastic band calculation*

The NEB calculation is one of the methods for finding a saddle point (near a saddle point) and the minimum energy path between reactant and product in chemical reaction [10,11]. In the NEB calculation, the saddle point search is performed by optimizing a number of intermediate images, R_i along the reaction path. Here, the intermediate images, R_i are generated as the interpolating structures between the reactant structure and the product one, and connected as an elastic band. Practically, the elastic band optimization by the NEB calculation is performed by summing the spring forces along the band between the intermediate images, and the projection component of the force due to the potential perpendicular of the band. Therefore, the total force acting on an intermediate image is expressed by using the energy of the system,

$$F_i = F_i^s |_{||} - \nabla E(R_i)|_{\perp} .$$

Then the true force is given by,

$$\nabla E(R_i)|_{\perp} = \nabla E(R_i) - \nabla E(R_i) \cdot \tau_i ,$$

where τ_i is the normalized local tangent at i-th image. The spring force is

$$F_i^s |_{||} = k(|R_{i+1} - R_i| - |R_i - R_{i-1}|)\tau_i ,$$

where k is the spring constant. The movements of the intermediate images according the forces in above equations are performed by using an optimization algorithm such as a projected velocity Verlet algorithm [10,11]. In recent study, a lot of optimization algorithm have been tested [12].

2.3. *Computational details*

Three drug molecules were selected as the test molecules in this study, as shown in Figure 1. These selected drug molecules are mainly metabolized by CYP 1A2 as the oxidation in Phase I reaction, and are well known experimentally that the metabolic site is majorly one in each molecule. These molecules has also been used well for evaluating the site of metabolic reaction of a molecule theoretically. We also used these molecules in our previous study [3].

In the assessment and the evaluation of the Fukui reaction indices by quantum chemical calculation, we used the DFT calculations with B3LYP exchange-correlation function and 3-21G basis set (denoted as B3LYP/3-21G). The initial structures were relaxed using the GAFF of the classical force field by AMBER software [13]. The Fukui reaction indices were estimated by using five population analysis methods: Mulliken, Löwdin, Hirshfelt, Iterative-Hirshfelt, and Natural population analysis. In this study, we only evaluated the Fukui reaction indices f^+ (the nucleophilic attack) because the present test molecules show the oxidation reaction, that is, the nucleophilic reaction. These calculations were performed by Gaussian09 software [14]. We also checked the case of the QM optimized structures by B3LYP/6-31G(d,p).

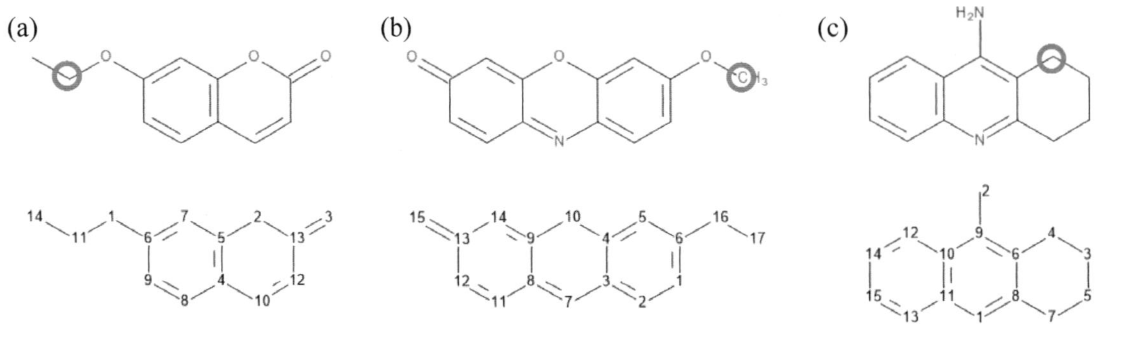

Figure 1. Structures of test molecules (a) 7-Ethoxycoumarin, (b) 7-Methoxyresorufin, (c) Tacrine. The circle with pink color shows the SOM of the molecules. The lowers shows the atom ID for each molecules.

In the evaluation of the transition energy from the hydrogen transition path with the orientation of drug molecule (7-Ethoxycoumarin) by the NEB method, the initial structures were prepared as follows. The CYP structure from X-ray structure (PDB ID: 2HI4) was used. Firstly, the Fe-Porphyrin part in CYP structure was replaced with the oxygen added FeO-Porphyrin that reproduces a reactant in metabolic reaction, as shown in our previous study [3]. Then the binding structures of the drug molecule on the CYP structure were obtained by the molecular docking calculations using the GOLD software [15] with the scoring function of ChemScore. At this time, thirty docking poses per a drug molecule were considered. After the molecular docking calculations, the structure relaxation of only drug molecule was performed by using the GAFF classical force field in AMBER software [13]. We selected three initial orientations of a drug molecule by considering the nearest distances between the SOM of drug molecule and the FeO-Porphyrin part in CYP. Then we prepared for the drug molecule and the FeO-Porphyrin part with hydrogen termination as the cluster model, which corresponds to the reactant structure (the reaction coordinate, rc = 0) in the NEB calculation. For the product structure (the reaction coordinate, rc = 1) in the NEB calculation, we just moved the hydrogen atom of the SOM atom in drug molecule to the oxygen atom of FeO-Porphyrin part. The distance between the hydrogen and the oxygen atom was set to 0.95 Å. The NEB calculations were performed by using the nine intermediated images of the system including the reactant and the product, and the dumped Verlet optimization algorism [10-12] with the B3LYP/3-21G level of theory. In this study, we set the maximum value of force of 0.09 hartree/bohr or less as the optimization convergence condition. We only analyzed the case of 7-Ethoxycoumarin molecule in this study. The NEB calculations were performed by NWChem software [16].

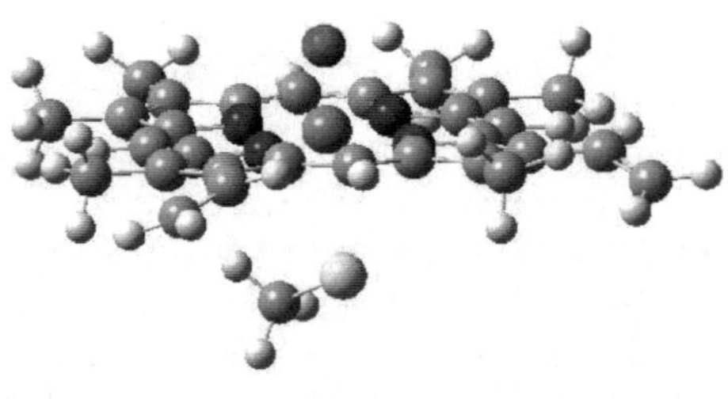

Figure 2. Structure of FeO-Porphyrin part of CYP (PDB ID: 2HI4). The spheres show carbon (gray), nitrogen (blue), oxygen (red), sulfur (yellow), iron (purple) and hydrogen (white).

In this study, although we limited our computational conditions in order to know the behavior the computational reaction indices and the hydrogen transition energies for the orientation of drug molecule, we will improve the computational accuracy using the higher quality of theory. Now we are going on further studies.

3. Reactivity analysis by Fukui reaction indices

First, we verified the numerical values of the Fukui reaction indices by five population analysis methods. Figure 3(a) shows the calculated Fukui reaction indices f^+ of the test molecule, 7-Ethoxycoumarin. Each value by five population analyses showed almost same tendency except for some parts. In order to specify the numerical features of these five population analysis methods, we performed the cluster analysis with a familiar Ward method. Figure 3(b) shows the dendrogram by the cluster analysis. As seen in Figure 3(b), Löwdin and Natural population analysis can be categorized as similar behavior, while Mulliken and Iterative-Hirshfelt population analysis seem to be close behavior numerically. From the cluster analysis of three test molecules, the similar tendency was confirmed. For the Fukui indices by five population analysis methods by the B3LYP/6-31G(d,p), the cluster analysis was almost similar result. We could mention that the Fukui reaction indices has less dependency of the basis set.

Next we evaluated the prediction of SOM by the Fukui reaction indices. As stated in the previous section, the values by the Fukui reaction indices mean the intensity of reactivity of a molecule. In the case of 7-Ethoxycoumarin illustrated in Figure 3, the SOM of 7-Ethoxycoumarin is the carbon atom of atom ID, 11 (see in Figure 2(a)). From Figure 3(a), the reactivity of the atom ID, 11 by the Fukui reaction indices showed the very lower and negative value in some population analysis. We can also see the other lower values of the atom ID, 4 and 14 with the same tendency. From the values of the Fukui reaction indices, the atoms with the low Fukui values are very stable for the nucleophilic reaction. We also checked the Fukui reaction indices f^- of this molecule, which mean the reactivity for the electrophilic attack. The result is shown in Figure 4. As seen in Figure 4, it is found that the atoms with the low Fukui values are the atom ID, 11 and 14. From these two Fukui reaction indices, we can see that the atoms of the atom ID, 11 and 14 are very stable for the nucleophilic and electrophilic reactions. We also obtained the similar results from other two test molecules (not shown in this study). It is considered that the Fukui reaction indices could be applicable for finding the potential of the SOM in drug molecule, although it could not specify the SOM. From the analysis of the Fukui reaction indices using more molecules, we would also derive newly findings that in the metabolic reaction the FeO-Porphyrin attacks the relatively stable atom in drug molecule and the reaction is progressing.

(a)

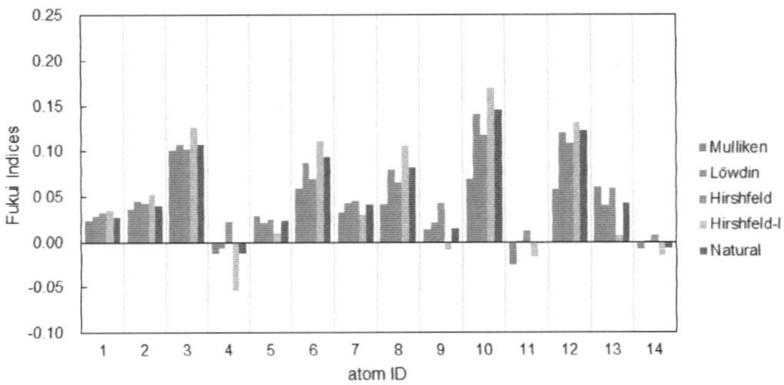

(b)

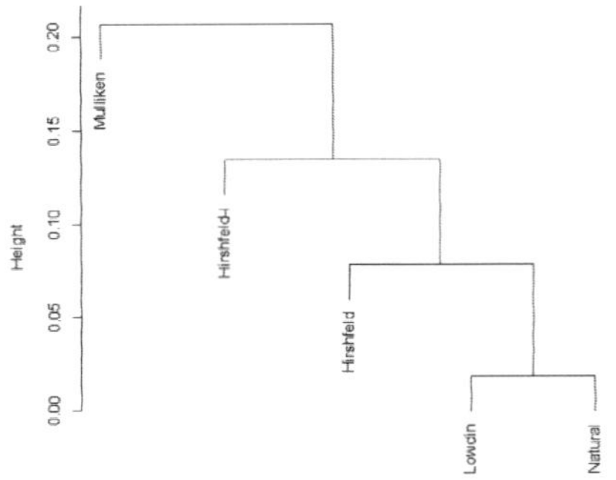

Figure 3. Reactivity analysis of 7-Ethoxycoumarin; (a) the calculated Fukui reaction indices f^+ using five population analysis methods, (b) the classification of five population analysis methods by the cluster analysis with Ward's method.

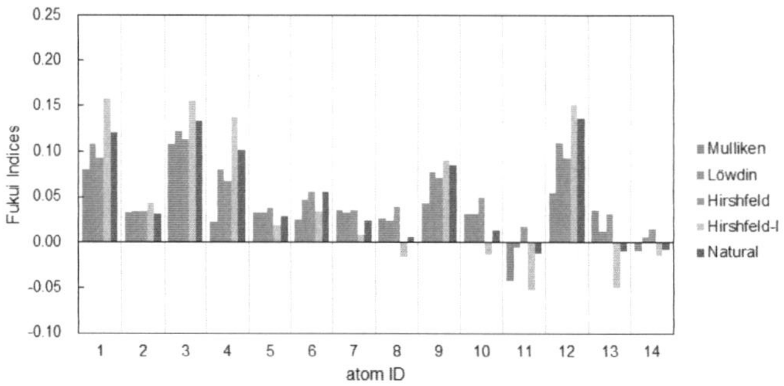

Figure 4. Reactivity analysis of 7-Ethoxycoumarin. The calculated Fukui reaction indices f^- using five population analysis methods.

4. Reaction path analysis by NEB calculation

In order to evaluate the hydrogen transition energy of 7-Ethoxycoumarin molecule, three orientations of the molecule by molecular docking calculations were selected from the order of the nearest distance between the SOM of the molecule and the oxygen in FeO-Porphyrin (g1 < g2 < g3), as shown in Figure 5. As stated in the previous section, the reactant structure and the product structure were prepared as the cluster models, which corresponds to the first image and the last image in the NEB calculations (see Figure 5(a)).

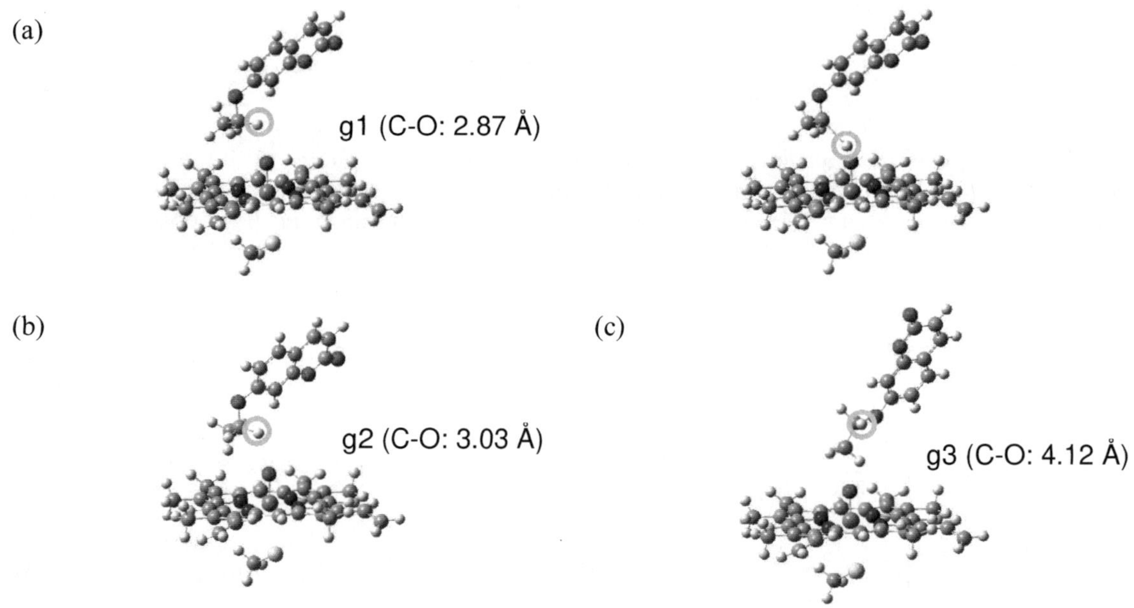

Figure 5. Three initial structures of 7-Ethoxycoumarin in the NEB calculation (denoted as g1, g2 and g3); (a) the distance between carbon atom of the molecule and the oxygen atom of FeO-Porphyrin, C-O: 2.87 Å. Hydrogen atom moves form the reactant structure (left) to the product structure (right), (b) C-O: 3.03 Å, (c) C-O: 4.12 Å. The circle (orange) show the moving hydrogen atom.

IOP Conf. Series: Journal of Physics: Conf. Series **1290** (2019) 012021 doi:10.1088/1742-6596/1290/1/012021

Figure 6 shows the reaction paths of the hydrogen transition reaction for three orientation of 7-Ethoxycoumarin molecule. The reaction coordinates, rc = 0 and rc = 1 correspond to the reactant and the product structures, respectively. The seven points between rc = 0 and rc = 1 show the structures of the intermediated images. In Figure 6, the energy values are plotted with the energy of g1 orientation as a reference value. Table 1 also show the hydrogen transition energies of each orientation as reference value at rc = 0 of each orientation.

For the initial structures at the reaction coordinate, rc = 0 in Figure 6, the nearest orientation to FeO-Porphyrin, g1 (0.0 kcal/mol) is slightly unstable, compared with the g2 (-4.8 kcal/mol) and g3 (-4.2 kcal/mol) orientations energetically. However, along the reaction coordinate to the product structure, the transition energy from the g1 orientation showed the lowest one, 37.5 kcal/mol at around rc = 0.75. The transition energy from the g2 orientation was 41.0 kcal/mol at around rc = 0.75. These transition energies by the NEB calculations were estimated less than 62.2 kcal/mol obtained by the empirical QM-based energy [3,4]. This implies that it is important to consider the good orientation of drug molecule in CYP metabolic reaction. On the other hand, the transition energy by g3 orientation shows the largest one, 114.3 kcal/mol at around rc = 0.625. The reason for this is that not only the distance between the hydrogen atom and the oxygen of FeO-Porphyrin is so long (C-O: 4.12 Å) but also the steric hindrance of other atoms in the molecule (the carbon of atom ID, 14 and the attached hydrogen atom, see Figure 5(c)) occurred on the transition path of the hydrogen atom. In our previous study, we have evaluated the distribution probability against the distance between atoms in a drug molecule and FeO-Porphyrin ring part of CYP by the combination of molecular docking and MD calculations. In this process, we have neglected the orientation of hydrogen atom of a drug molecules. By considering the orientation of hydrogen atom and/or the steric hindrance of other atoms in a drug molecule, we would be able to further improve our proposed SOM predicting scheme.

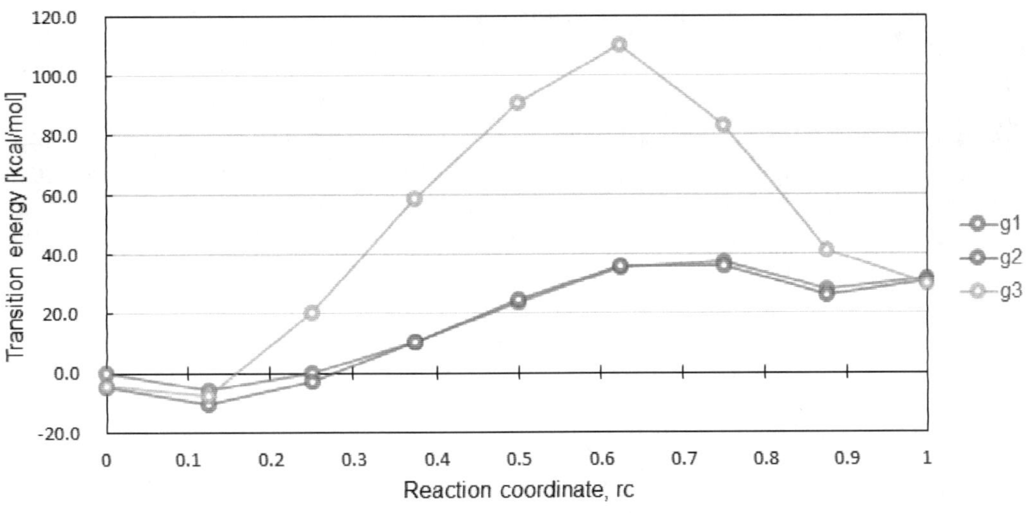

Figure 6. NEB reaction paths of hydrogen transition of 7-Ethoxycoumarin. The transition energies of the g2 and g3 orientation were converted based on the transition energy of the g1 orientation. The reaction coordinates, rc =0 and rc = 1 correspond to the reactant and product structures, respectively.

IOP Conf. Series: Journal of Physics: Conf. Series **1290** (2019) 012021 doi:10.1088/1742-6596/1290/1/012021

Table 1. Hydrogen transition energies of each orientation (referenced at rc = 0 of each orientation).

Orientation	Reaction coordinate (rc)	Transition Energy (kcal/mol)
g1	0.75	37.5
g2	0.75	41.0
g3	0.625	114.3

5. Conclusion

In this study, we performed a quantum chemical analysis for the chemical reactivity of drug molecules in CYP metabolic reaction motivated on two insights. In the reactivity analysis by the Fukui reaction indices (Fukui function), we first assessed the numerical values obtained by five population analysis methods, such as Mulliken, Löwdin, Hirschfeld, Iterative-Hirschfeld, and Natural population analysis. From the cluster analysis, we could categorize the groups of Mulliken and Iterative-Hirschfeld, and Löwdin and Natural population analysis. We also confirmed the less basis set dependency of the Fukui reaction indices, different from the conventional population analysis. Based on the findings, we evaluated the reaction potentials for the site of metabolic reaction using three test molecules. Although the Fukui reaction indices could not predict the SOMs completely, we could read the tendency that in the metabolic reaction the FeO-Porphyrin attacks the relatively stable atom of a drug molecule. Furthermore, in the analysis of the minimum energy path for the CYP metabolic reaction, we performed the NEB calculations for the hydrogen transition reaction in the first step of the CYP metabolic reaction. Although we evaluated only one test molecule case, we found that the minimum energy path of hydrogen transition would depend on the orientations of a drug molecule, that is, the orientation of hydrogen atom and the steric hindrance of other atoms. We expect that these findings would be useful for improving our prediction scheme for the site of metabolic reaction.

Acknowledgements

This work was partly supported by the JSPS KAKENHI projects: Grant Numbers, 26860084, 16KT0168 to T. O. and 16K05648 H. S.. The calculations were performed in part on the RIKEN HOKUSAI supercomputer. The authors also acknowledge Prof. Y. Okamoto (Nagoya Univ.) and Prof. H. Akai (Univ. of Tokyo) for critical comments and fruitful discussion.

References

[1] For example, S. Shaik, S. Cohen, Y. Wang, H. Chen, D. Kumar, W. Thiel, Chem. Rev. 110, 949-1027, (2010).

[2] For example, J. Kirchmair, A. H. Göller, D. Lang, J. Kunze, B. Testa, I. D. Wilson, R. C. Glen, G. Schneider, Nature Reviews Drug Discovery, 14, 387-404 (2015).

[3] H. Saito, T. Otsuka, N. Okimoto, M. Taiji et al., to be appreaed.

[4] P. Rydberg, D. E. Gloriam, J. Zaretzki, C. Breneman, L. Olsen, ACS Med. Chem. Lett., 1, 96-100 (2010).

[5] R. G. Parr, W. Yang, Density-functional theory of atoms and molecules, Oxford University Press, 1989.

[6] W. Yang, W. J. Mortier, J. Am. Chem. Soc., 108, 5708-5711 (1986).

[7] K. Fukui, T. Yonezawa, H. Shingu, J. Chem. Phys., 20, 772 (1952).

[8] K. Fukui, T. Yonezawa, C. Nagata, H. Shingu, J. Chem. Phys., 22, 1433 (1954).

[9] For example, P. Bultink, R. Carbó-Dorca, W. Langenaeker, J. Phys. Chem., 118, 4349-4356 (2003).
[10] G. Henkelman, B. P. Uberuaga, H. Jónsson, J. Chem. Phys., 113, 9901 (2000).
[11] G. Henkelman, H. Jónsson, J. Chem. Phys., 113, 9978 (2000).
[12] D. Sheppard, R. Terrell, G. Henkelman, J. Chem. Phys., 128, 134106 (2008).
[13] D. A. Case et al., AMBER 12; University of California: San Francisco, 2012.
[14] M. J. Frisch et al., Gaussian 09, revision A.08; Gaussian, Inc.: Wallingford, CT, 2009.
[15] G. Jones, P. Willett, R. C. Glen, A. R. Leach and R. Taylor, J. Mol. Biol., 267, 727-748, (1997).
[16] M. Valiev, E.J. Bylaska, N. Govind, K. Kowalski, T.P. Straatsma, H.J.J. van Dam, D. Wang, J. Nieplocha, E. Apra, T.L. Windus, W.A. de Jong, Comput. Phys. Commun. 181, 1477 (2010).

IOP Conf. Series: Journal of Physics: Conf. Series **1290** (2019) 012022 doi:10.1088/1742-6596/1290/1/012022

A new formulation of heat dissipation in a rocking Büttiker-Landauer ratchet model

Karina I Mazzitello[1,5], José L Iguain[2,5], Yi Jiang[3], Fereydoon Family[4] and C Miguel Arizmendi[1]

[1] Instituto de Investigaciones Científicas y Tecnológicas en Electrónica, Universidad Nacional de Mar del Plata, Argentina
[2] Instituto de Investigaciones Físicas de Mar del Plata, Universidad Nacional de Mar del Plata, Argentina
[3] Department of Mathematics and Statistics, Georgia State University, Atlanta, GA 30303,USA
[4] Department of Physics, Emory University, Atlanta, GA 30322, USA

E-mail: arizmend@fi.mdp.edu.ar

Abstract. Thermal ratchets achieve net particle transport through rectification of thermal fluctuations, which arise from one or more heat baths in the system. We propose a new formulation of heat dissipation from the ratchet to the thermal baths, using a rocking Büttiker-Landauer ratchet model. We found that heat transport between the ratchet and the heat baths is related to the effective temperature through the generalization of the fluctuation-dissipation theorem for systems far from equilibrium. We showed that the net heat transport between the ratchet and the heat baths is different from Fourier's law and is the sum of two terms which are proportional to the nth power of the difference between the effective temperature of ratchet and the temperature of the baths. The power n depends only on the temperature of the bath, while the thermal conductivity also depends on the ratchet potential. These findings suggest that anomalous heat dissipation can be a non-equilibrium measure for systems far from equilibrium.

1. Introduction

Fluctuations-induced transport has been extensively studied both experimentally and theoretically [1, 2, 3]. This process appears in simple stochastic models, such as thermal ratchets in which some of the energy in a nonequilibrium bath is transformed into work or into a current of particles at the expense of increased entropy [4, 5]. Ratchets or Brownian motors are kept away from equilibrium by external fluctuations or temperature gradients, which make them completely different from Carnot engines. Studies on the thermodynamics of these ratchet transport have been mainly focused on efficiency [6, 7, 8]. In this study, we focus on heat dissipation between the ratchet and the heat baths, because high efficiency may be achieved with considerable heat dissipation losses. Another reason for our focus on heat dissipation is that the entropy plays such an important role in fluctuations-induced transport processes that they have been associated with Maxwell's demon performance [9]. Heat transport is usually associated to the empirical Fourier's law, which states that the time rate of heat transfer through a material is proportional to the negative gradient in the temperature and to the area, at right angles

CONICET researcher

to that gradient, through which the heat flows. The Fourier's law has been experimentally examined in virtually all 3D thermal conductors. Nevertheless, anomalous heat conduction was found in one dimensional (1D) momentum-conserving model systems, such as Fermi-Pasta-Ulam (FPU)- lattices [10] and the 1D diatomic Toda lattices [11]. Similar anomalous heat conduction behavior has also been predicted in momentum-conserving physical materials, such as in carbon nanotubes [12], silicon nanowires [13] and in polymer chains [14]. Experimental evidence of the violation of Fourier's law in nanotube thermal conductors has been obtained in [15]. On the other hand, normal heat conduction obeying Fourier's law has been obtained for 1D nonlinear lattice systems where the total momentum is not conserved, such as 1D nonlinear Frenkel-Kontorova [16] lattices and ϕ^4 lattices [17].

We propose in this work to relate the process of heat transport to the effective temperature through the generalization of the fluctuation-dissipation theorem on far from equilibrium systems. The effective temperature is defined by the comparison between induced and spontaneous fluctuations. Recently Harada [18] has investigated the energetic efficiency of a rocking ratchet at a nonequilibrium steady state as a function of the effective temperature. In that work, a rise of the effective temperature with respect to the bath temperature has been found that causes a decrease in the efficiency of the ratchet through irreversible heat dissipation to the bath. The effective temperature of these systems gives us relevant information about the distribution of the input energy and how much work can be extracted from the free energy.

We study the heat transfer on a thermal Büttiker-Landauer (BL) [19, 20] ratchet, with a rocking driving force and transport reversal. To calculate the effective temperature [21], we focus on the spontaneous dynamic fluctuations and the response (induced) fluctuations to a weak constant force on the ratchet. The current reversal is due to the presence of two heat reservoirs in the B-L ratchet model [22]. The rest of the paper is organized as follows. We describe the model in section 2 in detail. In section 3, the results are divided into two parts: the analysis of the mean current, power and effective temperature (subsection 3.1) and the heat conduction of the ratchet to the reservoirs (subsection 3.2). Finally, section 4 consists of concluding remarks.

2. The model

We consider an overdamped Brownian particle in a sawtooth potential subjected to an unbiased external force that pushes it left and right periodically (rocking ratchet). This model has been first proposed by Büttiker [19] and Landauer [20] to describe a molecular motor in biological systems [23]. The stochastic differential equation (Langevin equation) for such a particle is given by

$$\gamma \frac{dx}{dt} = -\frac{dV(x)}{dx} + \sqrt{2\gamma k_B T(x)}\xi(t) + F(t), \tag{1}$$

where the left-hand side describes a frictional force experienced by the particle when it is moving relative to its environment, with γ being the drag coefficient. The first term of the right-hand side is the force due to the sawtooth potential $V(x)$, with period L such that $V(x) = V(x+L)$ described by

$$V(x) = \begin{cases} \frac{V_0}{L_1}x, & 0 < x < L_1 \\ \\ \frac{V_0}{L_2}(L-x), & L_1 < x < L \end{cases} \tag{2}$$

where V_0 is the barrier height and L_1 and L_2 are the projections over the x-axis of both sides of the sawtooth as shown in Fig. 1(a). The second term of the right-hand side of Eq. (1) describes the fluctuations of the ratchet with the surrounding through $\xi(t)$, a zero-mean, delta-correlated, Gaussian white noise i.e., $\langle \xi(t) \rangle = 0$ and $\langle \xi(t)\xi(t') \rangle = \delta(t-t')$. Here, $\langle ... \rangle$ denotes an ensemble average over the distribution of the fluctuating force $\xi(t)$. These fluctuations are imposed onto the particle by the interaction with two heat baths at different temperatures, one for each side of the potential given by

$$T(x) = \begin{cases} T+\delta, & 0 < x < L_1 \\ T, & L_1 < x < L \end{cases} \tag{3}$$

where $T(x)$ has the same period as the potential $V(x)$, that is $T(x) = T(x+L)$.

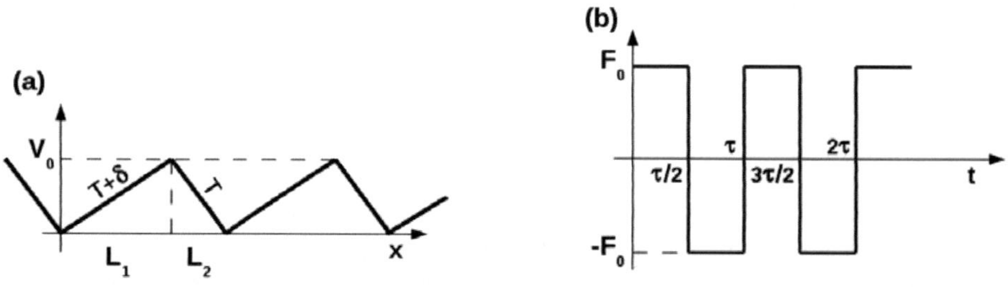

Figure 1. (a) Asymmetric potential $V(x)$ with period length $L = L_1 + L_2$ ($V(x+L) = V(x)$) amplitude V_0 and asymmetry $\Delta = L_1 - L_2$; $T+\delta$ and T indicate the temperature profiles. (b) Driving force $F(t)$ that pushes the particle left and right periodically, with temporal period τ and amplitude F_0.

Ultimately, $F(t)$ in Eq. (1) is an external driving force with a temporal period τ and amplitude F_0, that satisfies the condition $\int_0^\tau F(t)dt = 0$ (see Fig. 1(b)).

For simplicity and without loss of generality, we assume the drag coefficient $\gamma = 1$ and measure the energy in units of k_B ($k_B = 1$) and length in units of L ($L = 1$), respectively.

3. Results

3.1. Current, input power and effective temperature for the rocking B-L ratchet

According to the Einstein relation the ratio between independent measurements of fluctuations and the response to a weak external force on a particle such as, the diffusivity D and the mobility μ is equal to the thermal temperature T when the system is kept close to thermal equilibrium ($D/\mu = T$). The mobility and the fluctuations must be related, because both come from the same origin, the interaction of the particle with its surroundings.

The Einstein relation is a form of the fluctuation-dissipation theorem, strictly valid for an equilibrium thermal system or close to it. However, many out-of-equilibrium systems have shown that this relation gives rise to a well-defined effective temperature which is different from the bath temperature and governs the heat flow [24, 25, 26]. Indeed, in [21] have introduced the notion of effective temperature T_{eff} for glassy systems from a modification of the fluctuation-dissipation theorem (FDT) as

$$\frac{\chi}{MSD} = \frac{1}{2T_{eff}}, \tag{4}$$

where MSD and χ are the mean-squared displacement of the particle and the response function, respectively. The response function is the average displacement of the particle due to an weak constant force divided by the amplitude of this force. Thus, the expressions for MSD and χ are given by

$$\begin{aligned} MSD &= \left\langle |x(t) - \langle x(t)\rangle|^2 \right\rangle \\ \chi &= \langle |x_h(t) - x_{h=0}(t)| \rangle / h \end{aligned} \tag{5}$$

where h is the amplitude of the force.

Modified and ordinary FDT are equivalent, for systems in equilibrium (see Eq. (4)). Indeed, $MSD = 2Dt$ for a Brownian particle at a temperature T. On the other hand, if we pull gently on the particle with a constant force, its surrounding responds with a viscous, dissipative force and its average displacement divided by the force after a large time is $\chi = \mu t$. Then, the ratio $\chi/MSD = \mu/2D$. From the Einstein relation and Eq. (4), we arrive at $T_{eff} = T$, in thermal equilibrium. In general, for an out-of- equilibrium system in contact with a thermal bath, $T_{eff} \geq T$, when the system reaches a steady state [27]. However, for a very confined system, $T_{eff} < T$. In this model, confinement means that the particle cannot get out from the two nearest barriers of the ratchet potential. We will only focus on the case where the Brownian particle is not confined, that is when the particle goes through different barriers.

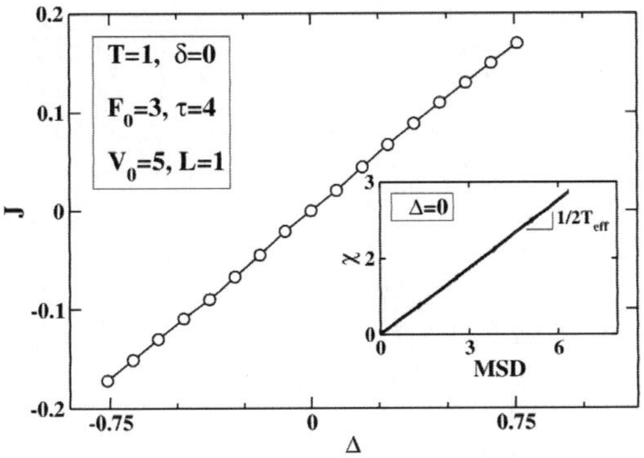

Figure 2. Current J as a function of the asymmetry parameter Δ of the potential for a single heat bath at both sides of it. The inset shows χ versus C, using the same parameter values as in the main plot and for $\Delta = 0$. All results in this paper are obtained for $\tau = 4$, chosen under the criterion described in the text above Eq. 6.

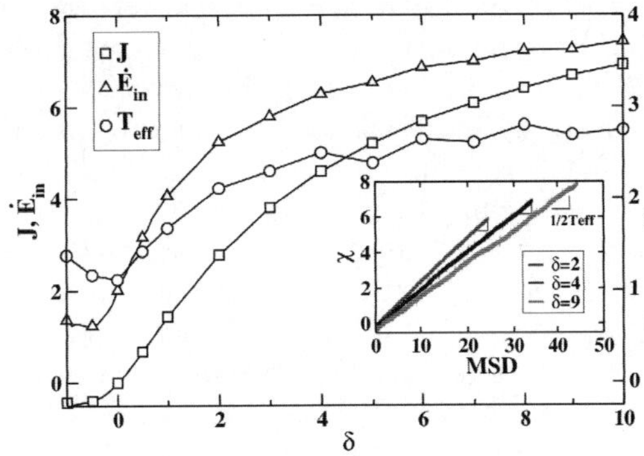

Figure 3. Current J, input power $\dot{E}_{in}$ and effective temperature T_{eff} versus temperature difference δ between the sides of a symmetric potential $\Delta = 0$, for $T = 1$, $F_0 = 3$ and $V_0 = 5$. (The left (right) scale corresponds to J and $\dot{E}_{in}$ (T_{eff})). Error bars are approximately the same size as the symbols except for T_{eff}, which they increase with the absolute value of δ (two error bars are shown as references). The inset shows χ versus C, using the same parameter values as in the main plot and for $\delta = 2, 4$ and 9.

In order to gain some insight on the fluctuations-induced transport of the model, in addition to effective temperature, we also compute the average current J and the input mean energy per unit of time $\dot{E}_{in}$, as functions of the different parameters of the model. The current J and the input mean power $\dot{E}_{in}$ have been obtained in [22, 28]. The average current over a fluctuation period τ of $F(t)$ is $J = \frac{1}{\tau}\int_0^\tau J(t)dt$. If the period τ of the external force is longer than any other time scales of the system, there exists a quasi-steady state which leads to

$$\begin{aligned} J &= \frac{1}{\tau}\left(\int_0^{\tau/2} J(t)dt + \int_{\tau/2}^{\tau} J(t)dt\right) \\ &= \frac{1}{2}\left(J_+ + J_-\right) \end{aligned} \tag{6}$$

where J_+ (J_-) is the current induced by the force $F(t)$ when it is equal to F_0 ($-F_0$) (see Fig.1(b)).

The input mean energy per unit of time or power $\dot{E}_{in}$ comes from the fluctuations generated by the external driving force $F(t)$ to the ratchet and is given by [8]:

$$\dot{E}_{in} = \frac{1}{\tau}\int_0^{\tau} J(t)F(t)dt = \frac{F_0}{2}\left(J_+ - J_-\right). \tag{7}$$

Figure 2 shows the current J in the quasi-steady state as a function of the asymmetry parameter Δ of the potential, for a single heat reservoir ($\delta = 0$). J is an anti-symmetric function as expected, negative for $\Delta < 0$ and positive for $\Delta > 0$. When $\Delta = 0$, the current $J = 0$ however, the input mean energy per unit of time $\dot{E}_{in} \neq 0$. Moreover, we found constant $\dot{E}_{in}$ for every value of the asymmetry parameter Δ. Therefore, the energy consumption per unit of time is independent of the current.

The inset of Fig. 2 shows the response function χ versus the mean-squared displacement MSD, using the same parameter values as in the main plot and for $\Delta = 0$. χ and MSD are calculated from Eq. (5). To compute χ, we simulate two replicas of the system, with the perturbative force $h = 10^{-3}$ applied to one of them. The replicas evolve using the same random numbers. Then, the results are obtained averaging over $1.5 - 3 \times 10^5$ replicas. According to Eq. (4), the slope of χ vs. MSD in the inset of Fig. 2 is equal to $1/2T_{eff}$, when the ratchet reaches the steady state. Thus, we obtain $T_{eff} = 1.12 \pm 0.01$, for $\Delta = 0$, $T = 1$, $\delta = 0$, $V_0 = 5$ and $F_0 = 3$. Moreover, we find that T_{eff} is independent of the asymmetry parameter Δ of the potential, with one bath. We will return to this result in the next section.

Figure 3 shows the current J, the input power $\dot{E}_{in}$ and the effective temperature T_{eff} versus the temperature difference δ between the sides of a symmetric potential $\Delta = 0$, for $T = 1$, $F_0 = 3$ and $V_0 = 5$. The temperature difference δ controls the magnitude of these three variables and the direction in the case of the current ($\dot{E}_{in}$ and T_{eff} are by definition always positives). When $\delta = 0$, there is no current and both the input power and effective temperature are non-zero. Then, as δ increases, the three variables increase indefinitely, as shown in Fig. 3. Note that, T_{eff} appears to converge to an asymptotic value; however, it keeps increasing as δ increases.

The inset of Fig. 3 shows the response function χ versus the mean-squared displacement MSD, using the same parameter values as in the main plot and for three positive values of $\delta = 2$, 4 and 9. As δ increases, the straight line slopes in the inset decrease, revealing the increase of T_{eff} (see Eq. (4)).

Finally, we study the current J, the input power $\dot{E}_{in}$ and the effective temperature T_{eff} as a function of temperature T, a combination of Δ and δ. As we mentioned previously, the current J for this model has been extensively studied in [22, 28]. Positive current, single or double reversals have been obtained for different combinations of Δ and δ. Two particular cases, $\Delta = 0.1, -0.8$, and fixed $\delta = 0.1$ with direct and single reversal current, respectively are shown in Fig 4(a). For $\Delta = 0.1$, $J > 0$, the motor always moves to the right (see Fig. 1(a)) following the regular behavior of a rocking ratchet, while for $\Delta = -0.8$, J may reverse its direction. Despite the differences between their currents, the input power $\dot{E}_{in}$ is the same for both cases as shown in Fig 4(a). $\dot{E}_{in}$ increases with T, up to a limit value in which the

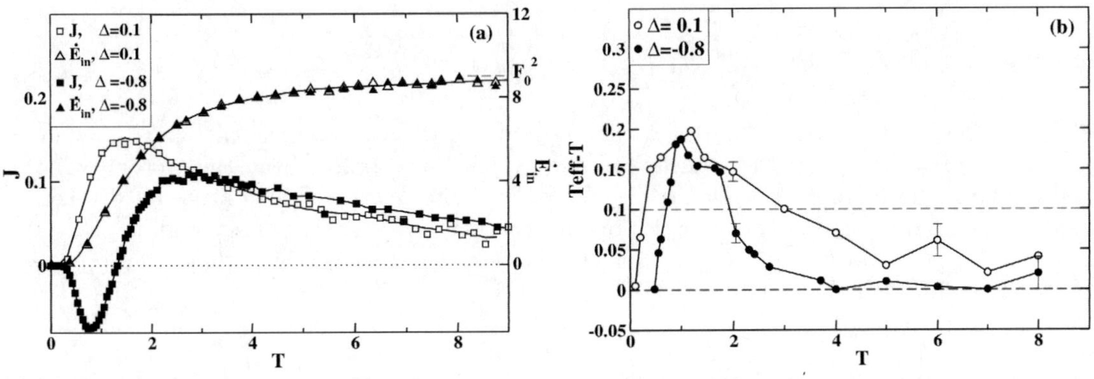

Figure 4. J, $\dot{E}_{in}$ and (b) $(T_{eff} - T)$ as functions of T, for $\delta = 0.1$, $F_0 = 3$, $V_0 = 5$, and two values of the asymmetry parameter Δ. The left (right) scale in (a) corresponds to J ($\dot{E}_{in}$). F_0^2 denotes the maximum asymptotic of $\dot{E}_{in}$. When $\Delta = 0.1$, $J > 0$, the ratchet behaves as a refrigerator, while when $\Delta = -0.8$, the ratchet behaves as heat machine (see section 3.2.2 for further details). The bottom (up) dashed line in (b) represents the cold (hot) bath at T ($T+\delta$). Error bars are approximately the same size as the symbols in (a) and increase with the value of T in (b) (two error bars are shown as references for $\Delta = 0.1$ and -0.8, respectively).

motor moves as an overdamped Brownian particle periodically pushed to the left and right by the force F_0 (see Fig. 1(b)). Certainly, when the sawtooth potential does not offer resistance to the ratchet movement, $J = 0$ and $\dot{E}_{in} = F_0^2$.

Plots of the difference $(T_{eff} - T)$ versus T, for the two case studies, $\Delta = 0.1, -0.8$, and fixed $\delta = 0.1$ are shown in Fig. 4(b). In general, T_{eff} can be between the bath temperatures or can be greater than both of them (see $(T_{eff} - T)$ data between the dashed lines or above them in Fig. 4(b), where the bottom and upper dashed lines represent the cold and hot baths at T and $T + \delta$, respectively). As we have mentioned before, the motor always transfers a net heat in a steady state. The heat conduction of a system far from equilibrium mostly does not obey a linear transfer law with the temperature gradient. In the next section we will study the heat conduction starting with the simplest case in which there is only one thermal reservoir exchanging heat with the motor and continuing on the more general case of two thermal reservoirs. We will show that the Brownian particle can absorb heat from the cold reservoir and transfer heat to the hot one o vice versa, even when the effective temperature of the Brownian particle is greater than both reservoir temperatures, working as refrigerator or heat engine, respectively. In fact, the system is far from equilibrium continuously excited by the external agent $F(t)$, which may result in a heat flux against the temperature gradient concerned.

Note that, $\dot{E}_{in}$ is the same function of T for both Δ values, while T_{eff} is always greater for $\Delta = 0.1$ than for $\Delta = -0.8$ (see Figs. 4(a) and (b)). This last result is due to the difference of the relative contact time between the motor and the reservoirs. Indeed, in a steady state, the motor spends a longer time on the long side of the potential. Thus, for $\Delta = 0.1$, the motor spends more time in contact with the hot bath than the cold one and vice versa, for $\Delta = -0.8$, reaching a higher effective temperature in the first case.

3.2. Heat conduction for the rocking B-L ratchet

In general, a Brownian motor exchanges energy with three types of systems: the thermal baths, a load that transports, and the potential or external agent that is the source of nonequilibrium [7, 8, 29]. The flow of energy is presented in Fig. 5. The first law of thermodynamics leads to

IOP Conf. Series: Journal of Physics: Conf. Series **1290** (2019) 012022 doi:10.1088/1742-6596/1290/1/012022

$$\dot{E}_{in} = \dot{W} + \dot{Q}_T + \dot{Q}_{T+\delta}, \tag{8}$$

where $\dot{E}_{in}$ is the input power, $\dot{W}$ is the work done by the ratchet against the load per unit of time and $\dot{Q}_T$ and $\dot{Q}_{T+\delta}$ are the transfers of energy per unit of time from the motor to the thermal baths at the temperature T and $T + \delta$, respectively. All these magnitudes are taken averaging over a fluctuation period τ (see previous section). In the stationary regime, $\dot{E}_{in}$ is given by the Eq. (7) and $\dot{Q}_T$ and $\dot{Q}_{T+\delta}$ can be positives or negatives, depending on the heat transferred to or from the bath, respectively.

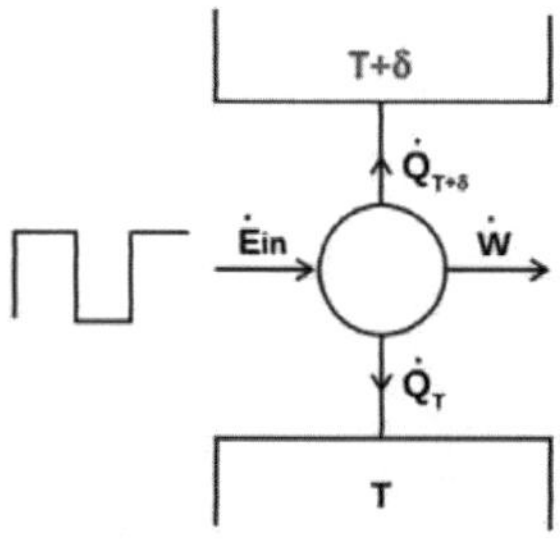

Figure 5. Energy flows in the motor (circle) subjected to a load and an external agent (square wave) in contact with two thermal baths at temperatures T and $T + \delta$.

In many experimental systems the heat transport responds to a linear expression with the temperature gradient, known as Fourier's law. We propose here, that the heat transport is a power-law of the temperature difference between the motor and the thermal bath,

$$\dot{Q}_T = c|T_{eff} - T|^{n_T}, \tag{9}$$

with c the thermal conduction coefficient and n_T an exponent that only depends on T.

Next, we will test this expression operating with one heat bath and between two different heat baths and also its valid range.

3.2.1. Operating with one heat bath In order to determinate the behavior of the heat transport with the temperature gradient, we begin with the simplest case: a particle without load moving in a symmetric potential, with only one heat reservoir (i. e. $\dot{W} = 0$, $\Delta = 0$ and $\delta = 0$). In this case, the particle transfers heat to the reservoir and taking into account Eqs. (7) and (8), the heat flux is given by $\dot{Q}_T = \frac{F_0}{2}(J_+ - J_-)$. Fig. 6 (a) shows $\dot{Q}_T$ as a function of $(T_{eff} - T)$, in log-log scales, for three different temperatures. The heat flux from the particle to the reservoir follows a power-law as predicts Eq. (9) with an exponent n_T decreasing with the bath temperature T (n_T is provided in the inset of Fig. 6(a), for different T). This exponent is independent of the barrier height, while the coefficient c in Eq. (9) depends on T and V_0 as shown in Fig. 6(b). The Eq. (9) is valid for low T and $F_0 \leq 2V_0$, that is when the B-L ratchet is working (i.e. $F_0 \leq |dV/dx|$, for all x). The range of T within which the Eq. (9) is valid is limited by the condition that the B-L ratchet is working.

3.2.2. Operating between two different heat baths When the system is working with two different heat baths the net heat transferred from a Brownian particle to two reservoirs at T and $T+\delta$ is

$$\begin{aligned} \left(\dot{Q}_T + \dot{Q}_{T+\delta}\right) &= c_1 \;\; |T_{eff} - T|^{n_{T+\delta}} + \dot{Q}_V \\ &+ c_2 \;\; |T_{eff} - T|^{n_T} - \dot{Q}_V, \end{aligned} \tag{10}$$

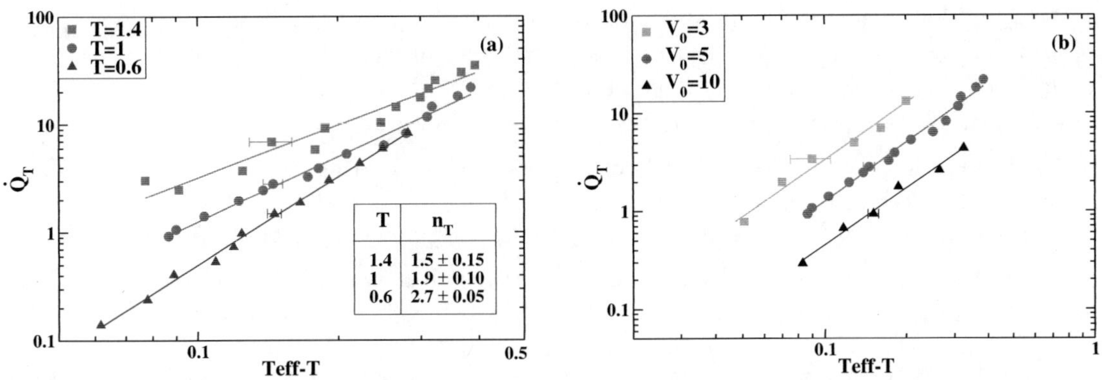

Figure 6. $\dot{Q}_T$ against $(T_{eff} - T)$, in log-log scales, for a ratchet operating with one heat bath, $\delta = 0$, $\Delta = 0$ and different values of V_0 and T: $V_0 = 5$ (a) and $T = 1$ (b). The inset in panel (a) shows the exponents n_T obtained from the slopes of straight lines. The vertical error bars are approximately the same size as the symbols while the horizontal ones increase with the value of T in (a) and decease with the value of V_0 in (b) (three error bars are shown as references in each panel).

where c_1 and c_2 depend on the bath temperatures and the sawtooth potential parameters, $\dot{Q}_V$ is the heat flux due to the ratchet force at the $T + \delta$ side and $-\dot{Q}_V$ is the heat flux due to the ratchet force at the T side. These two heat fluxes cancel each other to obtain the net heat transfer [29].

Fig. 7 shows the net heat flux $\left(\dot{Q}_T + \dot{Q}_{T+\delta}\right)$ from a motor without load to two thermal reservoirs versus $(T_{eff} - T)$, in log-log scales, for $T = 0.9$ and $T + \delta = 1$. The solid line corresponds to the best fit of the data using Eq. (10), with the exponents $n_{0.9} = 2.5$ and $n_1 = 1.9$. A good agreement between simulated data and Eq. (10) is achieved. It must be pointed out that these same exponents are obtained for a ratchet operating with one heat bath at the temperatures $T = 0.9$ and 1, respectively (for the sake of space, only the exponent n_1 is shown in Fig. 6(a)).

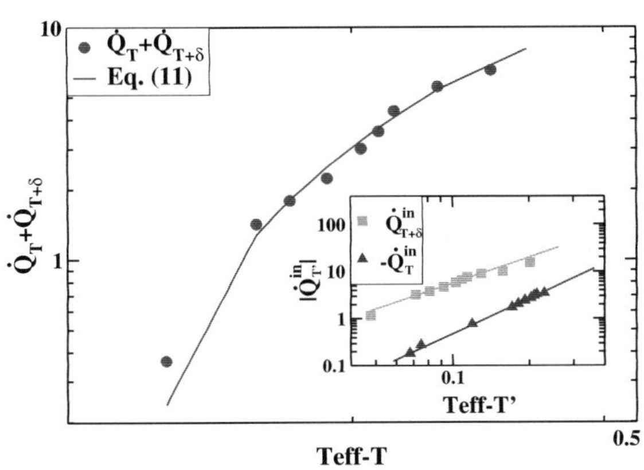

Figure 7. $\dot{Q}_T + \dot{Q}_{T+\delta}$ against $(T_{eff} - T)$, in log-log scales, for a ratchet operating between two different heat baths, $T = 0.9$, $\delta = 0.1$, $\Delta = -0.8$ and $V_0 = 5$. Symbols correspond to model results and solid line corresponds to the best fit of the results using Eq. (10) and subject to the exponents $n_{0.9} = 2.5$ and $n_1 = 1.9$. The inset shows $|\dot{Q}^{in}_{T'}|$ versus $(T_{eff} - T')$, in log-log scales, for $T' = 0.9$ and $T' = 1$. Symbols are obtained using the currents induced in each side of the sawtooth potential (see text). The slope of the blue (red) straight line is equal to $n_{0.9}$ (n_1).

The net flux $(\dot{Q}_T + \dot{Q}_{T+\delta})$ is obtained using the induced currents in each side of the sawtooth potential. According to Eq. (8), the input power due to the external agent $\dot{E}_{in}$ is directly the sum of the heat fluxes from the motor without load to the thermal reservoirs. On the other hand, the integral of Eq. (7) can be splitted into two parts related to the external driving force on each side of the sawtooth potential and thus arrive to the expression $\dot{Q}^{in}_{T'} = \frac{F_0}{\tau}\left(\tau^{T'}_{+} J^{T'}_{+} - \tau^{T'}_{-} J^{T'}_{-}\right)$, where $T' = T, T+\delta$, $J^{T'}_{+}$ ($J^{T'}_{-}$) is the current in the side of the sawtooth potential at temperature T' induced by the external agent $F(t)$ when it is equal to F_0 ($-F_0$) and $\tau^{T'}_{+}$ ($\tau^{T'}_{-}$) is the time it takes for the particle to travel on the side of the sawtooth potential at T' while $F(t) = F_0$ ($-F_0$). Note that, the currents $J_\pm$ are recovered through the weighted means $J_\pm = \frac{\tau^{T}_{\pm}}{\tau} J^{T}_{\pm} + \frac{\tau^{T+\delta}_{\pm}}{\tau} J^{T+\delta}_{\pm}$.

The inset of Fig. 7 shows $|\dot{Q}^{in}_{T'}|$ calculated from the induced currents versus $(T_{eff} - T')$, in log-log scales, for the ratchet operating between two different heat baths at $T' = 0.9$ and $T' = 1$ (see previous paragraph). This figure confirms that, indeed, the heat fluxes related to the external driving force follow a power-law to each side of the sawtooth potential namely, $\dot{Q}^{in}_{T'} = \frac{F_0}{\tau}\left(\tau^{T'}_{+} J^{T'}_{+} - \tau^{T'}_{-} J^{T'}_{-}\right) = c'|T_{eff} - T'|^{n_{T'}}$, with $T' = T, T+\delta$ and the sum $\left(\dot{Q}_T + \dot{Q}_{T+\delta}\right)$ is given by Eq. (10). This expression is valid for two reservoirs at temperatures close to each other and under the conditions of Eq. (9).

Note that, $\dot{Q}^{in}_{T=0.9} < 0$ while $\dot{Q}^{in}_{T+\delta=1} > 0$. Nevertheless, when the heat transfer related to the ratchet force is considered, the ratchet absorbs heat from the reservoir at $T + \delta = 1$ and transfers heat to the other one at $T = 0.9$, working as a heat engine. The ratchet can also work as a refrigerator or transferring heat to both reservoirs. In Fig. 4(b) the different behaviors of the ratchet depend on the asymmetry parameter Δ of the ratchet: Heat engine for $\Delta = -0.8$ and refrigerator for $\Delta = 0.1$.

4. Conclusions

In this work, we study numerically the heat transfer from a Büttiker-Landauer rocking ratchet to the heat reservoirs, and find that the net heat transfer to both heat sources depends on the difference of the effective temperature T_{eff} and each reservoir temperature T. We examine different setting of the parameters that govern the ratchet transport. The B-L ratchet in an asymmetric potential with an external driving force with two thermal baths can reach a T_{eff} that is between the bath temperatures or greater than both of them. This higher effective temperature is possible because of the external agent that sustains the system far from equilibrium. Moreover, the ratchet motor at T_{eff} can absorb heat from a reservoir at temperature $T < T_{eff}$ or transfer heat to a reservoir at $T > T_{eff}$. Interestingly, the directions of the heat fluxes between the ratchet and the two reservoirs depend on the asymmetry parameter of the potential, independent of the value of T_{eff}. The net heat transfer to the heat reservoirs is the sum of the heat fluxes related to the external force for both sources $\dot{Q}^{in}_{T}$ and $\dot{Q}^{in}_{T+\delta}$. The directions of these heat fluxes are independent of T_{eff}, which only regulates their magnitudes. We found that the magnitudes of these heat transports from the motor to the reservoirs are proportional to the n_T power of the difference between T_{eff} of the motor and the temperature of the bath concerned. The power n_T depends only on the temperature of the bath, while the thermal conductivity coefficient also depends on the ratchet potential (the barrier height of the potential and its asymmetry for the case of two reservoirs). The exponent n_T of this power-law increases as the bath temperature decreases, reaching values much greater than one. Thus, the net heat conduction, far from being linear with the temperature gradient as is the phenomenological Fourier's law is a non-equilibrium measure provided by the value of the exponent. Finally, this expression for the net heat conduction of the ratchet still holds true when the temperatures of the thermal baths are low enough and close to each other, and when amplitudes of external energy input are lower than or in the same order of magnitude as the barrier height of the potential.

IOP Conf. Series: Journal of Physics: Conf. Series **1290** (2019) 012022 doi:10.1088/1742-6596/1290/1/012022

References

[1] Hänggi P and Marchesoni F 2009 *Rev. Mod. Phys.* **81** 387–442
[2] Jülicher F and Ajdari A Prost J 1997 *Rev. Mod. Phys.* **69** 1269–1281
[3] Reimann P 2002 *Phys. Rep.* **361** 57–265
[4] Milonas M 1995 *Phys. Rev. Lett.* **74** 10
[5] Sanchez J R, Family F and Arizmendi C M 1998 *Phys. Lett. A* **249** 281
[6] Asfaw M 2014 *Phys. Rev. E* **89** 012143
[7] Sekimoto K 1997 *Phys. Soc. Jpn.* **66**
[8] Parrondo J M R and De Cisneros B J 2002 *Appl. Phys. A* **75** 179–191
[9] Leff H S and (Eds) R A F 2002 *Maxwell's Demon 2 Entropy, Classical and Quantum Information, Computing* (Institute of Physics Publishing)
[10] Lepri S, Livi R and Politi A 1997 *Phys. Rev. Lett.* **78** 1896
[11] Hatano T 1999 *Phys. Rev. E* **59** R1
[12] Zhang Gand Li B 2005 *J. Chem. Phys.* **123** 114714
[13] Yang N, Zhang G and Li B 2010 *Nano Today* **5** 85
[14] Liu J and Yang R 2012 *Phys. Rev. B* **86** 104307
[15] Chang C W, Okawa D, Garcis H, Majumdar A and Zettl A 2008 *Phys. Rev. Lett.* **101** 075903
[16] Hu B, Li B and Zhao H 1998 *Phys. Rev. E* **57** 2992
[17] Hu B, Li B and Zhao H 2000 *Phys. Rev. E* **61** 3828
[18] Harada T and Yoshikawa K 2004 *Phys. Rev. E* **69** 031113
[19] Büttiker M 1987 *Z. Phys. B* **68** 161
[20] Landauer R 1988 *J. Stat. Phys.* **53** 233
[21] Cugliandolo L F, Kurchan J and Peliti L 1997 *Phys. Rev. E* **55** 3898–3914
[22] Ai B, Wang L and Liu L 2005 *Phys. Rev. E* **72** 031101
[23] Oster G and Wang H 2002 *Molecular Motors* (Weinheim: Wiley-VCH) pp. 207-228
[24] Kurchan J 2001 *Jamming and Rheology: Constrained Dynamics on Microscopic and Macroscopic Scales* (Taylor and Francis, London)
[25] Makse H A and Kurchan J 2002 *Nat.* **415** 614–617
[26] Loi D, Mossa S and Cugliandolo L F 2008 *Phys. Rev. E* **77** 051111
[27] Harada T 2003 *e-print cond-mat/0310547* 1–4
[28] Zeng C H and Wang H 2012 *Chin. Phys. B* **21** 050502
[29] Parrondo J M R, Jiménez de Cisneros B and Brito R 2000 *In: Stochastic Processes in Physics: Chemistry, and Biology* (Springer, Berlin) lect. Notes Phys. 527

IOP Conf. Series: Journal of Physics: Conf. Series **1290** (2019) 012023 doi:10.1088/1742-6596/1290/1/012023

Two-dimensional solitary wave solution to the quadratic-cubic nonlinear Schrödinger equation

Sarun Phibanchon[1], **Yuttakarn Rattanachai**[2]

[1] Faculty of Education, Burapha University
169 Longhardbangsaen road, Sansuk, Muang, Chonburi, 20131, Thailand
[2] Faculty of Sciences and Liberal Arts, Rajamangala University of Technology Isan
744 Saranarai Road, Muang, Nakorn Ratchasima, 30000, Thailand

E-mail: sarunp@go.buu.ac.th

Abstract. The Madelung fluid transformation is applied to find the link between the modified Korteweg-de Vries and the quadratic-cubic nonlinear Schrödinger equation. The two-dimensional solitary wave solution of the quadratic-cubic nonlinear Schrödinger equation will be determined by the Petviashvili method. This solution will be used for the initial condition for the time evolution to study the stability analysis. The spectral method is applied for the time evolution.

1. Introduction

One of possible solutions of the nonlinear wave equation, which retain its shape as a result of the balance between dispersion and non-linearity, is known as a solitary wave. For the one-dimensional spatial dimension, there are two major types that the solitary wave solutions can be found. One is called the Korteweg-de Vries (KdV) equation,

$$\phi_t + \phi\phi_x + \phi_{xxx} = 0. \tag{1}$$

where the subscripts x and t denote the differentiation with respect to space and time, respectively. It's originally derived for a shallow water wave, (see e.g. [1] for further details). This equation can also be found in plasma in which describe a weakly nonlinear ion-acoustic waves [2]. Another type is well-known as the cubic nonlinear Schrödinger (cNLS) equation,

$$\mathrm{i}\phi_t + \phi_{xx} + |\phi|^2\phi = 0. \tag{2}$$

which is the crucial equation to describe a traveling soliton along fiber optics [3, 4]. However, there is a transformation between (1) and (2) via the Madelung's fluid transformation [5]. The idea of this transformation is the use of the following wave form,

$$\phi(x,t) = \sqrt{\rho(x,t)}\mathrm{e}^{\mathrm{i}\Theta(x,t)}, \tag{3}$$

where $\rho = |\phi|^2$ is the Madelung's fluid density and Θ is a wave phase. For the modified nonlinear Schrödinger equation, we consider the quadratic-cubic nonlinear Schrödinger (qc-NLS) equation,

$$\mathrm{i}\phi_t + \phi_{xx} + \phi_{yy} + |\phi|\phi + |\phi|^2\phi = 0, \tag{4}$$

IOP Conf. Series: Journal of Physics: Conf. Series **1290** (2019) 012023 doi:10.1088/1742-6596/1290/1/012023

and the modified KdV equation, which is expected to be related to (4), can be written as [6, 7, 8, 9],

$$\rho_t + \rho\rho_x + \rho^{1/2}\rho_x + \rho_{xxx} = 0. \tag{5}$$

This equation also governs a weakly nonlinear ion-acoustic wave as well as nonlinear dust ion-acoustic waves in plasma with some electrons are trapped on the ion-acoustic waves while some moves freely. Thc traveling wave solutions of KdV and cNLS are not difficult to determine but these might be more difficult for the higher dimensional solutions. The bending soliton solution of Zakharov-Kutnetzsov (ZK) equation with the long-wavelength perturbation [10, 11],

$$\phi_t + \phi\phi_x + \phi_{xxx} + \phi_{yyx} = 0,$$

is other methods to find the higher soliton solution. As well as the perturbed soliton solution of modifiled Kadomtsev-Petiashvili (KP) equation [12],

$$(\phi_t + \phi^{1/2}\phi_x + \phi_{xxx})_x = \phi_{yy},$$

which gives the 2D lump solution. We then find the higher soliton solution from the perturbed planed soliton but it sometimes might not be stable. However, if there is a suitable method to obtain 2D soliton solution directly, the higher soliton solution of other types of the nonlinear wave equation will be more studied. Before we move to the numerical method for the 2D solution, we will consider the Madelung's fluid transformation for the connection between (4) and (5).

2. Madelung's fluid transformation

We will repeat the original approach [5, 13] for (4). The main result of this transformation is to obtain the governed fluid equations. We first substitute (3) into (4). There are real and imaginary parts, the real part can be written as

$$\frac{1}{2}\rho_t + (\rho V)_x = 0. \tag{6}$$

This equation can be related to the continuity equation. For the imaginary part, we have

$$V_t + 2VV_x = \frac{\partial}{\partial x}\rho^{1/2} + \frac{\partial}{\partial x}\rho + \frac{\partial}{\partial x}\left[\frac{1}{\sqrt{\rho}}\frac{\partial^2}{\partial x^2}\sqrt{\rho}\right], \tag{7}$$

where $V(x,t) = \frac{\partial\Theta}{\partial x}$ denotes the current velocity. This equation can be referred as the momentum equation. To solve for ρ, the real part (6) will be multiplied by V,

$$\rho\left(\frac{\partial}{\partial t}V + 2V\frac{\partial}{\partial x}V\right) = -V\frac{\partial\rho}{\partial t} + \rho\frac{\partial V}{\partial t} - 2V^2\frac{\partial\rho}{\partial x}. \tag{8}$$

We will combine the momentum equation,(7), and the modified real part,(8), and integrate with respect to time,

$$-V\frac{\partial\rho}{\partial t} + \rho\frac{\partial V}{\partial t} + 4\left[-c(t) + \frac{\partial}{\partial t}\int V dx\right]\frac{\partial}{\partial x}\rho$$
$$= \left(\rho\frac{\partial}{\partial x}\sqrt{\rho} + 2\sqrt{\rho}\frac{\partial}{\partial x}\rho\right) + 3\rho\frac{\partial}{\partial x}\rho + \frac{\rho_{xxx}}{2}$$

where $c(t)$ is a function of time. To obtain the soliton solution, the current velocity is assumed to be a constant, $V = V_0$,

$$\Theta = V_0 x - 2c_0 t \quad \text{and} \quad c(t) = c_0$$

IOP Conf. Series: Journal of Physics: Conf. Series **1290** (2019) 012023 doi:10.1088/1742-6596/1290/1/012023

The traveling wave equation can be written as

$$-V_0\frac{\partial}{\partial t}\rho - 4c_0\frac{\partial}{\partial x}\rho = \left(\rho\frac{\partial}{\partial x}\sqrt{\rho} + 2\sqrt{\rho}\frac{\partial}{\partial x}\rho\right) + 3\rho\frac{\partial}{\partial x}\rho + \frac{\rho_{xxx}}{2} \tag{9}$$

If a new variable is introduced as $\xi = x - 2V_0 t$, the stationary equation of (9) can be found as

$$\left(4V_0^2 - 8c_0\right)\frac{\mathrm{d}}{\mathrm{d}\xi}\rho = 3\sqrt{\rho}\frac{\mathrm{d}}{\mathrm{d}\xi}\rho + 6\rho\frac{\mathrm{d}}{\mathrm{d}\xi}\rho + \frac{\mathrm{d}^3}{\mathrm{d}\xi^3}\rho,$$

which is the modified KdV equation, (4). We will consider the numerical 2D solution of (4) by applying the iterative method.

3. Petviashvili method

The ground state of the nonlinear Schrödinger equation with the power-law potential had been proposed by Petviashvili [14] and revised by Lakoba and Yang [15]. The convergence of the method for the homogeneous equation has been shown by Pelinovsky and Stepanyants [16]. To determine the solution, we consider

$$\phi(x, y, t) = u(x, y)\mathrm{e}^{\mathrm{i}\mu t} \tag{10}$$

where $u(x, y) > 0$ and μ is the propagation constant. After substitute (10) into (4), we obtain

$$-\mu u + u_{xx} + u_{yy} + u^2 + u^3 = 0.$$

A new variable will be introduced as

$$Mu = (\mu - \partial_{xx} - \partial_{yy})u = u^2 + u^3. \tag{11}$$

To determine the steady solution, (11) will be iterated until it meets the steady solution, namely,

$$u_{i+1} = M^{-1}(u_i^2 + u_i^3), \tag{12}$$

where i denotes a number of iteration from zero to any integer number. The result of (12) might be either zero or infinity. The key idea of Petviashvili method is to find the stabilized factor which maintain the result of the iteration without diverge to zero or infinity. We can now rewrite (12) as

$$u_{i+1} = S_i^{\gamma} M^{-1}(u_i^2 + u_i^3),$$

where γ is a constant and this stabilizing factor can be obtained by inner product of the previous value, u_i,

$$S_i = \frac{< u_i, Mu_i >}{< u_i, u_i^2 + u_i^3 >}.$$

To determine γ, we consider the exact solution $u(x, y) = O(1)$. At some iterations, the solution becomes $u_i(x, y) = O(\epsilon)$, where $\epsilon \leqslant 1$ or $\epsilon \geqslant 1$. The order of magnitude of the stabilizing factor can be represented as

$$S_i = O(\epsilon^{-1}).$$

The order of magnitude of $(u(x, y)$ becomes

$$u_{i+1} = O(\epsilon^{-\gamma+2}). \tag{13}$$

This implies that $\gamma = 2$. The Gaussian function will be used as the initial condition,

$$u_0(x, y) = 5e^{-(x^2+y^2)},$$

where the length in x and y spaces are $L_x = L_y = [-5, 5]$ and spacing for each space is 128 ($N_x = N_y = 128$). We choose $\mu = 3/2$ for the examination. Figure 1 shows how the initial Gaussian function approaches to its steady state solution for each step of iteration. It looks like the Gaussian function expands until it meets the steady state. To be confirm this structure we consider the second example of the initial condition,

$$u_0(x, y) = 5\,\mathrm{sech}^2(\sqrt{x^2 + y^2}) \cos x \sin y.$$

The result is shown in Figure 2, there are a few steps between Figure 2(a)-(b), before it becomes the Gaussian-like function.

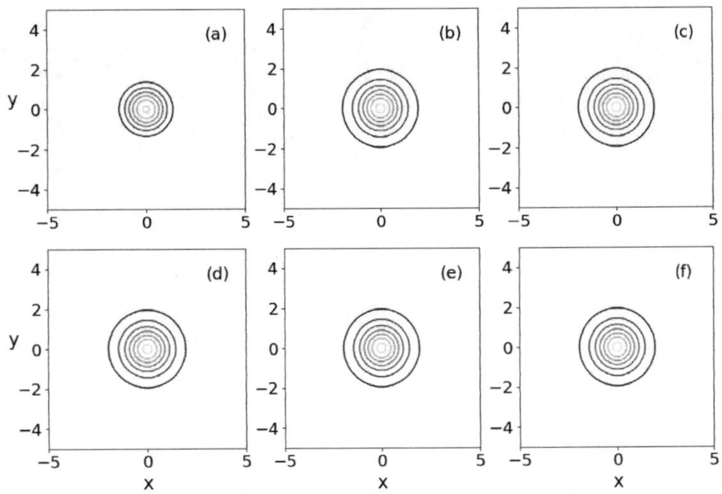

Figure 1. Result of each iteration: (a) i = 0, (b) i=10, (c) i = 20, (d) i = 30, (e) i = 40, (f) i = 52.

We will use these results as the initial condition for the time evolution. We expect to get the same shape for the evolution.

4. Time evolution of a vortex soliton

We apply the Fourier transform to the spatial coordinates of (4),

$$\frac{d\phi}{dt} = -\mathrm{i}\left(F^{-1}\left(\left(\xi^2 + \chi^2\right) F(\phi)\right) - |\phi|\phi - |\phi|^2\phi\right)$$

where N_x and N_y are the number of mesh points in the x and y direction. $(x_l, y_m) = (lL_x/N_x, mL_y/N_y)$, in which L_x and L_y are the lengths of the domain in the x and y directions. $(\xi_p, \chi_q) = 2\pi(p/L_x, q/L_y)$ for $p = 0, \ldots, N_x - 1$ and $q = 0, \ldots, N_y - 1$. The Runge-Kutta method [17] is applied to the time derivative, while the initial condition is the final step of the Petviashili method. The time evolution of the stationary state will be shown in Figure 3. The initial condition is calculated for the stationary state, there are no any movement during the simulation. This result confirms that the results from the iterative method are the possible 2D solution of (4).

IOP Conf. Series: Journal of Physics: Conf. Series **1290** (2019) 012023 doi:10.1088/1742-6596/1290/1/012023

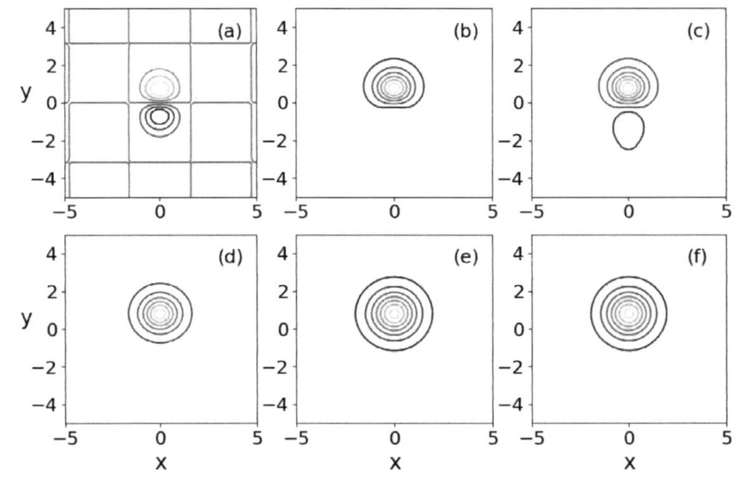

Figure 2. Result of each iteration: (a) i = 0, (b) i=1, (c) i = 2, (d) i = 3, (e) i = 30, (f) i = 44.

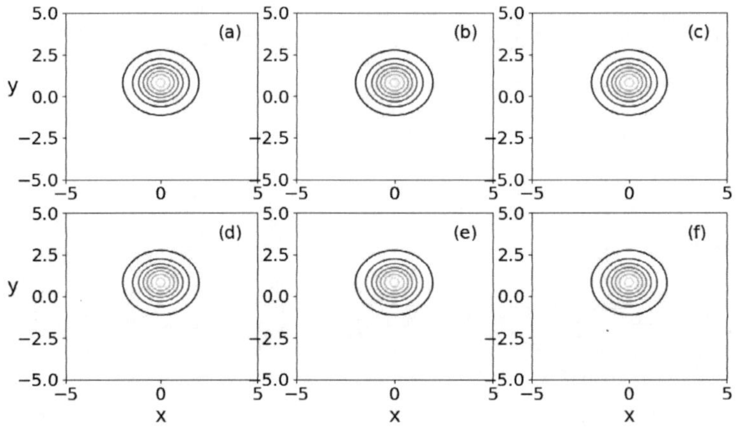

Figure 3. Time step evolution: (a) t = 0, (b) t=2, (c) t = 4, (d) t = 6, (e) t = 8, (f) t = 10.

5. Conclusion

The 2D solitary wave solution of the cqNLS can be obtained by the Petviashvili method. The time evolution is shown that the results are correct. We used the one example of the computational methods to calculate the vortex-like solution of the nonlinear wave equation. It's also a good start to study some phenomena of the vortex solitons or some excited states of others nonlinear wave equations.

6. Acknowledgments

SP wishes to acknowledge support from Faculty of Education, Burapha University.

7. References

[1] Infeld E and Rowlands G 2000 *Nonlinear Waves, Solitons and Chaos* 2nd ed (Cambridge: Cambridge University Press)
[2] Washimi H and Taniuti T 1966 *Phys. Rev. Lett.* **17** 996–8
[3] Agrawal G 2001 *Nonlinear fiber optics* 3rd ed (Academic Press)
[4] Biswas A and Konar S 2007 *Introduction to non-Kerr Law Optical Solitons* (Boca Raton: Chapman and Hall)
[5] Fedele R and Schamel H 2002 *Eur. J. Phys. B* **27** 313–320
[6] Allen M A, Phibanchon S and Rowlands G 2007 *JPP* **73** 215–29
[7] Das G C, Sarma J, Gao Y T and Uberoi C 2000 *Phys. Plasmas* **7** 2374–80
[8] Shukla P K and Bharuthram R 1986 *Phys. Rev. A* **34** 4457–9
[9] Rattanachai Y and Phibanchon S 2018 *Journal of Physics: Conference Series* **1136** 012009
[10] Frycz P and Infeld E 1989 *Phys. Rev. Lett.* **63** 384–5
[11] Frycz P and Infeld E 1990 *Phys. Rev. A* **41** 3375–6
[12] Phibanchon S and Allen M A 2007 Time evolution of perturbed solitons of modified Kadomtsev-Petviashvili equations *The 2007 International Conference on Computational Science and its Applications (ICCSA 2007)* (Los Alamitos, CA, USA: IEEE Computer Society) pp 20–3
[13] Fedele R, Schamel H and Shukla P K 2002 *Phys. Scr.* **T98** 18–23
[14] Petviashvili V I 1976 *Sov. J. Plasma Phys.* **2** 257–8
[15] Lakoba T and Yang J 2007 *Journal of Computational Physics* **226** 1668–1692
[16] Pelinovsky D E and Stepanyants Y A 2004 *SIAM J. Numer. Anal.* **42** 1110–1127
[17] Press W H, Teukolsky S A, Vetterling W T and Flannery B P 1992 *Numerical Recipes in C* 2nd ed (Cambridge: Cambridge University Press)

IOP Conf. Series: Journal of Physics: Conf. Series **1290** (2019) 012024 doi:10.1088/1742-6596/1290/1/012024

Adomian Decomposition Method for the solitary wave solution to the modified Korteweg-de Vries equation

Songvudhi Chimchinda[1] **and Sarun Phibanchon**[2]
[1] Physics Department, Burapha University
169 Longhardbangsaen road, Sansuk, Muang, Chonburi, 20131, Thailand
[2] Faculty of Education, Burapha University
169 Longhardbangsaen road, Sansuk, Muang, Chonburi, 20131, Thailand

E-mail: songvudhi@buu.ac.th

Abstract. The Adomian decomposition method (ADM), is the one of the semi-analytical method, will be applied to study the time evolution of the soliton solution to the modified Korteweg-de Vries equation. We also provide the conservation laws to verify the dark solution obtained from this method. The ADM gives a good result for the approximation solution of the weakly nonlinear wave equation.

1. Introduction

Solitary waves are well known as the solution of the nonlinear wave equations. These kind of waves retain their shape as a result of the balance between nonlinearity and dispersion. The most familiar form of nonlinear wave equation to produce the solitary wave, is called Korteweg-de Vries (KdV) equation,

$$\phi_t + \phi\phi_x + \phi_{xxx} = 0,$$

where ϕ subscripts x and t denote differentiation with respect to space and time, respectively, which originally derived for water, (see e.g. [1] for further details). ϕ represents the amplitude of the wave. Although originally derived for water, the KdV equation occurs in many other contexts. It governs weakly nonlinear ion-acoustic waves in plasma when the electrons have a Maxwellian distribution [2]. In this case, ϕ is the electrostatic potential. Schamel proposed allowing for the trapping of some of the electrons on ion-acoustic waves [3]. The free and trapped electrons have different temperatures, although both still have Maxwellian distributions. This leads a modified equation for ion-acoustic waves,

$$\phi_t + \phi^{1/2}\phi_x + \phi_{xxx} = 0, \tag{1}$$

The traveling wave solution of (1) is not difficult to find [4], but to express both of spatial and temporal coordinates might be more difficult. As well as, (1) is not an integrable equation [5], the inverse scattering method [6] can not be applied to obtain the solution. To obtain the soliton solution of (1), we then consider method known as the Adomian decomposition method (ADM) [7, 8, 9]. The initial condition for (1) will be used as,

$$\phi_0(x) = (30\eta^2)^2 \operatorname{sech}^4 \eta(x - x_0), \tag{2}$$

IOP Conf. Series: Journal of Physics: Conf. Series **1290** (2019) 012024 doi:10.1088/1742-6596/1290/1/012024

η is a constant and x_0 denotes the initial position of the solitary wave. We also derive the conservation laws for the modified KdV equation where the solution must be satisfied to confirm the valid of the solution.

2. Adomian decomposition method for the modified KdV equation

The ADM has been extensively used for semi-analytical solution of the nonlinear wave equations such as nonlinear Schrödinger equation and its modified form [10, 11, 12]. The idea of this method is to rewrite the nonlinear terms into the Adomian polynomials. We now consider the nonlinear evolution equation as follows

$$\phi_t + \hat{L}\phi + \hat{N}\phi = 0,$$

where $\hat{L}$ is the linear differential operator and $\hat{N}$ denotes the nonlinear differential operator. The solution will be written in the form of a power series for N terms, namely,

$$\phi(x,t) = \sum_{n=0}^{N} u_n(x,t), \tag{3}$$

where $u_n(x,t)$ represents the unknown function. For the nonlinear operator, the expression for the power series will be shown as

$$\hat{N}\phi = \sum_{n=0}^{N} A_n$$

where

$$A_n = \frac{1}{n!}\frac{d^n}{d\lambda^n}\left[\hat{N}\left(\sum_{i=0}^{\infty}\lambda^i u_i(x,t)\right)\right]_{\lambda\to 0}.$$

The solution of (3)can be then determined as

$$\sum_{n=0}^{N} u_n(x,t) = L_t^{-1}\left[\hat{L}\left(\sum_{n=0}^{N} u_n\right)\right] - L_t^{-1}\sum_{n=0}^{N} A_n$$

We will applied this method to find the solitary wave solution.

3. Conservation laws for modified KdV equation

The conserved quantities will be derived which they do not vary with time. A conservation law can be expressed as

$$\frac{dQ}{dt} = 0,$$

where Q is a conserved quantity. Rearrange (1) to give

$$\phi_t = -\phi^{1/2}\phi_x - \phi_{xxx},$$

and integrate with respect to x, we then obtain

$$\frac{\partial}{\partial t}\int_{-\infty}^{\infty}\phi dx = -\frac{2\phi^{3/2}}{3}\Bigg|_{-\infty}^{\infty} = 0,$$

ϕ, ϕ_x and ϕ_{xx} have the same value as $x \to \pm\infty$. It is sometimes referred as the 'Mass'

$$\int_{-\infty}^{\infty}\phi dx = \mathrm{M}. \tag{4}$$

Another conserved quantity is easily to obtain by multiplying both sides with ϕ before integrating. This leads to

$$\frac{\partial}{\partial t}\int_{-\infty}^{\infty}\frac{\phi^2}{2}dx = -\frac{2}{5}\left(\phi^{5/2}\Big|_{\infty}^{\infty}\right) + \left(\frac{\phi_x^2}{2}\Big|_{\infty}^{\infty}\right) = 0.$$

This is referred as the 'Momentum',

$$\int_{-\infty}^{\infty}\frac{\phi^2}{2}dx = P. \tag{5}$$

The last conserved quantity, we first differentiate (1) with respect to x and multiply both sides with ϕ_x and then integrate both sides with respect to x,

$$\frac{\partial}{\partial t}\int_{-\infty}^{\infty}\frac{\phi^2}{2}dx = \int_{-\infty}^{\infty}\phi^{1/2}\phi_x\phi_{xx}dx \tag{6}$$

To determine the right-hand side of (6), we then multiply (1) with $\phi^{3/2}$ and integrate with respect to x,

$$\frac{\partial}{\partial t}\int_{-\infty}^{\infty}\frac{2\phi^{5/2}}{5}dx = \frac{3}{2}\int_{-\infty}^{\infty}\phi^{1/2}\phi_x\phi_{xx}dx \tag{7}$$

From (6) and (7), we have the conserved quantity called 'Energy',

$$\int_{-\infty}^{\infty}\frac{3\phi^2}{4} - \frac{2\phi^{5/2}}{5}dx = E. \tag{8}$$

To examine some results, we consider $\eta = 0.15$, and the initial condition at the origin ($\xi_0 = 0$), can be shown in Figure. 1

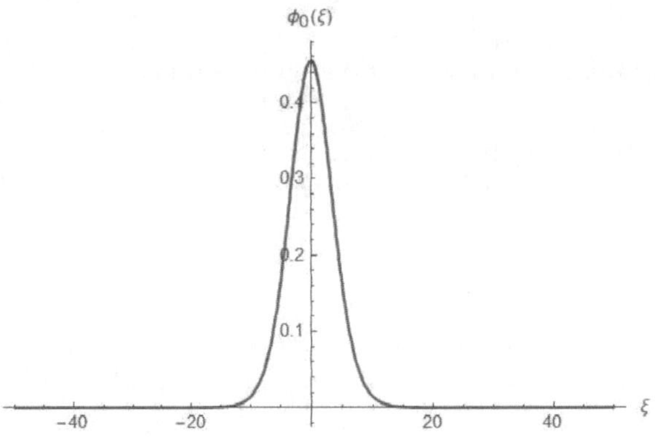

Figure 1. shows the initial profile

We corrected up to 3 terms, ($N = 3$), where the initial condition is written as

$$u_0 = 0.455\,\text{sech}^4(0.15x),$$

while u_1, u_2 and u_3 are determined as follows,

$$u_1 = t\tanh(0.15x)(0.098\,\text{sech}^6(0.15x) + 0.098\,\text{sech}^4(0.15x)\tanh^3(0.15x))$$

$$u_2 = t^2(-0.027\,\text{sech}^{10}(0.15x) + 0.0053\,\text{sech}^8(0.15x)\tanh^2(0.15x) + 0.0186\,\text{sech}^6(0.15x)\tanh^4(0.15x) + 0.106\,\text{sech}^4(0.15x)\tanh^6(0.15x))$$

$$u_3 = t^3(\text{csch}^7(0.3x)\sinh^8(0.15x)(\text{csch}^4(0.3x)\sinh^4(0.15x))^{3/2} (-14.693 - 2.938\sinh^2(0.15x) + 38.202\sinh^4(0.15x) + 26.447\sinh^6(0.15x)) + \text{sech}^4(0.15x)\tanh(0.15x)(0.001124\,\text{sech}^8(0.15x) - 0.000884\,\text{sech}^6(0.15x)\tanh^2(0.15x) - 0.00437\,\text{sech}^4(0.15x)\tanh^4(0.15x) - 0.0016\,\text{sech}^2(0.15x)\tanh^6(0.15x) + 0.000765\tanh^8(0.15x)))$$

The time evolution of the solution is shown in Figure. 2. The structure of the solution is nearly coherent or traveling without changing its shape.

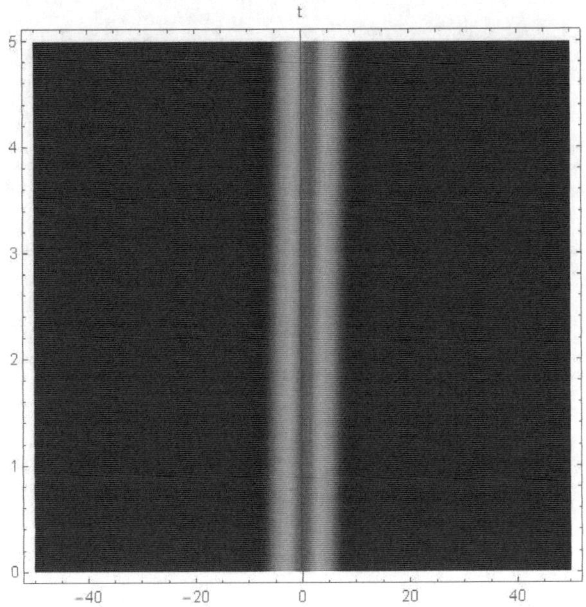

Figure 2. Time evolution of (1) with $0 <= N <= 3$ and $t \in [0, 5]$.

However,we can provide the Matheamtica file by direct contact the author (SC).

The conservation laws, moreover, are given in Table 1, where these parameters are conserved during the time evolution.

These results might be improved by adding a few terms for the approximation solution ($N > 3$) but this also increases a computational time.

4. Conclusion

The ADM has been quite successfully applied to calculate the solitary wave solution for the modified KdV equation. The good thing about this method is directly obtain the solution without using any assumption to the equation. We are strongly believed to the solitary solution by considering the conservation laws. However, if the nonlinear term contains a strong nonlinearity or some messy forms to the nonlinear operator, ADM will take some computational time to process.

time	Mass (4)	Momentum (5)	Energy (8)
0	4.05	1.26534	-5.69401
1	4.05	1.26533	-5.69394
2	4.05	1.26519	-5.69299
3	4.05	1.26465	-5.68917
4	4.05	1.26335	-5.68004
5	4.05	1.26102	-5.66397

Table 1. Computed quantities M, P and E for modified KdV which are satisfied the ADM

5. Acknowledgments

SP wishes to acknowledge support from NRCT.

6. References

[1] Infeld E and Rowlands G 2000 *Nonlinear Waves, Solitons and Chaos* 2nd ed (Cambridge: Cambridge University Press)
[2] Washimi H and Taniuti T 1966 *Phys. Rev. Lett.* **17** 996–8
[3] Schamel H 1972 *Plasma Phys.* **14** 905–24
[4] Biswas A 2009 *Communications in Nonlinear Science and Numerical Simulation* **14** 1830–1833
[5] Ramani A and Grammaticos B 1991 *J. Phys. A: Math. Gen.* **24** 1969–70
[6] Ablowitz M J and Segur H 1981 *Solitons and the Inverse Scattering Transform* (Philadelphia: SIAM)
[7] Adomian G 1988 *Journal of Mathematical Analysis and Applications* **135** 501–544
[8] Adomian G 1990 *Mathematical and Computer Modelling* **13** 17–43
[9] Cherruault Y and Adomian G 1993 *Mathematical and Computer Modelling* **18** 103–106
[10] Biswas A and Milovic D 2009 *CNSNS* **14** 1993–8
[11] Asma M, Othman W, Wong B and Biswas A 2018 *Optik* **164** 632–641
[12] González-Gaxiola O and Biswas A 2018 *Optical and Quantum Electronics* **50**

IOP Conf. Series: Journal of Physics: Conf. Series **1290** (2019) 012025 doi:10.1088/1742-6596/1290/1/012025

Propagation of dust ion acoustic solitary waves in dusty plasma with Boltzmann electrons

S. Das

Central Institute of Technology Kokrajhar
Kokrajhar, BTAD, Assam, Pin-783370, India

e-mail: s.das@cit.ac.in

Abstract. In this multispecies plasma model, consisting of negative mobile dusts, non-thermal ions and Boltzmann electrons, dust-ion acoustic solitary waves are studied through reductive perturbative technique by deriving corresponding Korteweg-de Vries (KdV) equation. The number of dust charge contained in a dust particle (Z_d) and the streaming speeds of mobile dusts (u_{d0}) and ions (u_{i0}) are observed to play very important role to form dust-ion acoustic compressive and rarefactive KdV solitons. Remarkably, some initial streaming of dust (u_{d0}) are found to be instrumental for both compressive and rarefactive KdV solitons in a short range of Z_d separating both kinds by some asymptotic lines. Also, the presence of low dust charges and lower ion streaming, compressive and rarefactive KdV solitons of either concave or convex characters are shown to reflect. For the higher streaming of mobile dusts, the amplitudes of the rarefactive KdV solitons characteristically changes from higher to lower showing convex character for this plasma model. A rigorous theoretical investigation has been made to show how the number of dust charge contained in a dust particle (Z_d) drastically change the amplitudes of the compressive and rarefactive KdV solitons.

1. Introduction

The Ion-Acoustic Soliton (IAS) is one of the beautiful manifestations of nonlinear structures of nature and its study have started since seventy of the last century with very simple stage of two component pasmas. The study of solitary wave was started in 1985 by Korteweg and de Vries [1] and then Washimi and Taniuti [2] in 1966.

In last few decades, IASW have been studied by many workers by taking two and multispecies of the plasmas (not reported here). But gradually the study of multispecies plasmas have been extended to insertion of various characteristics effects viz. magnetized or unmagnetized, plasmas of either cold or warm, relativistic or non-relativistic effects, consideration of pressure variations, quantum effects etc. Dust particles are omnipresent and are invariably immersed in the ambient astrophysical and space plasma. The existence and the effects of dusty plasmas are nicely depcted in planetary rings, earth's magnetosphere, stars' neighbourhood, Cometary tails, Asteroid zones, Circumstellar disks, dark molecular clouds, interstellar clouds, nebulae etc. [3-8]. Also, because of the tremendous applications of Dust –Ion Acoustic Waves (DIAW) in the various theoretical and laboratory activities, DIAW has become very popular in the field of plasma physics [9-11].

Properties of solitary waves are greatly affected [12-16] by the presence of dust particles in space. In absence of magnetic field, static dust grains in plasma can generate extremely low frequency dust acoustic (DA) waves [17] and can modify the existing wave spectra [18]. Investigation of DIAWs under

IOP Conf. Series: Journal of Physics: Conf. Series **1290** (2019) 012025 doi:10.1088/1742-6596/1290/1/012025

various plasma compositions is studied by many authors [6, 19-21]. In a dusty plasma with inertial charged dust and Boltzmann distributed electrons and ions, it have been studied [17] that DIAWs propagates linearly as a normal mode and nonlinearly as subsonic solitons of either positive or negative electrostatic potentials. Again, DA solitary wave have been studied in a dusty plasma by taking non-thermal parameter [22] in the nonthermal distribution of ions as variable. Creation of spatial inhomogeneity is observed [23, 24] for the distribution of immobile dust particles. It is also shown that an unmagnetized dusty plasma with the effects of vortex-like and nonthermal ion, large amplitude rarefactive as well as compressive DA solitons exists [25]. In an unmagnetized multicomponent plasma, existence of solitary waves are investigated with pressure variations in ions, negative ions and isothermal electrons [26]. For stationery dust, DIA solitary waves are investigated for a plasma consisting ions and electrons with their pressure (or temperature) variations for both lower and higher order situations followed by a comparative study [27]. Formation of DIA compressive and rarefactive solitons are also shown to exist for a relativistic plasma with stationery [28] and mobile [29] dust. Again, DIA waves are studied for non-thermal Cairns like distribution of ions in plasma [30]. Dust acoustic (DA) waves have been investigated foe a dusty plasma showing the implicit role of Cairns distributed cold ions [31]. The existence of both positive and negative solitary structures in some parameter space have been investigated with sufficient non thermality in the ions and negative charges on the dust along with electrons [32] following Cairns distribution [33].

In this paper, the existence of both compressive and rarefactive DIA KdV solitons are shown for the multispecies plasma model consisting of negative mobile dust, non-thermal ions and Boltzmann electrons. Also, the characteristic role (in the formation of DIA solitary waves) of several plasma parameters especially number of dust charges (Z_d) and streaming speeds of dusts (u_{d0}) and ions (u_{i0}) are reported in this paper. The paper is organized as 'Introduction' in section 1, 'Basic equations governing the dynamics of motion' in section 2, 'derivation of Korteweg-de Vries (KdV) equation' in section 3, 'Solution of KdV equation' in section 4, 'Results and discussion' in section 5, followed by the 'References'.

2. Basic equations governing the dynamics of motion

In this model of multispecies plasma, to investigate the propagation of solitary waves, we consider negative mobile dusts, non-thermal ions and Boltzmann electrons as follows

For dust particles,

$$\frac{\partial n_d}{\partial t}+\frac{\partial}{\partial x}(n_d u_d)=0 \tag{1}$$

$$\left(\frac{\partial}{\partial t}+u_d\frac{\partial}{\partial x}\right)u_d=Z_d\frac{\partial \varphi}{\partial x} \tag{2}$$

for ions,

$$\frac{\partial n_i}{\partial t}+\frac{\partial}{\partial x}(n_i u_i)=0 \tag{3}$$

$$\left(\frac{\partial}{\partial t}+u_i\frac{\partial}{\partial x}\right)u_i=-\frac{1}{Q}\frac{\partial \varphi}{\partial x},\ \text{where } Q=\frac{n_{do}}{n_{i0}} \tag{4}$$

for electrons

$$n_e=e^{\varphi} \tag{5}$$

and the Poisson equation

$$\frac{\partial^2 \varphi}{\partial x^2}=Z_d n_d+n_e-n_i \tag{6}$$

where u_d, u_i ; m_d, m_i and n_d, n_i are velocities, masses and densities the dust and ions respectively and Z_d is the number of charges contained in a dust grain and n_e is the density of electrons.

3. Derivation of the Korteweg-de Vries (KdV) equation

In order to derive the KdV equation from the set of equations (1)-(6), we use the stretched variables in terms of the smallness parameter ε as

$$\xi = \varepsilon^{\frac{1}{2}}(x - Vt),\ \tau = \varepsilon^{\frac{3}{2}}Vt \tag{7}$$

with the phase velocity V of the waves. We expand the flow variables asymptotically about the equilibrium state in terms of the smallness parameter ε as

$$\left.\begin{array}{l} n_d = n_{d0} + \varepsilon n_{d1} + \varepsilon^2 n_{d2} + \ldots \\ n_i = n_{i0} + \varepsilon n_{i1} + \varepsilon^2 n_{i2} + \ldots \\ n_e = n_{e0} + \varepsilon n_{e1} + \varepsilon^2 n_{e2} + \ldots \\ u_d = u_{d0} + \varepsilon u_{d1} + \varepsilon^2 u_{d2} + \ldots \\ u_e = u_{e0} + \varepsilon u_{e1} + \varepsilon^2 u_{e2} + \ldots \\ \varphi = \varepsilon\varphi_1 + \varepsilon^2\varphi_2 + \ldots \end{array}\right\} \tag{8}$$

Using the standard perturbation method, with the use of transformation (7), expansions (8) in the normalized set of equations (1)-(6), we determine the ε-order equations. From the coefficients of ε-order equations, on integration and using the boundary conditions

$$n_{i1} = 0, n_{e1} = 0; v_{i1} = 0, v_{e1} = 0 \text{ and } \varphi_1 = 0 \text{ at } |\xi| \to \infty \tag{9}$$

the first-order perturbed quantities are obtained as

$$\left.\begin{array}{l} n_{d1} = -\dfrac{n_{d0}Z_d}{(u_{d0} - V)^2},\ u_{d1} = \dfrac{Z_d}{u_{d0} - V},\ n_{i1} = \dfrac{n_{i0}}{(u_{i0} - V)^2 Q}, \\ u_{i1} = -\dfrac{1}{Q(u_{i0} - V)},\ n_{e1} = 1 \end{array}\right\} \tag{10}$$

Also, the use of stretched variables (8) and the expansions (9) in equation (6) following equations yield

$$1 - Z_d\sigma = \frac{1}{n_{i0}},\ \text{where } \sigma = \frac{n_{do}}{n_{i0}} \tag{11}$$

and $Z_d n_{d1} + n_{e1} - n_{i1} = 0$ (12)

We use the first-order quantities from (10) in (12), to get the phase velocity (V) equation as

$$1 - Z_d\sigma - \frac{1}{Q(u_{i0} - V)^2} - \frac{Z_d^2\sigma}{(u_{d0} - V)^2} = 0 \tag{13}$$

Again, from the coefficients of ε^2-order equations, the following equations are obtained

$$B_d\frac{\partial n_{d2}}{\partial\xi} + n_{do}\frac{\partial u_{d2}}{\partial\xi} + \frac{\partial}{\partial\xi}(n_{d1}u_{d1}) + V\frac{\partial n_{d1}}{\partial\tau} = 0 \tag{14}$$

$$B_d\frac{\partial u_{d2}}{\partial\xi} + n_{do}\frac{\partial u_{d2}}{\partial\xi} + u_{d1}\frac{\partial u_{d1}}{\partial\xi} + V\frac{\partial u_{d1}}{\partial\tau} - Z_d\frac{\partial\varphi_2}{\partial\xi} = 0 \tag{15}$$

$$B_i \frac{\partial n_{i2}}{\partial \xi} + n_{io} \frac{\partial u_{i2}}{\partial \xi} + \frac{\partial}{\partial \xi}(n_{i1} u_{i1}) + V \frac{\partial n_{i1}}{\partial \tau} = 0 \tag{16}$$

$$QB_i \frac{\partial u_{i2}}{\partial \xi} + n_{io} \frac{\partial u_{d2}}{\partial \xi} + Qu_{i1} \frac{\partial u_{i1}}{\partial \xi} + QV \frac{\partial u_{i1}}{\partial \tau} + \frac{\partial \varphi_2}{\partial \xi} = 0 \tag{17}$$

$$n_{e2} = \varphi_2 + \frac{\varphi_1^{\,2}}{2} \tag{18}$$

$$\frac{\partial^2 \varphi_1}{\partial \xi^2} = Z_d n_{d2} + n_{e2} - n_{i2} \tag{19}$$

From the equations (14)-(18), we find the values of $\frac{\partial n_{d2}}{\partial \xi}$, $\frac{\partial n_{e2}}{\partial \xi}$, $\frac{\partial n_{i2}}{\partial \xi}$ and using equation (19), the KdV equation is obtained as

$$\frac{\partial \varphi_1}{\partial \tau} + p\varphi_1 \frac{\partial \varphi_1}{\partial \xi} + q \frac{\partial^3 \varphi_1}{\partial x^3} = 0 \tag{20}$$

where $p = \frac{(1-Z_d\sigma)B_i^{\,4}B_d^{\,4}Q^2 - 3B_d^{\,4} + 3\sigma B_i^{\,4}Q^2 Z_d^{\,4}}{2VB_iB_dQ(B_d^{\,3} + \sigma Z_d^{\,2} B_i^{\,3} Q)}$, $q = \frac{(1-Z_d\sigma)B_i^{\,3}B_d^{\,3}Q - 3B_d^{\,4} + 3\sigma B_i^{\,4}Q^2 Z_d^{\,4}}{2V\left\{(1-Z_d\sigma)B_iB_d^{\,3} + \sigma Z_d^{\,2} B_i^{\,3} Q)\right\}}$

$$B_i = u_{i0} - V, \quad B_d = u_{d0} - V$$

4. Solution of Korteweg-de Vries (KdV) equation

To derive the solution of the KdV equation (20), we introduce the transformation $\eta = \xi - C_1\tau$ where C_1 is the soliton speed in the linear η-space and using the boundary conditions $\varphi_1 = \frac{\partial \varphi_1}{\partial \tau} = \frac{\partial^2 \varphi_1}{\partial \tau^2} = 0$ as $|\eta| \to \infty$, the solution is obtained as

$$\varphi_1 = \varphi_0 \operatorname{sec}h^2\left(\frac{\eta}{\delta}\right)$$

where φ_0 and δ are the amplitude and width of the solitary wave respectively and are given by $\varphi_0 = \frac{3C_1}{p}$ and $\sigma = \sqrt{\frac{4q}{C_1}}$.

5. Results and discussion

In this model of multispecies dusty plasma with negative mobile dusts, non-thermal ions and Boltzmann electrons, dust-ion acoustic (DIA) compressive and rarefactive KdV solitons are shown to exist in the plasma. Throughout the consideration of the paper, $Q = 0.54 \times 10^{-3}$ is taken to trace the curves.

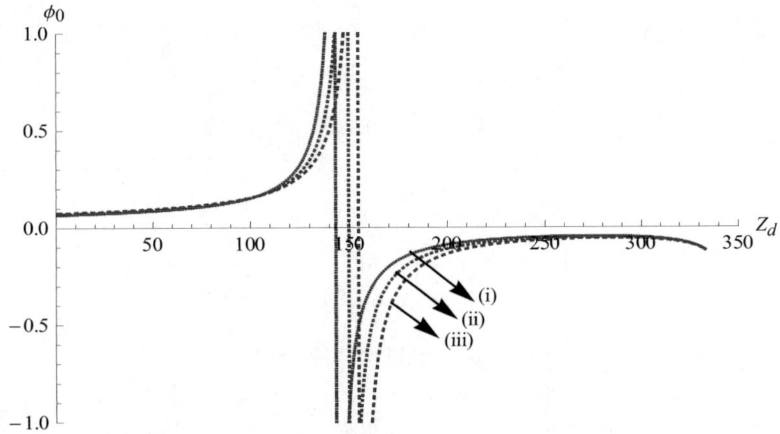

Figure 1: Amplitudes of compressive and rarefactive KdV solitons versus dust charges (Z_d) for fixed $u_{d0} = 4$, $\sigma = 0.001$ and u_{i0}=4 (i), 5 (ii), 6 (iii).

The smaller values of initial streamings of ions and dusts $[u_{d0} = 4, u_{i0} = 4\ (i), 5\ (ii), 6\ (iii)]$ are found instrumental for the generation of compressive and rarefactive DIA KdV solitons simultaneously (Fig. 1) with $\sigma = 0.001$. Starting with the linear increase of amplitudes of compressive DIA KdV solitons, suddenly shows a rapid and convex increase within a very small interval of Z_d but their rarefactive counter parts sharply decrease concavely for each ion streaming (u_{i0}) in a similar but opposite pattern. After attaining a fixed and certain range of Z_d, called critical range (e.g., $Z_{dr} \approx 130 - 150$) of dust charges, both the kinds (compressive and rarefactive) of DIA solitons are separated by disjoint lines like asymptotes. Within this critical range of dust charge (Z_{dr}), the initial streamings of ions and mobile dusts become almost equal (Fig. 1). But outside this critical range, for $u_{d0} \leq u_{i0}$ (or $u_{i0} < u_{d0}$) cause to generate compressive (rarefactive) DIA solitons of low and high amplitudes together. Immediate left/right of the critical range of dust charge (Z_{dr}) the compressive/rarefactive DIA solitons of much higher amplitudes form within a span of very small interval of Z_d. From very far away from the critical range of dust charge (Z_{dr}), for higher number of dust charges (Z_d) and smaller initial streamings of dust (u_{d0}) and ions (u_{i0}), the amplitudes of the compressive solitons become very low whereas because of much greater number of dust charges compared to u_{d0} and u_{i0} starting from very near of critical range of dust charge (Z_{dr}) to very far, the amplitudes of rarefactive DIA solitons follow similar character.

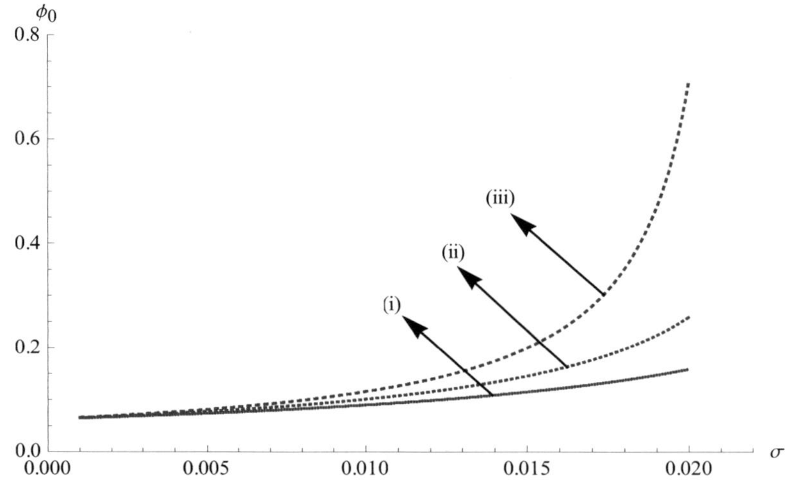

Figure 2: Amplitudes of compressive KdV solitons versus ion to dust density ratio (σ) for fixed u_{i0}=2, u_{d0}=0.5 and Z_d=20 (i), 25 (ii), 30 (iii).

The amplitudes of compressive DIA KdV solitons for very very small ion to dust density ratio (σ) starting with a reasonable (very small) amplitudes for fixed but smaller initial streamings of ion and dust viz. $u_{i0} = 2, u_{d0} = 0.5$ respectively corresponding to the small number of dust charge $Z_d = 20$ increases slowly (Fig. 2). But with the increase of number of dust charge [$Z_d = 25\ (i), 30\ (ii)$], the amplitudes of the DIA KdV solitons increases convexly (Fig. 2) but sharply for a smaller range of σ for the same values of other parameters.

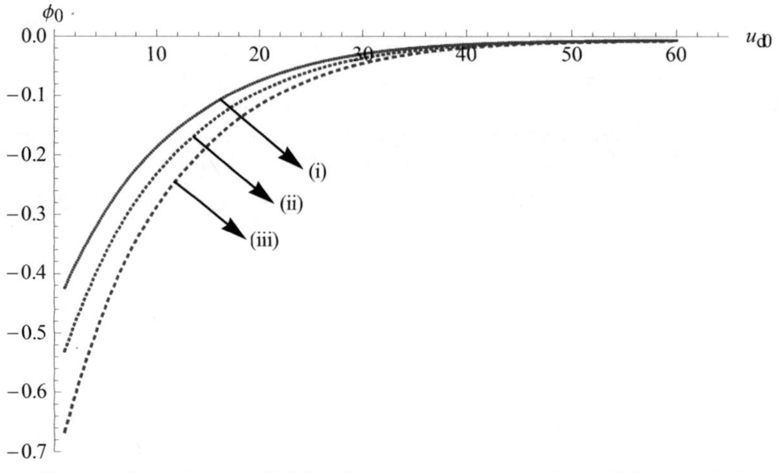

Figure 3: Amplitudes of rarefactive KdV solitons versus streaming of dust (u_{d0}) for fixed Z_d=300, $\sigma = 0.001$ and u_{i0}=3(i), 5(ii), 7(iii).

On the other hand, the growth of the amplitudes of rarefactive DIA KdV solitons decreases concavely (Fig. 3) relative to the variation of streamings of dusts for fixed values of $Z_d = 300, \sigma = 0.001$ and

$u_{i0} = 3(i), 5(ii)$and$7(iii)$. Initially, positive amplitudes of the rarefactive DIA KdV solitons are higher for very small initial streamings of dust ($u_{d0} < 2$) but with the increase of dust charge u_{d0}(Fig. 3).

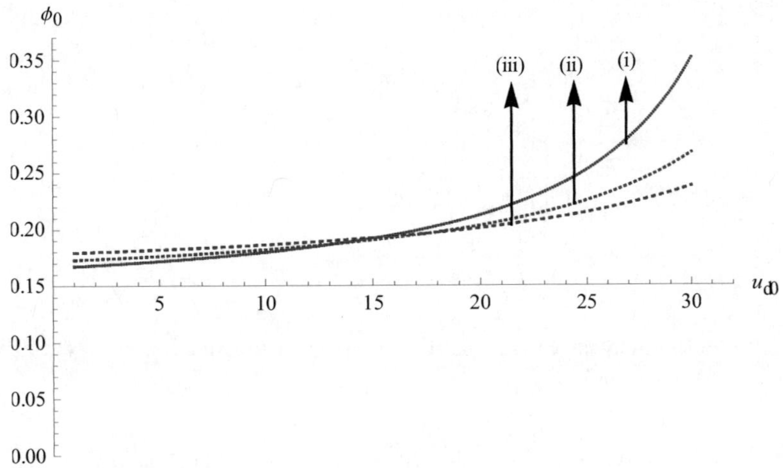

Figure 4: Amplitudes of compressive KdV solitons versus streaming of dust (u_{d0}) for fixed Z_d=200, $\sigma = 0.001$ and u_{i0}=35 (i), 40 (ii), 45 (iii).

Contrary to this, the smaller number of negative dust charges $Z_d = 200$ (compared to Fig. 3) into the plasma but greater initial streaming of ions $u_{i0} = 35\ (i), 40\ (ii), 45\ (iii)$ and $\sigma = 0.001$, the amplitudes of compressive DIA KdV solitons slowly increases with the smaller increase of initial streamings of dust i.e., $u_{d0} < 15$ for other parametric values of the plasma parameters as above. At some critical initial streamings of dusts $u_{dc} \approx 15$, the amplitudes of compressive DIA KdV solitons become same for all the three values of ion streamings. After the critical u_{dc}, the amplitudes of compressive DIA KdV solitons sharply increases (Fig. 4) with the increase of dust initial streaming (u_{d0}).

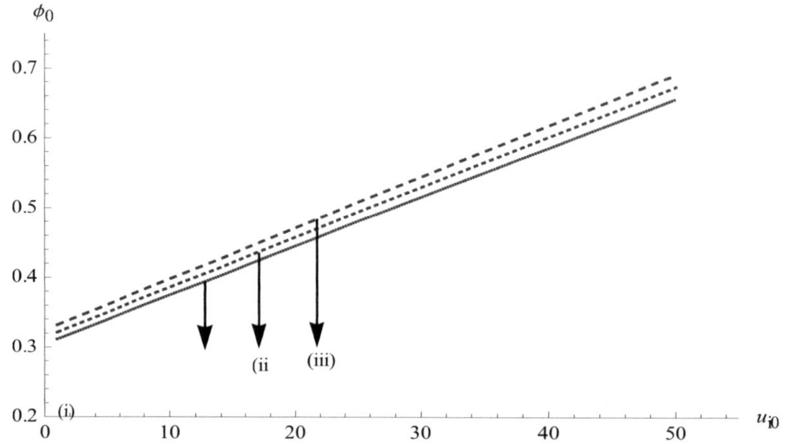

Figure 5: Amplitudes of compressive KdV solitons versus ion streaming u_{i0} for fixed Z_d=10, u_{d0}=0.1 and $\sigma = 0.001\ (i), 0.002\ (ii), 0.003\ (iii)$.

The amplitudes of compressive DIA KdV solitons increases almost linearly (Fig. 5) as initial streamings of ion (u_{i0}) increases for small $Z_d = 10$ very smaller initial streaming of dust $u_{d0} = 0.1$ and $\sigma = 0.001$. Of course, smaller the values of dust to ion density ratio (σ), smaller is the amplitude of DIA KdV solitons.

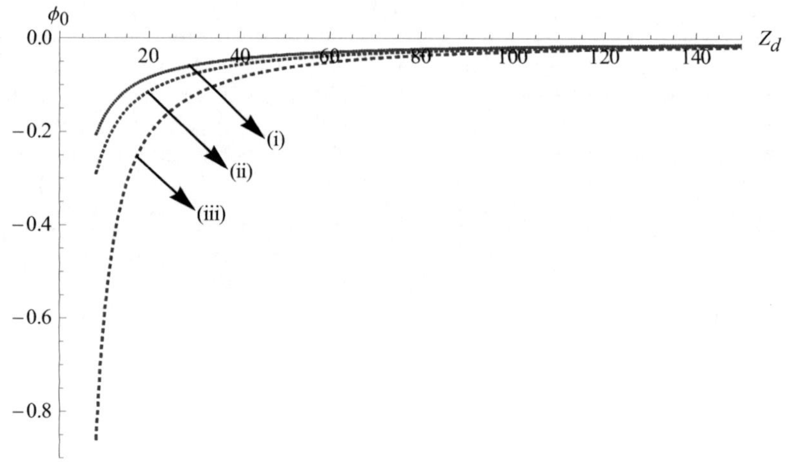

Figure 6: Amplitudes of rarefactive KdV solitons versus Z_d for fixed u_{d0}=50, $\sigma = 0.001$ and u_{i0}=3 (i), 5 (ii), 7 (iii).

Smaller number of dust charges (Z_d), the amplitudes of compressive DIA KdV solitons decreases sharply from higher to lower in a very short range of Z_d and then convexly tend to almost zero with the other plasma parameters as $u_{d0} = 50, \sigma = 0.001$. With the increase of initial streaming of ions [(i.e., $u_{i0} = 3\ (i)$ to $5\ (ii)$] keeping the other parametric values viz. $u_{d0} = 50, \sigma = 0.001$, the rate of decrease of amplitudes of DIA KdV solitons become more sharp from higher to lower and the interval of Z_d becomes more contracted (Fig. 6).

Frequent bursts of solar winds and magnetic storms effects the plasma scenario and as a result huge nonlinearity is created in space plasma. These nonlinearity create hazards to space vehicles and probes. Our present study of DIA solitary waves with the plasma parameters (mentioned in this paper with different situation) may be helpful (to some extent) to meet the hazards in dusty space plasma.

References

[1] Korteweg D J, de Vries G 1895 *Phil.Mag.* **39** 422
[2] Washimi H and Taniuti T 1966 *Phys. Rev. Lett.* **17** 996
[3] Goertz C K 1989 *Rev. Geophys.* **27** 271
[4] Mendis D A and Rosenberg M 1944 *Annu. Rev. Astron. Astrophys.* **32** 419
[5] de Angelis U 1992 Phys. Scripta **45** 465
[6] Shukla P K and Mamun A A 2002 *Introduction to Dusty Plasma Physics* (Institute of Physics Publishing Ltd., Bristol)
[7] Verheest F 2000 *Waves in Dusty Plasma Physics* (Kluwer AcademicPublishers, Dordrecht)
[8] Mendis D A, "Summary: Dust Plasma Interactions in Space", in Proc. *4th Int. Conf. Phys. Dusty Plasmas* 2005 AIP Conf., **799** 583
[9] Barkain A, D' Angelo N and Merlino R L 1996 *Planet Space Sci.* **44** 239
[10] Barkain A, Merlino R L and D' Angelo N 1995 *Phys. Plasmas* **2** 3563
[11] Nakamura Y, Bailung H, Shukla P K 1999 *Phys. Rev. Lett.* **83** 1602
[12] Shukla P K and Silin V P 1992 *Phys. Scr.* **45** 508
[13] Mamun A A 1999 *Astrophys. Space Sci.* **268** 443
[14] Ghosh S, Sarkar S, Khan M and Gupta M R 2000 *Phys. Lett. A* **274** 162
[15] Nakamura Y and Sarma A 2001 *Phys. Plasmas* **8** 3921
[16] Shahmansouri M and Tribeche M 2014 *Commun. Theor. Phys.* **61** 377
[17] Rao N N, Shukla P K, Yu M Y 1990 *Planet Space Sci.* **38** 543
[18] Bliokh P V and Yaroshenko V V 1985 *Sov. Astron.* **29** 330
[19] Verheest F 1992 *Planet Space Sci.* **40** 1
[20] Duan W S, Xong X R, Shi Y R and Sun J A 2003 *Chaos Soliton Fract.* ***16*** 767
[21] Wang Y Y and Jhang J F 2006 *Phys. Lett. A* **352** 155
[22] Pakzad H R 2009 *Astrophys. Space Sci.* **324** 41
[23] Whipple E C, Northop T G and Mendis D A 1985 *J. Geophys. Res.* **90** 7405
[24] de Angelis U, Formisano V and Giordano M 1988 *J. Plasma Phys.* **40** 399
[25] Mamun A A, Cairns R A and D'Angelo N 1996 *Phys. Plasmas* **3(7)** 2610
[26] Kalita B C and Barman S N 1995 *J. Phys. Soc. Japan* **64** 784
[27] Kalita B C and Das Samiran 2017 *J. Plasma Phys.* **83(5)** 905830502
[28] Kalita B C and Das Samiran 2017 *IEEE Trans. Plasma Sci.* **46(4)** 790
[29] B C Kalita and Samiran Das 2014 *Astrophys. Space Sci.* **352 (2)** 585
[30] Asgari H, Muniandy S V and Wong C S 2013 *Phys. Plasmas* **20** 023705
[31] Kalita B C and Kalita Rekha 2015 *J. Plasma Phys.* **82(2)** 905820201
[32] Verheest F and Pillay S R 2008 *Phys. Plasmas* **15** 013703
[33] Cairns R A, Mamun A A, Bingham R, Dendy R, Bostrom R, Nairns C M C and Shukla P K 1995 *Geophys. Res. Lett.* **22** 2709

IOP Conf. Series: Journal of Physics: Conf. Series **1290** (2019) 012026 doi:10.1088/1742-6596/1290/1/012026

On the carbuncle instability of the HLLE-type solvers

Zhijun Shen, Jian Ren, Xia Cui

Laboratory of Computational Physics, Institute of Applied Physics and Computational Mathematics, P. O. Box 8009-26, Beijing 100088, China

E-mail: shen_zhijun@iapcm.ac.cn, ren_jian@iapcm.ac.cn, cui_xia@iapcm.ac.cn

Abstract. The carbuncle phenomenon is a numerical instability that affects the numerical capturing of shock waves when low-dissipative upwind scheme is used. This paper investigates shock instabilities of the HLLE-type methods for the Euler equations under the strong shock interaction, where the HLLE-type methods include the HLLE, HLLC, HLLEM, HLLCM and HLLEC Riemann solvers with specific wavespeed estimates. Based on a matrix stability analysis for two dimensional steady shocks, a new factor to influence carbuncle phenomenon is pointed out and the choice of the signal velocity plays an important role. A numerical flux function with wave velocity estimates which can crisply resolve shocks seems to be vulnerable to the shock anomalies even if the numerical fluxes to be regarded to be free from the carbuncle phenomenon. A suggestion to the choice of the wave speed is proposed when calculating strong shock wave problems.

1. Introduction

Shock-capturing Godunov schemes have become a mainstay in aerospace calculations and also in a range of other fields such as astrophysics, space physics, fusion and several others. This is due to their low numerical dissipation, their high level of robustness, and their ability to exactly capture discontinuities. However, such upwind Riemann solvers have their own peculiar flaws. One of the well-known flaws is the carbuncle phenomenon.

The carbuncle phenomenon is a numerical instability that affects the numerical capturing of shock waves. It was first reported by Peery and Imlay [11] when they calculated a high speed flow interacting with a blunt body. Quirk [12] found that the carbuncle instability was also related to the even-odd decoupling problem when he simulated planar moving shocks. Since the discovery of this pathology, one may associate the carbuncle instability with several other instabilities involving the propagation of shocks on a multidimensional computational mesh. For example, a shock moving on a multidimensional mesh with skewed mesh lines [3] or uneven mesh spacing [13] [20] will also show deficiencies. In all such circumstances, overcoming the carbuncle instability also contributes to overcome all these other deficiencies associated with strong shocks moving on multi-dimensional meshes.

In the procedure of seeking the cause of shock instability, the first stability analysis was offered by Quirk [12] who realized that the instability was triggered by an exact representation of the linearly degenerate intermediate waves. Riemann solvers like the HLLE solver, which does not resolve intermediate waves, are always stable, while low dissipative Riemann solvers like the Roe-type solver were unstable unless a large dose of dissipation was introduced by using Harten's fix [7]. Quirk constructed a hybrid method to cure shock instability. It switches to a more dissipative Riemann solver in the vicinity of shock wave to efficiently dissipate intermediate

IOP Conf. Series: Journal of Physics: Conf. Series **1290** (2019) 012026 doi:10.1088/1742-6596/1290/1/012026

waves, and restores low dissipative scheme, like Roe Riemann solver, on other domains. Such healing has a hypotheses that the high dissipative scheme is always stable in calculations.

Sanders et al.[14] noticed that the dimension by dimension extension of one-dimensional upwind schemes to the multidimensional equations of gas dynamics often leads to poor results when computing strong shocks. They showed that this failure is an instability which is the result of inadequate crossflow dissipation implied by strictly upwind schemes. Further studies show that the carbuncle phenomenon mainly arises from the perpendicular cell face of shock wave [9],[15] [19]. Adding appropriate amount of dissipation on these faces is conductive to damp out the carbuncle phenomenon. Related numerical algorithms involve the modification of linear shear wave. Two concrete Riemann solvers are the HLLCM scheme [16] and HLLEC scheme [19]. They originate from the unstable HLLC [18] and the HLLEM flux functions [5], but illustrate very good stability since shear waves in original schemes are smeared or deleted.

Only a few works realized that the normal dissipation is also source of instability. Kitamura et al. [8] made some comparisons to a slew of Riemann solvers and found that "there are at least two kinds of shock instabilities at work: one is one-dimensional (1-D) mode and the other is multidimensional mode". If a scheme is unstable in 1-D, then it remains unstable in two dimensional calculation. However, there are also some puzzling behaviors for the above conclusion. Actually, some schemes in [8] (EC-Roe(α = 0.2),AUSMPW+, RoeM2) are 1D unstable but 2D stable (see Table 2 in [8]). In [19], the HLLEC flux has similar performance, see Table 3 in [19]. Therefore corresponding connection between 1D and 2D instability is not clear.

In this paper, we discuss the stability of HLLE-type numerical fluxes, including HLLC, HLLCM, HLLEM and HLLEC. One of the common characters of these numerical fluxes is that the right-going and left-going wave velocities need to be given artificially. There are many methods to estimate the speeds. The most popular approach is to estimate them directly. Davis [2] proposed an exceedingly simple algorithm by comparing the maximum and minimum characteristic velocities of both states. He further suggested that one could estimate the shock velocities from an intermediate state based on the Roe average. This was implemented as a wavespeed algorithm due to Einfeldt et al.'s work[5]. For the HLLC flux, Batten et al. [1] showed that Einfeldt et al.'s estimates can resolve contact waves exactly and guarantee the method is positively conservative. Especially, the algorithm "has proved extremely robust and can yields the exact velocity for isolated shocks", and thus is recommended for practical computations [1]. The classical HLLEM adopts Einfeldt et al.'s wave algorithm naturally, and it also possesses above discontinuity resolving properties after selecting the appropriate contact velocity, see [10].

However, the HLLC and HLLEM Riemann solvers were found to be unstable to the carbuncle instability due to their full family of waves construction. While the HLLCM flux [16] is a modification of the HLLC flux and the HLLEC [19] is one of the HLLEM flux. The main modification focus on smearing or deleting the linear shear wave. According to the analysis in [16] and [19], the HLLCM method and HLLEC method do not suffer from carbuncle instability. In particular, in almost all literatures, the HLLE scheme is believed to be carbuncle free.

In this paper, we will report some numerical experiments with strong steady shock waves, in which instability phenomena occur for the three so-called 'stable' numerical fluxes: the HLLE, HLLCM and HLLEC schemes equipped with Einfeldt et al.'s wavespeed algorithm. As a simple cure, Davis's wavespeed algorithm is adopted and greatly improves the stability of numerical calculations. A matrix stability analysis is put forward to validate the finding.

The outline of the paper is as follows. In Section 2, the governing equations and some Riemann solvers are introduced. An unstable phenomenon is described in Section 3, the computing effects using two different wavespeed estimates are illustrated in the section. In Section 4, a linear matrix stability analysis is proposed, and the stability of two choices is described. In particular, an extreme counterexample is given to illustrate the limitation of the two classical wavespeed algorithms. Finally, the conclusions are summarized in Section 5.

2. Governing equations and numerical methods

The governing equations for inviscid flow in two dimensions are as follows:

$$\frac{\partial \mathbf{U}}{\partial t} + \frac{\partial \mathbf{F}(\mathbf{U})}{\partial x} + \frac{\partial \mathbf{G}(\mathbf{U})}{\partial y} = 0, \tag{1}$$

where the state vector and flux vectors are

$$\mathbf{U} = \begin{bmatrix} \rho \\ \rho u \\ \rho v \\ \rho E \end{bmatrix}, \qquad \mathbf{F}(\mathbf{U}) = \begin{bmatrix} \rho u \\ \rho u^2 + p \\ \rho u v \\ \rho E u + p u \end{bmatrix}, \qquad \mathbf{G}(\mathbf{U}) = \begin{bmatrix} \rho v \\ \rho u v \\ \rho v^2 + p \\ \rho E v + p v \end{bmatrix},$$

where ρ, p, E are the fluid density, pressure and total energy respectively, and $\mathbf{u} = (u, v)$ is the fluid velocity. The equation of state is in the form

$$p = (\gamma - 1)\rho e = (\gamma - 1)\rho \left[E - \frac{1}{2}(u^2 + v^2) \right],$$

where γ is the specific heat ratio and e is the specific internal energy.

For simplicity assume that the boundaries of the computational domain are aligned with the coordinate directions x and y. Consider a typical finite volume or computational cell $\Omega_{i,j}$ of dimensions $\Delta x \times \Delta y$. To each intercell boundary there corresponds a numerical flux.

A semi-discrete finite volume method to solve (1) reads

$$\frac{d\mathbf{U}_{i,j}}{dt} = -\frac{\mathbf{H}_{i+1/2,j} - \mathbf{H}_{i-1/2,j}}{\Delta x} - \frac{\mathbf{H}_{i,j+1/2} - \mathbf{H}_{i,j-1/2}}{\Delta y}, \tag{2}$$

where the numerical fluxes $\mathbf{H}_{i+1/2,j}$ and $\mathbf{H}_{i,j+1/2}$ are approximate intercell ones of $\mathbf{F}(x_{i+1/2}, y_j)$ and $\mathbf{G}(x_i, y_{j+1/2})$ respectively. They can be obtained by many methods. In this paper, we concentrate on the HLLE-type Riemann solvers.

2.1. Some approximate Riemann solvers

When discretizing the equations (1), Riemann solver is an important ingredient in numerical methods. The HLL Riemann solver [6] is one of lots of Riemann solvers [17]. It approximates the solution of Riemann problem with two waves propagating at speeds of S_L and S_R. They are the lower and upper bounds for the physical signal speeds with which the information of the initial discontinuity is transported. Given the initial conserved states $\mathbf{U}_L$ and $\mathbf{U}_R$ of a Riemann problem, the numerical flux of the HLL approximate Riemann solver can be expressed as

$$\mathbf{H}_{HLL} = \frac{S_R^+ \mathbf{F}_L - S_L^- \mathbf{F}_R + S_R^+ S_L^- (\mathbf{U}_R - \mathbf{U}_L)}{S_R^+ - S_L^-}, \tag{3}$$

with $S_L^- = \min(S_L, 0)$ and $S_R^+ = \max(S_R, 0)$. $\mathbf{F}_L$ and $\mathbf{F}_R$ are fluxes on the left and right sides.

To satisfy the entropy and the positivity conditions, Einfeldt et al.[5] suggested adequate bounds by making use of the Roe-averaged eigenvalues,

$$S_L = \min(u_L - a_L, \hat{u} - \hat{a}), \quad S_R = \max(u_R + a_R, \hat{u} + \hat{a}), \tag{4}$$

where a_L and a_R are the sound speeds of the left and the right states. The superscript $\hat{}$ denotes the Roe-averaged values throughout this paper. With the wavespeed estimates, the HLL flux is also called HLLE scheme. It needs to point out that the HLLE flux has too much dissipation and has difficulty to simulate practical problems. Many modifications have done to retain the shear wave and entropy wave.

2.2. HLLEM and HLLEC schemes

The HLLEM scheme [5] is an modified scheme of the HLLE flux, which improves the resolution of contact discontinuity by reusing the information of contact discontinuity in terms of modifying the intermediate state. The numerical flux function of the HLLEM scheme can be written by

$$\mathbf{H}_{HLLEM} = \mathbf{H}_{HLLE} - \frac{S_R^+ S_L^-}{S_R^+ - S_L^-}(\delta_2 \hat{\alpha}_2 \hat{\mathbf{r}}_2 + \delta_3 \hat{\alpha}_3 \hat{\mathbf{r}}_3), \tag{5}$$

where $\hat{\mathbf{r}}_k$ correspond the right eigenvectors of the flux Jacobian $\hat{\mathbf{A}}$ evaluated at the intermediate states $\hat{\mathbf{U}}$. The wave strengths $\hat{\alpha}_k$ are the approximate values of the projection from $\mathbf{U}_R - \mathbf{U}_L$ onto $\hat{\mathbf{r}}_k$, i.e.,

$$\mathbf{U}_R - \mathbf{U}_L = \sum_{k=1}^{4} \hat{\alpha}_k \hat{\mathbf{r}}_k.$$

δ_k are the anti-diffusion coefficients defined by Park et al. [10],

$$\delta_2 = \delta_3 = \frac{\hat{a}}{|\hat{u}| + \hat{a}}, \tag{6}$$

which has resolve the stationary contact discontinuity exactly.

However, two main problems have been reported for this scheme. One is multidimensional shock instability and another is nonexistence of strong receding flows. The HLLEC flux is a modification of the HLLEM one proposed by Xie et al. [19],

$$\mathbf{H}_{HLLEC} = \mathbf{H}_{HLLE} - \frac{S_R^+ S_L^-}{S_R^+ - S_L^-} \delta_2 \hat{\alpha}_2 \hat{\mathbf{r}}_2. \tag{7}$$

In this formula, the shear wave $\delta_3 = 0$ has been dropped off but the contact wave keeps unchanged. In [19], the HLLEC flux is regarded as free from shock instability.

2.3. HLLC-type schemes

The HLLC method [18] is considered to be the most preferred finite difference method because of its simplicity and ability to capture discontinuities accurately. In the Riemann solver, four constant states, which can be denoted from left to right $\mathbf{U}_L, \mathbf{U}_L^*, \mathbf{U}_R^*, \mathbf{U}_R$, are separated by signal velocities S_L and S_R and contact velocity S_*. The flux function can be expressed by

$$\mathbf{H}_{HLLC} = \begin{cases} \mathbf{F}_L, & \text{if} \quad S_L \geq 0, \\ \mathbf{F}_L + S_L(\mathbf{U}^*_{HLLC,L} - \mathbf{U}_L), & \text{if} \quad S_L < 0 \leq S_*, \\ \mathbf{F}_R + S_R(\mathbf{U}^*_{HLLC,R} - \mathbf{U}_R), & \text{if} \quad S_* < 0 < S_R, \\ \mathbf{F}_R, & \text{if} \quad S_R \leq 0, \end{cases} \tag{8}$$

where the conserved vectors in the star region are

$$\mathbf{U}^*_{HLLC,K} = \rho_K^* \begin{bmatrix} 1 \\ S_* \\ v_K \\ E_K^* \end{bmatrix} = \rho_K^* \begin{bmatrix} 1 \\ S_* \\ v_K \\ E_K^{*,1d} + \frac{1}{2} v_K^2 \end{bmatrix}, \quad \text{for} \quad K = L, R, \tag{9}$$

in which the total energy is decomposed into sums of a pure one-dimensional normal quantity $E_K^{*,1d}$ and a tangent kinetic energy $v_K^2/2$. Introduce following notations

$$\alpha_L = \rho_L(S_L - u_L), \quad \alpha_R = \rho_R(S_R - u_R),$$

then the contact velocity, density and total energy in the star region can be written as

$$S_* = \frac{\alpha_R u_R - \alpha_L u_L + p_L - p_R}{\alpha_R - \alpha_L}, \tag{10}$$

$$\rho_K^* = \frac{\alpha_K}{S_K - S_*}, \quad E_K^{*,1d} = e_K + (S_* - u_K)\left(S_* + \frac{p_K}{\alpha_K}\right), \quad K = L, R. \tag{11}$$

In order to enhance the stability, authors in [16] introduce the shear viscosity to the HLLC scheme. The modified algorithm is called the HLLCM Riemann solver and has similar flux form with (9),

$$\mathbf{H}_{HLLCM} = \begin{cases} \mathbf{F}_L, & \text{if} \quad S_L \geq 0, \\ \mathbf{F}_L + S_L(\mathbf{U}^*_{HLLCM,L} - \mathbf{U}_L), & \text{if} \quad S_L < 0 \leq S_*, \\ \mathbf{F}_R + S_R(\mathbf{U}^*_{HLLCM,R} - \mathbf{U}_R), & \text{if} \quad S_* < 0 < S_R, \\ \mathbf{F}_R, & \text{if} \quad S_R \leq 0. \end{cases} \tag{12}$$

Here, the conserved vectors in star states are

$$\mathbf{U}^*_{HLLCM,K} = \rho_K^* \begin{bmatrix} 1 \\ S_* \\ v^* \\ E^*_{HLLCM,K} \end{bmatrix} = \rho_K^* \begin{bmatrix} 1 \\ S_* \\ \frac{\alpha_R v_R - \alpha_L v_L}{\alpha_R - \alpha_L} \\ E_K^{*,1d} + \frac{1}{2}\frac{\alpha_R v_R^2 - \alpha_L v_L^2}{\alpha_R - \alpha_L} \end{bmatrix}, \quad K = L, R. \tag{13}$$

The main difference between the HLLC flux and the HLLCM flux is in the tangent velocity. The change of total energy is along with the variation of tangent velocity.

3. An instability phenomenon and cure

Shock instability of a numerical scheme means the shock front might be destroyed by a very slight perturbation. Such perturbation might come from grid, initial values, boundary conditions, even just rounding errors. In what follows, we use numerical experiments to illustrate some solution behaviors of these HLLE-type schemes.

The raw problem is a steady normal shock wave. The initial data are given by the exact Rankine-Hugoniot solution in x-direction. The upstream and downstream states are

$$\mathbf{W}_L = (\rho_L, u_L, v_L, p_L) = (1, \sqrt{\gamma}M_a, 0, 1), \quad x < 0.1, \tag{14}$$

and the downstream state is

$$\mathbf{W}_R = (\rho_R, u_R, v_R, p_R) = \left(f(M_a), \frac{\sqrt{\gamma}M_a}{f(M_a)}, 0, g(M_a)\right), x > 0.1, \tag{15}$$

where $M_a = u_L/a_L$ is the upstream Mach number of the incoming flow,

$$f(M_a) = \left(\frac{2}{\gamma+1}M_a^2 + \frac{\gamma-1}{\gamma+1}\right)^{-1}, \quad g(M_a) = \frac{2\gamma}{\gamma+1}M_a^2 - \frac{\gamma-1}{\gamma+1}.$$

The computational domain is on a rectangular one of $[0, 0.2] \times [0, 1]$ in the x-y plane, and the mesh has 20×100 uniform cells. The specific heat ratio $\gamma = 1.4$, Mach number $M_a = 20$. The signal velocity estimates adopt Einfeldt et al.'s method (4). A slight random perturbation of the relative order 10^{-15} is added to the downstream states. The upper and bottom boundary conditions are the outflow. The boundary conditions of the left and right side are the inflow and outflow respectively.

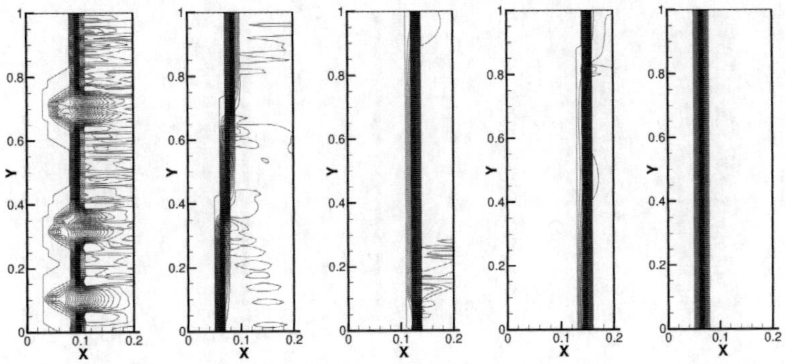

Figure 1. The density contours for 2D steady flow with Einfeldt et al.'s wave velocity estimates. The grid resolution is 20×100 cells. Thirty-one equally spaced contour lines from $\rho = 1$ to $\rho = 6$. From left to right: HLLEM, HLLC, HLLEC, HLLCM, HLLE. All of them are not stable.

For this one-dimensional shock problem, the shock wave front should remain stationary. The density contours using the schemes of HLLEM, HLLC, HLLEC, HLLCM and HLLE are displayed in Fig. 1. The shock positions in all schemes move and thus deviate from the correct steady state solution. Some move to the left and some to the right. Notice that this is a multi-dimensional phenomenon since such moving never been observed in a pure one-dimensional calculation(grid number is 1 in the y-direction). The carbuncle phenomenon has caused the normal shock to form a bulge which is unphysical.

Such solution behavior is a little surprising. It is well known that the HLLEM and HLLC flux are both able to exactly capture the contact and shear discontinuities, and thus they are susceptible to the carbuncle problem. But for the HLLEC, HLLCM and HLLE fluxes, the shear viscosities have been manipulated to the largest extend. These shear viscosities provide a kind of multi-dimensional dissipative mechanism on the face perpendicular to the shock, and they should suppress the drawbacks.

One knows that a good aspect ratio of the grid can alleviate the instability. If we add more cells in a direction normal to the shock, the anomalies should be cured partly. Fig. 2 illustrate the contour plots of density using the mesh with 40×100 cells. In this experiment, the first two fluxes still suffer from the carbuncle phenomena at once. For other schemes, the shock resolutions are enhanced and the sawtooth forms of the unstable mode disappear. However, the fluxes except the HLLE one still yield wrong shock velocity. For the HLLE flux, if we increase upstream mach number, for example, $M_a = 40$, the unstable phenomenon will appear again (not shown here). This numerical experiment shows that the refinement of the grid indeed plays an important role in alleviating numerical oscillation but does not cure the problem completely.

Clearly, the numerical experiments valid that there are at least two kinds of shock unstable mechanism in above numerical fluxes [8]. The primary one is due to lack of multi-dimensional dissipation in the shear direction of the shock. It can lead to the calculation collapse at once, just like the case in the HLLEM and HLLC schemes. The secondary might be from a kind of normal direction unstable mechanism of shock. We do not call it one dimensional unstable mode since such instability never been observed in pure one dimensional calculation. Such instability can be seen only when the multi-dimensional dissipation has been built in numerical flux, and results in wrong shock location, as shown in the fluxes of the HLLEC, HLLCM and HLLE.

In order to cure the second instability, we need add dissipation in one dimensional flux

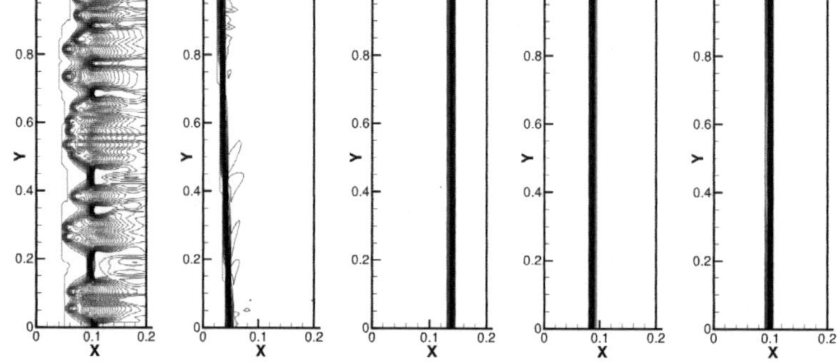

Figure 2. The density contours for 2D steady flow with Einfeldt et al.'s wave velocity estimates. The grid resolution is 40×100 cells. Thirty-one equally spaced contour lines from $\rho = 1$ to $\rho = 6$. From left to right: HLLEM, HLLC, HLLEC, HLLCM, HLLE.

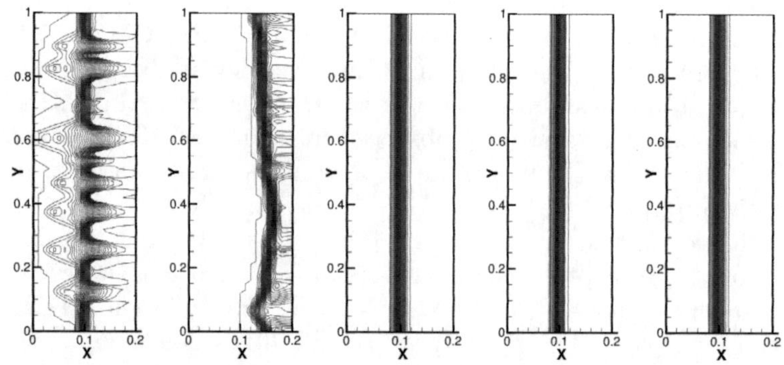

Figure 3. The density contours for 2D steady flow with Davis' wave velocity estimates. The grid resolution is 20×100 cells. Thirty-one equally spaced contour lines from $\rho = 1$ to $\rho = 6$. From left to right: HLLEM, HLLC, HLLEC, HLLCM, HLL(D). The last three fluxes are stable.

function. Here we adjust the signal velocities of the HLLC-type schemes.

Davis's wavespeeds algorithm [2] is expressed as

$$S_L = \min(u_L - a_L, u_R - a_R), \quad S_R = \max(u_L + c_L, u_R + c_R). \tag{16}$$

In general, the estimates are more diffusive than (4) since they can not distinguish an isolated shock.

The numerical results using Davis's estimates (16) are displayed in Fig. 3. The HLLC flux and HLLEM flux (we still use the same notations although different wavespeed estimates are adopted) suffer from catastrophic instability once more, while anomalous phenomena in the other fluxes disappear.

From above numerical experiments, we can draw some conclusions: the viscosity on the perpendicular cell face of shock wave is not high enough to damp out normal direction numerical instabilities. The grid refinement can modify bulge phenomenon but does not solve instability problem. A wavespeed algorithm which can crisply resolve shock seems to be vulnerable to the shock anomalies, and Davis's wave estimates have better performance in restraining instability.

4. A Linear Stability Analysis

In order to closer investigate the situation of shock instability, a kind of stability analysis method with shock structure [4] is used to illustrates stability mechanism of numerical scheme.

4.1. Matrix Stability Analysis

A matrix-based stability analysis [4] has been used to study the occurrence of unstable modes during the shock wave computation. Here we describe briefly the approach in [4] [15] for the convenience of discussion.

Similar to numerical experiments in preceding section. Calculations are performed on a 2D domain $[0,1] \times [0,1]$. The grid is composed of regular Cartesian cells without perturbation. The total cell number is $M = I \times J$.

For Einfeldt et al.'s algorithm, the initial steady shock states are (14) and (15) but with an equilibrium intermediate state,

$$\mathbf{W}_{i,j}^0 = \begin{cases} \mathbf{W}_L, & i < I/2, \quad j = 1, \cdots, J, \\ \mathbf{W}_m, & i = I/2, \quad j = 1, \cdots, J, \\ \mathbf{W}_R, & i > I/2, \quad j = 1, \cdots, J, \end{cases} \tag{17}$$

where the intermediate state $\mathbf{W}_m$ locates on the Hugoniot curve based on the downstream state. This means

$$\begin{aligned} \rho_m &= \alpha_\rho \rho_L + (1-\alpha_\rho)\rho_R, \quad p_m = \frac{(\gamma+1)\rho_m - (\gamma-1)\rho_R}{(\gamma+1)\rho_R - (\gamma-1)\rho_m} p_R, \\ u_m &= u_R + (p_R - p_m)\left[\frac{2/\rho_R}{(\gamma-1)p_R + (\gamma+1)p_m}\right]^{1/2}, \quad v_m = 0, \end{aligned}$$

where $0 < \alpha_\rho \leq 1$.

For the stability analysis of a steady field, we assume that

$$\mathbf{U}_{i,j} = \mathbf{U}_{i,j}^0 + \delta \mathbf{U}_{i,j}, \tag{18}$$

where $\mathbf{U}_{i,j}^0$ is the conserved vector of $\mathbf{W}_{i,j}^0$ and $\delta\mathbf{U}_{i,j}$ is a small perturbation vector.

Substituting (18) into (2), and the numerical flux functions are linearized around the steady mean value, then we finally get the error evolution of all M cells in the domain,

$$\frac{d}{dt}\begin{pmatrix} \delta\mathbf{U}_{1,1} \\ \vdots \\ \delta\mathbf{U}_{I,J} \end{pmatrix} = S \begin{pmatrix} \delta\mathbf{U}_{1,1} \\ \vdots \\ \delta\mathbf{U}_{I,J} \end{pmatrix}, \tag{19}$$

where S is the stability matrix. Details can be referred to [4] and [15]. When considering only the evolution of initial errors, the solution of the linear time invariant system (19) is

$$\begin{pmatrix} \delta\mathbf{U}_{1,1} \\ \vdots \\ \delta\mathbf{U}_{I,J}^0 \end{pmatrix} = \exp^{St} \begin{pmatrix} \delta\mathbf{U}_{1,1}^0 \\ \vdots \\ \delta\mathbf{U}_{I,J}^0 \end{pmatrix}. \tag{20}$$

The perturbation remains bounded only if the maximum of the real part of the eigenvalues of S is negative.

$$\max(Re(\lambda))) \leq 0. \tag{21}$$

IOP Conf. Series: Journal of Physics: Conf. Series **1290** (2019) 012026 doi:10.1088/1742-6596/1290/1/012026

Remark 1. Different from Einfeldt et al.'s wave algorithm, Davis's wave speed estimates can not provide a single point equilibrium state to form a stationary shock wave. We obtain the stationary viscous profile $\mathbf{U}_{i,j}^0$ by a pure one-dimensional numerical calculation, whose initial value still adopts $\mathbf{W}_{i,j}^0$ in (17).

4.2. Analysis results

After obtaining numerical steady shock, we may carry out stability analysis for all the HLLE-type Riemann solvers. At first, we consider Einfeldt et al.'s algorithm.

By varying a parameter α_ρ, we check the stability of these solutions to perturbations. Just like pointed out by Dumbser et al. in [4], when $\rho_m \to \rho_L(\alpha_\rho \to 1)$, instability might occur. We represent stability domain of a flux function as $(0, \alpha_{\max})$ by the parameter α_ρ. The upper bounds $\alpha_{\max}$ of stability region depend on different numerical flux functions and strong shocks. They are listed in the Table 1.

Table 1. The upper bounds $\alpha_{\max}$ of stability region for Einfeldt et al.'s algorithm.

Mach	HLLEM	HLLC	HLLEC	HLLCM	HLLE
10	0.382	0.382	0.816	0.816	0.935
20	0.375	0.375	0.810	0.810	0.928
30	0.374	0.374	0.809	0.809	0.927
...					
100	0.373	0.373	0.808	0.808	0.926

Table 1 indicates clearly all flux formulae have unstable regions to perturbations, including the HLLE scheme. The upper bounds $\alpha_{\max}$ of the HLLEM flux are identical to those of the HLLC flux for same shock strength, and $\alpha_{\max}$ of the HLLEC scheme are the same as the HLLCM one. The ranking order is listed according to linear stability,

$$\text{HLLEM} = \text{HLLC} < \text{HLLEC} = \text{HLLCM} < \text{HLLE}.$$

With the increase of shock strength, the values of $\alpha_{\max}$ decrease, and this means the unstable region increases. A surprising finding is that the upper bounds have minor variations with the increase of upstream Mach number. We conjecture that there exists a common critical density jump to trigger local instabilities for different shock strengths.

Secondly, we implement a stability analysis for the HLLEC and HLLCM fluxes with initial parameter $\alpha_\rho = 0.81$. The maximal real parts of eigenvalues of stability matrix of the two fluxes (vs. upstream Mach number) are displayed in Fig. 4. The analysis shows that Davis's wavespeed estimates are stable for any strength of shock wave while Einfeldt et al.'s wavespeed estimates begin to destabilize with the increase of the upstream Mach number. Hence Davis's algorithm has better stability than Einfeldt et al.'s algorithm.

More analysis have been preformed but do not show here. For example, we employ Davis's wavespeed algorithm in x-direction flux and Einfeldt et al.'s in y-direction flux, or alternately, use Einfeldt et al.'s wavespeed algorithm in x-direction flux and Davis's in y-direction flux, then the stability of the hybrid fluxes are always the same as the single flux equipped with x-direction wavespeed algorithm. These results validate that the main factor to influence stability of the HLLEC and HLLCM fluxes comes from the nonlinear acoustic wave in normal direction, rather than comes from the insufficient viscosity in linear wave of transverse direction.

The mechanism why Davis's algorithm can provide more stable calculation than Einfeldt et al.'s algorithm is still not clear, but we try to understand it. From the first analysis, we know that if the state of the intermediate point ρ_m is sufficiently close to the downstream state, then

IOP Conf. Series: Journal of Physics: Conf. Series **1290** (2019) 012026 doi:10.1088/1742-6596/1290/1/012026

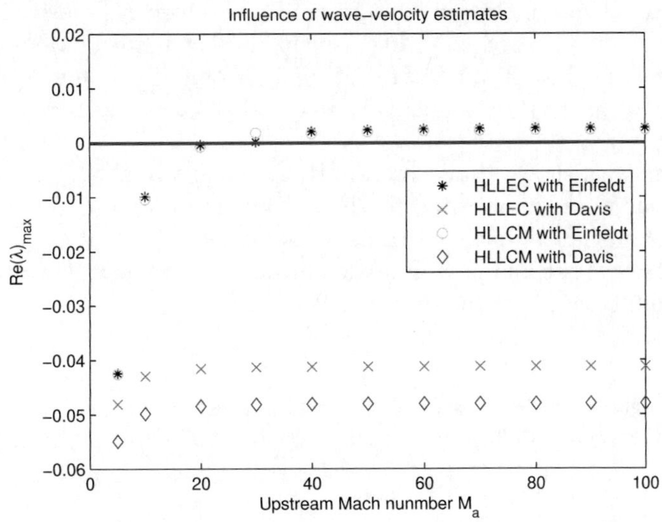

Figure 4. Maximal real part of eigenvalues of stability matrix vs. Mach number.

all Riemann solvers will be carbuncle free no matter how high the upstream Mach number. From the second analysis and numerical experiments, Davis's algorithm forces the intermediate point to be close to the downstream state, or deceases the density jump in the shock layer. The discussion of the concrete amount of the density jump in Davis's algorithm is beyond the scope of the article.

A natural question arises here: is Davis's wavespeed algorithm applicable for all strong shock waves? The reply depends on degree of density jump in a numerical shock layer. In an extreme case, the answer is no. In fact, Davis's wave speed estimates are not lower and upper bounds for the physical signal velocity, therefore it is possible to fail in shock calculations. It is worthwhile pointing out that Einfeldt et al. [5] stated incorrectly that (4) are lower and upper bounds for the physical signal velocity. Here we give an concrete case to demonstrate the limitation of two algorithms. Both of Einfeldt et al.'s algorithm and Davis's algorithm might underestimate the shock velocity.

Considering a Riemann problem with the following initial values,

$$\mathbf{W}_L = (\rho_L, u_L, v_L, p_L, \gamma_L) = (1, 1, 0, 0, 3),$$
$$\mathbf{W}_R = (\rho_R, u_R, v_R, p_R, \gamma_R) = (1, -1, 0, 0, 3).$$

In this collision problem, the shock strength is infinite and the shock speed is 2. Note that the initial sound speed (in pre-shock) is 0, thus $S_L = \min(u_L, u_R) = -1$, $S_R = \max(u_L, u_R) = 1$ according to Davis's wavespeed algorithm. In addition, Roe average states $\hat{u} = 0, \hat{c} = 0$, thus S_L an S_R in Einfeldt et al.'s estimates are identical with Davis's estimates, which are smaller than the physical shock wave velocity.

Except such extreme cases, Davis's choice is rather robust in practice. We recommend this method in simulations to strong shock wave problems.

5. Conclusions

In this paper, we discuss the stability of some HLLE-type schemes when calculating strong shock wave problems. In particularly, we pay attention to those schemes which claimed to avoid carbuncle phenomenon. The conclusion about "only the eigenvalue associated to the linear vorticity mode is responsible for the instability" seems not correct [4]. All HLLE-type Riemann solvers, including those are believed to be free carbuncle anomalies, suffer from severe instability. Nonlinear acoustic waves rather than linear vorticity and shear waves play an important role to damp out instability.

A linear stability analysis shows that Einfeldt et al.'s algorithm, which is able to capture an isolated shock exactly, is more susceptible to the carbuncle problem. While Davis's estimates have more dissipation and hence may overcome such numerical instability. We recommend to use the simple David's wave estimates when calculating strong shock wave problems, although the algorithm has its limitation in some extreme situations.

Acknowledgments
This project was supported by the National Natural Science Foundation of China (U1630249, 11871112, 11471048), the Science Challenge Project (No. JCKY2016212A502) and the Foundation of LCP.

References

[1] Batten P, Clarke N, Lambert C and Causon D M 1997 *SIAM J. Sci. Comput.* **18** 1553-70
[2] Davis S F 1988 *SIAM J. Sci. Stat. Comput.* **9** 445-73
[3] Dukowicz J K, Meltz B 1992 *J. Comput. Phys.* **99** 115-34
[4] Dumbser M, Morschetta J M, Gressier J 2004 *J. Comput. Phys.* **197** 647-70
[5] Einfeldt B, Munz C D, Roe P L, Sjogreen B 1991 *J. Comput. Phys.* **92** 273-95
[6] Harten A, Lax P D and van Leer B 1983 *SIAM Rev.* **25** 35-61
[7] Harten A 1983 *J. Comput. Phys.* **49** 357-93
[8] Kitamura K, Roe P L, and Ismail F 2009 *AIAA Journal,* **47** 44-53
[9] Pandolfi M and D'Ambrosio D 2001 *J. Comput. Phys.* **166** 271-301
[10] Park S H, Kwon J H 2003 *J. Comput. Phys.* **188** 524-42
[11] Peery K M and Imlay S T 1988 *AIAA Paper* 88-2924
[12] Quirk J 1994 *Int. J. Numer. Meth. Fluid.* **18** 555-74
[13] Rider W J 2000 *J. Comput. Phys.* **162** 395-410
[14] Sanders R, Morano E and Druguet M C 1998 *J. Comput. Phys.* **145** 511-37
[15] Shen Z J, Yan W and Yuan G W 2014 *Commun. Comput. Phys.* **15** 1320-42
[16] Shen Z J, Yan W and Yuan G W 2016 *J. Comput. Phys.* **309** 185-206
[17] Toro E F 1997 *Riemann Solvers and Numerical Methods for Fluid Dynamics* (Berlin: Springer)
[18] Toro E F, Spruce M, Speares W 1994 *Shock Wave* **4** 25-34
[19] Xie W J, Li W, Li H, Tian Z. Y, Pan S 2017 *J. Comput. Phys.* **350** 607-37
[20] Xu K, Hu J 1998 *J. Comput. Phys.* **142** 412-27

IOP Conf. Series: Journal of Physics: Conf. Series **1290** (2019) 012027 doi:10.1088/1742-6596/1290/1/012027

On the geometric structures in evolutionary games on square and triangular lattices

Evgeni Burovski[1,2], Aleksandr Malyutin[1,2], Lev Shchur[1,2]

[1] National Research University Higher School of Economics, 101000 Moscow, Russia
[2] Science Center in Chernogolovka, 142432 Chernogolovka, Russia

Abstract. We study a model of a spatial evolutionary game, based on the Prisoner's dilemma for two regular arrangements of players, on a square lattice and on a triangular lattice. We analyze steady state distributions of players which evolve from irregular, random initial configurations. We find significant differences between the square and triangular lattice, and we characterize the geometric structures which emerge on the triangular lattice.

1. Introduction

Methods of statistical physics, originally developed for describing collective phenomena in macroscopic collections of atoms and molecules, find their uses in studying a broad range of phenomena, which, traditionally, would be considered out of scope. Some problems with origins in social sciences can be cast in the language of the emergence of collective behavior from elementary interactions between individual agents. [1] A prominent one among these is the so-called evolution of cooperation [2], i.e., the emergence of cooperative behavior where individual interactions favor competition.

A prototypical model where cooperation emerges from pairwise interactions is provided by the classic Prisoner's dilemma game. Its spatially extended versions, where a macroscopic number of agents repeatedly interact via the rules of the Prisoner's dilemma contain surprisingly rich variety of behaviors, including scenarios where cooperation evolves in the long-time limit [3, 4].

A very simple yet highly non-trivial model, brought forward by Nowak and May [5, 6], defines the synchronous, discrete-time, deterministic evolution of an ensemble of agents, which are arranged in a regular pattern. Despite its apparent simplicity, the model displays surprisingly rich behavior and contains a series of dynamic regimes, separated by sharp transitions.

A natural question to ask is what is the role of the geometry of the arrangement of players. Original work [5, 7], and recent efforts [8, 9] mostly concentrated on the square grid geometry, where agents are located at the sites of a square lattice in two spatial dimensions. *A priori*, it is not immediately clear whether a different local connectivity has a qualitative effect: on one hand large-scale structures should be insensitive to the local details; on the other hand there are examples (e.g., variants of percolation problem, [10]) where processes on square and triangular lattice display significant differences.

In this paper, we thus compare the dynamics of the evolutionary game on the square and triangular lattices. The paper is organized as follows. In Section 2 we describe the model and review some previously known results. Section 3 contains results of simulations and data analysis. In Section 5 we discuss results and future work.

2. Model of spatial evolution game

A prototypical model of the game theory is the so-called *Prisoner's dilemma* (PD), played by two players in discrete time steps. In each round of the game, a player uses one of two possible strategies, *cooperate*, $\mathcal{C}$, or *defect*, $\mathcal{D}$, and receives a payoff which depends on the strategies of the player and its opponent [11]. We use the following payoff structure [5, 7]: (i) If two players are $\mathcal{D}$, they receive nothing; (ii) If both players $\mathcal{C}$, each of them receives a payoff of S, which we set to $S = 1$ without loss of generality; (iii) In the interaction of $\mathcal{C}$ and $\mathcal{D}$, the $\mathcal{D}$ receives a payoff $T > S$ and the $\mathcal{C}$ receives zero. This way, the payoff structure only depends on a single parameter, the payoff parameter $b = T/S$.

For the spatial version of the game [5, 7], we consider a collection of players arranged in the vertices of a regular grid. In each round, an agent plays pairwise games with its neighbors, and the total payoff of an agent is the sum of the pair-wise payoffs. The game is globally synchronous, so that the strategy of a player is only updated after all pairwise games are finished and all payoffs for all players are known.

Once all payoffs are known, players update their strategies: each agent adopts the strategy which led to the maximum payoff *among its neighbors*. This concludes the time step, and the game repeats with the new distribution of strategies.

We consider two arrangements of players. First, we consider a square grid, where players are arranged in the sites of a square lattice. In a round, an agent plays nine games: with itself and its eight neighbors (chess king's moves). We include self-interaction for consistency with earlier work [5, 7]. Second, we consider the triangular lattice arrangement. Since the triangular lattice can be thought of as a square lattice with additional diagonal bonds (SW and NE, but not SE and NW), a round of the game has an agent interacting with its six neighbors (and itself).

The evolution of the game can be thought of as a specific kind of a cellular automation [12]. We note however that in the language of cellular automata, these games have rather large transition tables: the state of a player at time step t depends on the states of its next-to-next-nearest neighbors at time step $t-1$. This way, the transition table for the square grid game has 2^{25} rules, and 2^{19} rules for the triangular grid.

The dynamics of the game is governed by the payoff ratio, b. The discrete structure of payoffs gives rise to a series of dynamic regimes separated by transitions at discrete values of b. Given an initial configuration, the dynamics of the game is exactly identical for all values of b between two consecutive transitions.

The transition points themselves are simply rational fractions with the numerator and denominator between one and the number of pairwise games played by an agent (i.e., the number of nearest neighbors on the lattice plus one for self-interaction.) For the square lattice, the candidate fractions between $1 < b < 3$ are

$$9/8, 8/7, 7/6, 6/5, 5/4, 9/7, 4/3, 7/5, 3/2, 8/5, 5/3, 7/4, 9/5, 2, 9/4, 7/3, 5/2, 8/3.$$

On the triangular lattice, the transitions are located at

$$7/6, 6/5, 5/4, 4/3, 7/5, 3/2, 5/3, 7/4, 2, 7/3, 5/2.$$

3. Dynamic regimes and steady states of the game

We consider a finite game field of the size $L \times L$ with periodic boundary conditions in both directions for both square and triangular lattices. We start from random configuration, where at $t = 0$ an agent is $\mathcal{C}$ with probability $p_0 = 0.1$. We directly simulate the game numerically. We let the system evolve for a certain number of burn-in steps (we typically take 1000 steps for the burn-in). Fig. 1 shows the steady state average density of cooperators, f_C, (i.e., the number of cooperators divided by the total number of sites, L^2) as a function of the payoff parameter b. Transitions are clearly manifest as sharp jumps in the density of the components. To gain insight into the nature of the transitions, it is instructive to consider small isolated objects, "islands" of one strategy immersed into the sea of the other strategy.

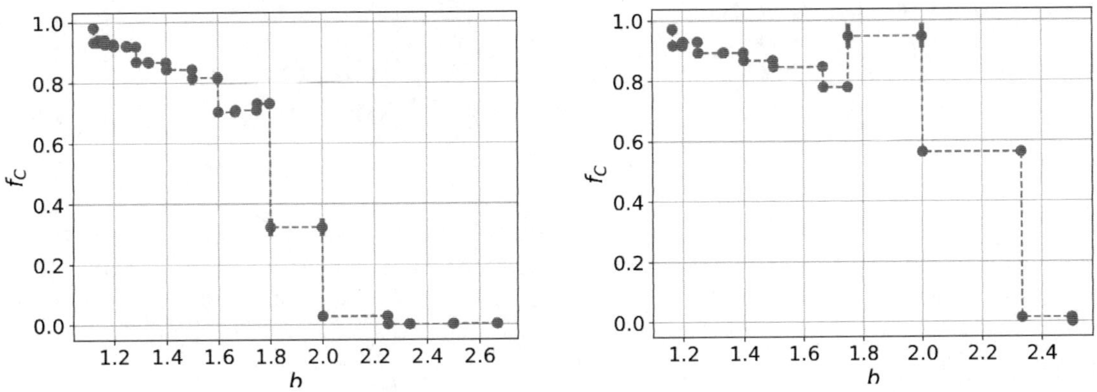

Figure 1. Concentration of cooperators $\mathcal{C}$, f_C, as a function of the payoff parameter b. **Left**: square lattice. **Right**: triangular lattice. Each point is an average over 10 measurements taken every 1000 steps after 1000 burn-in steps for 10 independent realizations of the initial conditions with the initial density of $\mathcal{C}$, $p_0 = 0.1$. Dashed lines are to guide an eye. The game field size is $L = 60$. See text for discussion.

3.1. *Square lattice*

This geometry was discussed in detail in [7], so here we only mention the key findings. For $b < 1$, cooperators always win. Defectors win unconditionally for $b > 3$. A single isolated defector is stable for $1 < b < 9/8$.

For $b > 9/8$, a variety of lattice animals is possible: there are various gliders, rotators, breathers and so on. Consider a straight $\mathcal{D}$ line one cell thick, separating two half-planes of $\mathcal{C}$. It grows in the perpendicular direction for $b > 3/2$, and becomes three sites thick at the next step. A straight line interface between $\mathcal{C}$ and $\mathcal{D}$ grows into the direction of C for $b > 3$, while cooperators invade the $\mathcal{D}$ region across the straight line interface for $b < 2$. This way, for $3/2 < b < 2$, a straight line of $\mathcal{D}$ alternates between thickness of one and three $\mathcal{D}$ with the period of 2 time steps.

A corner of $\mathcal{D}$ grows outwards into the $\mathcal{C}$ region for $b > 9/5$. Therefore, in the regime of $9/5 < b < 2$, a square 3×3 cluster of $\mathcal{D}$ grows from the corners and shrinks in the middle of each face. This way, for $9/5 < b < 2$, an isolated $\mathcal{D}$ generates a series of self-similar patterns, "evolutionary rugs", with the mean density $f_C = 12 \log 2 - 8$ [7].

3.2. *Triangular lattice*

The analysis of specific isolated objects is similar to that on the square lattice, Sec. 3.1. An isolated defector disappears after a single step for $b < 1$. For $1 < b < 7/6$ a single defector is stable. For $b > 7/6$ a single defector grows to a cluster of seven $\mathcal{D}$, see Fig. 2. Further evolution of the seven-$\mathcal{D}$ cluster depends on b: for $b < 6/3$, the cluster shrinks back to a single $\mathcal{D}$ at the center, so that we have a blinker of the size alternating between 1 and 7. For $b > 7/3$, it keeps growing, while for $2 < b < 7/3$ a seven-$\mathcal{D}$ cluster is stable.

A straight line of $\mathcal{D}$ of unit thickness disappears if $b < 5/4$. Otherwise, it remains stable for $5/4 < b < 7/4$, and grows for $b > 7/4$, see Fig. 2. The result of the latter case, a line of $\mathcal{D}$ of

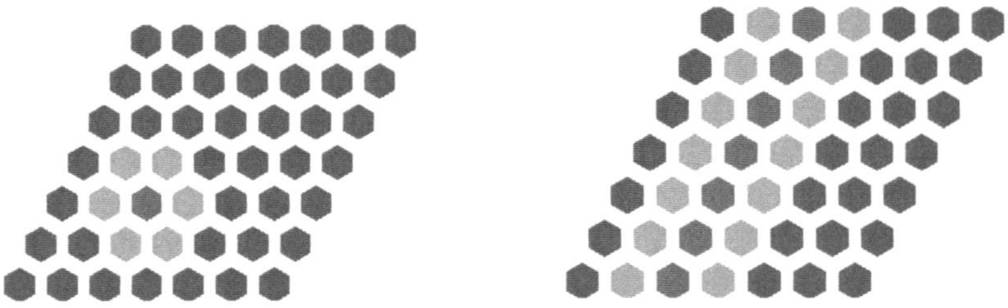

Figure 2. (Color online.) **Left:** The cluster of seven $\mathcal{D}$. In a single time step, an isolated $\mathcal{D}$ (red) grows to a cluster of seven sites (yellow) for $b > 7/6$. The cluster becomes a blinker with period of one time step for $7/6 < b < 2$, is stable for $2 < b < 7/3$ and grows without bound for $b > 7/3$. **Right:** Time evolution of a straight line of $\mathcal{D}$. See text for discussion. Color coding is consistent with Fig. 3

thickness three, remains stable for $5/2 < b < 7/2$, and shrinks back for the values of b below that range. Therefore, a straight line blinker with unit period exists in the range $7/4 < b < 5/2$.

3.3. Unstructured initial conditions: square lattice

Fig. 3 shows representative snapshots of the game field in the steady state which develops from a random initial configuration with $p_0 = 0.1$ for several values of the payoff parameter b. For $b < 9/5$, the system develops relatively static web-like structures of defectors. The most non-trivial regime is $9/5 < b < 2$. In this regime both kinds of strategies grow into regions of the opposite strategy, leading to the chaotic evolution where the state of the game changes significantly at the time scale of several time steps, and the mean density of cooperators fluctuates around the $f_C \approx 0.31$ (see Sec 3.1). [5].

3.4. Unstructured initial conditions: triangular lattice

Fig. 4 shows representative snapshots of the game field in the steady state which develops from a random initial configuration with $p_0 = 0.1$ for several values of the payoff parameter b. Typical steady state distributions of strategies strongly depend on the payoff parameter.

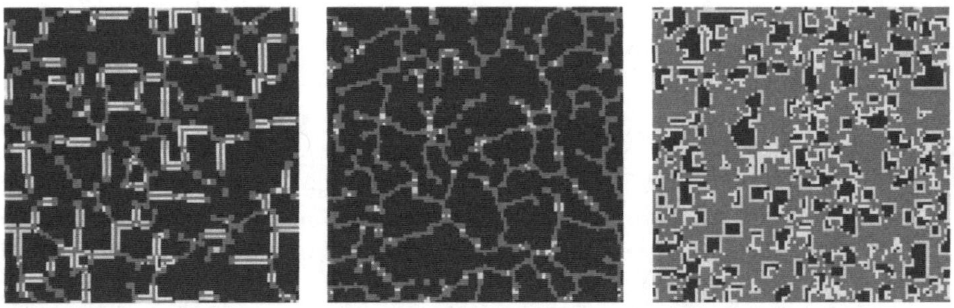

Figure 3. (Color online.) Representative snapshots of the game field for $b = 1.74$ (left), $b = 1.79$ (center) and $b = 1.81$ (right). The color coding is consistent with Ref. [5, 7]: blue is $\mathcal{C}$, red is $\mathcal{D}$, yellow is a $\mathcal{D}$ that was a $\mathcal{C}$ in the preceding round, and green is a $\mathcal{C}$ that was a $\mathcal{D}$ in the preceding round [8]. See text for discussion.

Figure 4. (Color online.) Representative snapshots of the triangular grid game field for $b \in (5/3, 7/4)$ (left), $b \in (7/4, 2)$ (center) and $b \in (2, 7/3)$ (right). Here the field size is $L = 60$. The color coding is consistent with Fig. 3. See text for discussion.

For $2 < b < 7/3$, the system develops an unstructured pattern of $\mathcal{C}$ and $\mathcal{D}$, which can be thought of as a random collection of seven-$\mathcal{D}$ clusters (cf Sec. 3.2), arranged in random ways, which are dictated by a specific realization of initial conditions. The pattern is mostly static, with small-scale oscillations at the boundaries.

Across the transition at $b = 2$, typical arrangements of the game field are very different. The field is dominated by cooperators, with relatively small isolated regions of $\mathcal{D}$ blinkers consisting of a static core $\mathcal{D}$ region, surrounded by a layer of "swingers", which frequently change their strategy.

For $5/3 < b < 7/4$, typical patters are also different. In this regime, defectors also arrange in static web-like "backbone" which is interconnected and spans the system size. (This is in contrast to regime of $7/4 < b < 2$, where $\mathcal{D}$ clusters are isolated.) Relatively small clusters of "swingers", where strategies oscillate with periods of the order of several time steps attach to this static web of $\mathcal{D}$.

4. Switching times

To further quantify the dynamics of the game, we study the mean time for for an agent to change their strategy. Specifically, we perform the following simulation: Starting from a random configuration, we evolve it for 10^4 steps (after 10^3 burn-in steps). For each player we record the time steps where they switched the strategy, and compute the mean time between switches. Results are summarized in Table 1. For each range of the payoff parameter b we show the number of players who switched their strategy at least once, n_{sw}, and the mean time between switches *for these players*, τ_{sw}.

For the square lattice, $b \in (9/5, 2)$ stands out: all L^2 players change their strategies with the time scale of around four steps. All other regimes are much more static, with a relatively small fraction of "swinging players" who change the strategy every time step on average.

On the triangular lattice, the special regime is $7/4 < b < 2$, where a sizable fraction of players engage in relatively slow processes, with time scales much larger then one. We note that the time scales are comparable to the simulation times, therefore we cannot yet rule out that what we observe is some very slow transient phenomenon. All other regimes are static, with only a small fraction of fast "swinging" players.

5. Conclusions and outlook

We study an evolutionary game based on the Prisoner's Dilemma with regular arrangements of agents in a plane. Despite its apparent simplicity, this deterministic, globally synchronous game displays surprisingly rich behavior in the long time limit. We compare the steady states of the game, for two ways of arranging the players, on the square grid and on the triangular

IOP Conf. Series: Journal of Physics: Conf. Series **1290** (2019) 012027 doi:10.1088/1742-6596/1290/1/012027

Triangular lattice			Square lattice		
b range	n_{sw}	τ_{sw}	b range	n_{sw}	τ_{sw}
5/4	92	1.0	4/3	99	1.1
4/3	12	1.0	7/5	7	1.5
7/5	153	1.0	3/2	233	1.5
3/2	64	1.1	8/5	148	1.1
5/3	57	1.7	5/3	105	1.1
7/4	348	590	7/4	13	1.0
2	35	1.3	9/5	900	4.1

Table 1. Switching times and numbers for the square and triangular lattice. The simulation is 10^4 steps on lattices with $L = 30$. n_{sw} is the number of players, which switched their strategy more then once, and τ_{sw} is the mean time between switches for these n_{sw} players. Each row shows results for the payoff parameter between the value in the row and the next row (e.g., "b range 5/4" means $5/4 < b < 4/3$). See text for discussion.

grid. The local structure of connections between neighboring players significantly affects the possible steady states: for the square grid, there exists a chaotic regime; on the triangular lattice, steady states are mostly static. However, these static patterns are significantly different for different values of the payoff parameter. It would be interesting further characterize these emergent geometric structures and their oscillatory patterns of evolution, which we intend to do in a future study.

Acknowledgments

This work was partially supported by RFBR grant 16-07-01122 (development of the simulation algorithms). A.M. acknowledges the support of the Academic Fund Program at the National Research University Higher School of Economics (HSE) in 2018-2019 (grant No 18-05-0024) and by the Russian Academic Excellence Project "5-100".

References

[1] M. Perc, J.J. Jordan, D.G. Rand, Z. Wang, S. Boccaletti, A. Szolnoki, *Statistical physics of human cooperation*, Phys. Rep. **687**, 1 (2017), and references therein.
[2] see, *e.g.*, R. Axelrod, *The Evolution of Cooperation*, Basic Books, 2006, and references therein.
[3] J. Maynard Smith, *Evolution and the Theory of Games*, Cambridge University Press, (1982).
[4] J.W. Weibull, *Evolutionary Game Theory*, MIT Press, (1995).
[5] M.A. Nowak and R.M. May, *Evolutionary games and spatial chaos*, Nature **359**, 826 (1992).
[6] M.A. Nowak, *Evolutionary Dynamics: Exploring the equations of life*, The Belknap Press, (2006).
[7] M.A. Nowak and R.M. May, *The spatial dilemmas of evolution*, Int. J. Bifurcation and Chaos **3**, 35 (1993).
[8] S. Kolotev, A. Malyutin, S. Krashakov, E. Burovski, and L. Shchur, *Dynamic fractals in spatial evolutionary games*, Physica A **499**, 142 (2018).
[9] S. Kolotev, A. Malyutin, E. Burovski, and L. Shchur,*Interfaces in evolutionary games*, J. Phys.: Conf. Ser. **955**, 012023 (2018).
[10] J. Adler, *Bootstrap percolation*, Physica A **171**, 453 (1991).
[11] see, *e.g.*, S. Tadelis, *Game Theory: An Introduction*, Princeton University Press, 2013, and references therein.
[12] see, *e.g.*, J.L. Shiff, *Cellular Automata: A Discrete View of the World*, Wiley-Interscience; 1 edition (January 6, 2008), and references therein.

IOP Conf. Series: Journal of Physics: Conf. Series **1290** (2019) 012028 doi:10.1088/1742-6596/1290/1/012028

Phase diagram of six-state clock model on rewired square lattices

Tasrief Surungan

Department of Physics, Hasanuddin University, Makassar, South Sulawesi, 90245, Indonesia

E-mail: tasrief@unhas.ac.id

Abstract. The six-state clock model (SSCM) on rewired square lattice is studied using Monte Carlo simulation with Wang-Landau algorithm. This is a discrete counterpart of the well-known XY model, the native host of a unique topological phase transition called Kosterlitz-Tholess (KT) transition. The model has two KT transitions, i.e., at temperature T_1 and T_2 , where $T_1 < T_2$. The first transition separates the lower temperature magnetic order and the quasi-long range order (QLRO) also known as KT phase; while the second transition separates the QLRO and the higher temperature paramagnetic phase. It has been established that the presence of KT phase is affected by the presence of randomness in the form of site and bond dilution. This intermediate phase is totally ruled out if bonds or sites of the lattice are no longer percolated. Here different type of randomness is probed, namely the rewired lattices, obtained by randomly adding one extra bond to each lattice site, and connect the site to one of its next-nearest neighbors. As a results, the average number of neighbors C increases. The increase of C affects the existence of KT phase. For each value of C, the KT temperatures, T_1 and T_2, were estimated from the plot of specific heats. Variation of KT temperatures for different values of C is observed, which is plotted with respect to each corresponding C to obtain the system phase diagram.

1. Introduction

Phase transitions are ubiquitous natural phenomena and appear in various realizations. Based on the properties of broken symmetry of the microscopic ingredient of the system, phase transitions are grouped into two main categories, i.e., the discontinous and continuous phase transitions [1]. The former type also called first order (FO) is characterized by the discontinuous change of the free energy of the system, exemplified by transition of water-ice, one of the well known PVT (Pressure-Volume-Temperature) systems [2]. In this type of transition, two phases co-exist, accompanied by the presence of latent heat. For the second type, the free energy of the system changes continuously and no co-exisiting phases observed during transition. A firm example of this is a spontaneous magnetization of the ferromagnetic material which changes from paramagnetic to an ordered magnetic phase. It occurs at Curie temperature (T_c), commonly referred to as critical point and phenomena nearby this point are referred to as critical phenomena.

The interesting aspect in studying critical phenomena is that the behavior of system (the critical properties) usually indicates universality, where different systems may exhibit identical behavior; characterized by a set of critical exponents. With universality, understanding properties of certain system enables one to comprehend other systems. Values of critical exponents are usually controlled by a limited number of variables, such as spin symmetry, coupling interaction and lattice structures. Randomness in the form of defect or impurity can be a part of these variables; therefore may change the universality class of certain system. Different set of critical exponents represent different type of universality class.

The effect of randomness, in a variety of forms such as site vacancies and diluted bonds, on phase transition has been intensively investigated. According to Harris criterion [3], randomness is relevant for the pure system experiencing second order phase transition with specific heat exponent $\alpha > 0$, and irrelevant for $\alpha < 0$; otherwise it is marginal (inconclusive). The two dimensional (2D) Ising model is a classic example of system experiencing second order phase transition with positive α. Various investigations related to the role of randomness on phase transition of this system had been reported, including the most recent studies probing the critical behaviour of the Ising model on rewired square lattice [4, 5].

Motivated by the studies above mentioned, here we investigate the phase transition of FM q-state clock model on rewired square lattices. The pure system of this model is well known to experience Kosterlitz-Thouless (KT) transition, which is a transition from a low temperature quasi long range order (QLRO) called KT phase to high temperature disordered phase. The QLRO (KT phase) which is a topological phase formed by vortex-antivortex pairs [6, 7], is the only type of order allowed by the Mermin-Hohenberg theorem for 2D systems [8]. Due to discrete symmetry of the clock model, additional KT transition occur at lower temperature, separating the ferromagnetic true long range order (TLRO) and the KT phase [9]. The number of vortex-antivortex pairs increases with the temperature until the system experiences the KT transition. The KT transition corresponds to the unbinding of vortex-antivortex pairs, which leads the system to a high-temperature disordered phase. Probing QLRO of planar spin model with randomness is an important subject. As this type of order is not a true long range order, one may question about its vulnerability against impurity. The continuous counterpart of clock model, the XY model (with dilution) found a relevance in the study of superconductivity, particularly for depicting the interaction between vortices and the spatial inhomogenity due to the impurities.

The effect of randomness in the form of site and bond dilution for system undergoing KT transition has been investigated and found that the KT phase persists so long as the lattice sites and bonds are percolated [10, 11, 12]. In this report, different type of randomness is studied, i.e., the rewired regular lattices. While number of neighbours of each site in a regular lattice is homogeneous, assigned by a definite integer called coordination number, it is in contrast for a rewired lattice where a fractional coordination number may be obtained, depending on the way of rewiring. Rewiring a regular lattices can be considered as increasing lattice spatial dimension, where several extra bonds are added to each lattice site. In the present study, only one extra bond is randomly added to each site, which connects the site to one of its diagonal neighbours. The concentration of bonds added determines the average number of bonds (ANOB). Therefore, a site (spin) may have up to seven bonds. ANOB which is assigned as C varies from 2.0 to 3.0, where $C = 2.0$ and $C = 3.0$ correspond respectively to unrewired and the fully rewired square lattice. One particular realization of these lattices is shown in Fig. 1.

This paper reports the study of six-state clock models on rewired square lattice by using Monte Carlo (MC) method with Wang-Landau algorithm. The subsequent sections of the paper is organized as follows: The model and method are explained in Section 2. Results and discussion are presented in Section 3. Section 4 is devoted to summary and concluding remarks.

2. Model and Method of Simulation

The 6-state clock model is written in the following Hamiltonian

$$H = -\sum_{\langle ij\rangle} J_{ij}\, \boldsymbol{S}_i \cdot \boldsymbol{S}_j = -\sum_{\langle ij\rangle} J_{ij} \cos(\theta_i - \theta_j), \tag{1}$$

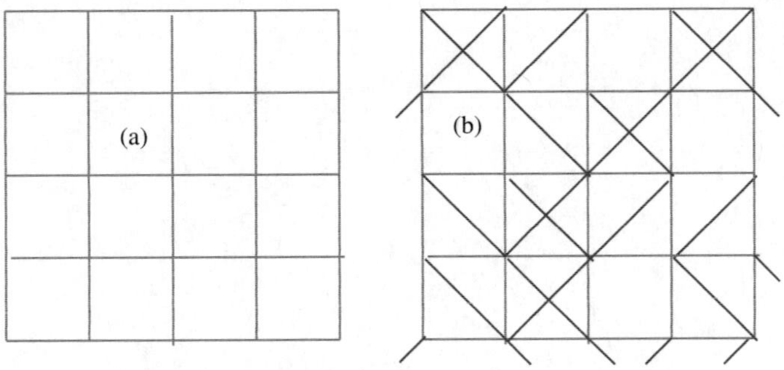

Figure 1. (Color online) (a) Unrewired square and (b) rewired square lattice where one bond is randomly added to each site.

where $\boldsymbol{S}_i$ and $\boldsymbol{S}_j$ are spins occupying respectively the site i-th and j-th of the lattice; and J_{ij} is the coupling interaction between them. For a FM six-state clock model, possible spin orientation takes discrete values, $\theta_i = 2\pi q/6$ with $q = 0, \cdots, 5$; and $J_{ij} = J < 0$. The summation is carried out over all directly connecting pairs of spins. For a regular lattice such as a square or a cubic lattice, all sites possess similar number of neighbours, corresponds to a fixed coordination number. We currently consider rewired square lattices with fractional connectivity, i.e., with ANOB C varies from 2.0 to 3.0. A rewired lattice is regarded as a quasi-regular structure as the translational symmetry is (partially) preserved, where the notion of spatial dimension remains. Our procedure of constructing the lattice is by randomly connecting each lattice site to one of its diagonal neighbors. There are 4 nearest neighbors of each site for the original square lattice while 7 neighbors in maximum for a rewired lattice. Therefore, a fractional ANOB may be obtained for a randomly rewired lattice.

Due to the complexity of the energy landscape, which may be difficult to tackle when using the the standard Metropolis MC algorithm, here we used the Wang-Landau (WL) algorithm[13]. It is an extended ensemble sampling algorithm where random walk is performed in the whole energy space. The central quantity to obtain is the energy density of states (DOS). This algorithm can accelerate the dynamics of the random walk so that the DOS $g(E)$ can be efficiently and correctly evaluated. This is done through the introduction of the transition probability defined as follows

$$P(E \to E') = \min\left[1, \frac{g(E)}{g(E')}\right] \tag{2}$$

where $g(E)$ is the DOS, which set at the beginning of the simulation to be $g(E) = 1$ for the whole states. The DOS is then evaluated through iterative random walk and its exact value is obtained when the histogram $h(E)$ of energy achieves flatness condition *i.e.* the histogram for all possible E is not less than some value of the average histogram, say, 0.80. The DOS and $h(E)$ are updated every time E is visited through the relation

$$\ln g(E_i) \to \ln g(E_i) + f_i \tag{3}$$
$$h(E_i) \to h(E_i) + 1 \tag{4}$$

where f_i is the modification parameter which is gradually reduced to zero when the DOS reaches the convergence.

It is to be noticed that using the WL algorithm, average energy as well as specific heat can be directly extracted from the DOS, both FM and AF systems. Thermal average of physical quantity Q is obtained from the standard relation

$$\langle Q \rangle_\beta = \frac{\int dE\ g(E)\ Q(E)\ \exp(-\beta E)}{\int dE\ g(E)\ \exp(-\beta E)} \tag{5}$$

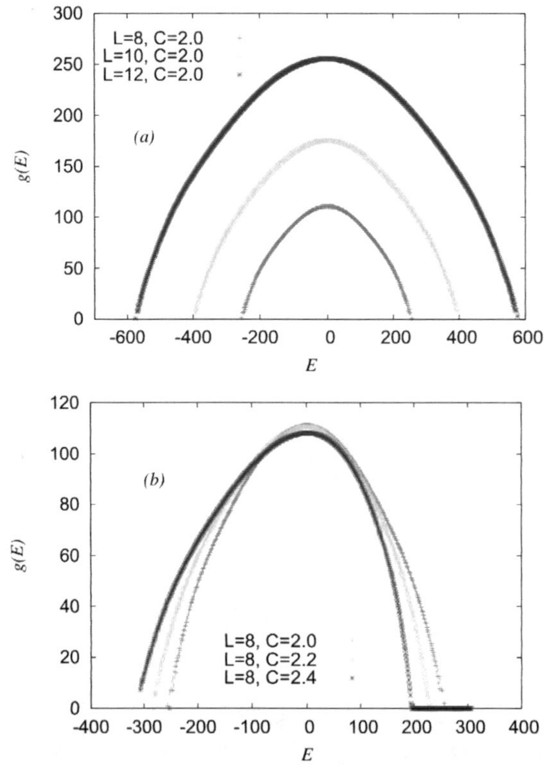

Figure 2. (Color online) Density of states of (a) pure system with several system sizes and (b) system size $L = 8$, with various bond concentrations.

where $\beta = 1/kT$, k and T respectively being the Boltzmann constant and temperature. The following is the detail steps of the algorithm:

(i) Set (a) random spin configuration and initial iteration factor $f_i = 1$; (b) Set the initial DOS and histogram as $\ln g(E_i) = 1$ and $h(E_i) = 0, \forall_i$.

(ii) Based on the transition probability (Eq. 2), update the spin configuration.

(iii) Update the DOS and the histogram as Eq. 3 and go to step (ii) if the flatness condition of the histogram is fulfilled.

(iv) Modify the iteration factor $f_i = f_i/2$, refine the DOS by using condition $\sum_{E_i} g(E_i) = 6^N$ (for the case of the six-state clock model); and reset the histogram $h(E_i) = 0$ before going back to step (ii).

(v) Do steps (i) to (iv) until the DOS is convergent, then measure physical quantity Q_i with a definite number of MCS's.

For the application of the WL algorithm to the large system size, we perform window break strategy, where the whole range of the energy is cut into several reasonable widths. The result from each window is later joined to obtain the resultant.

3. Results and Discussion

This paper studies the six-state clock model on rewired square lattices with various concentration of bonds C by using Wang-Landau algorithm of MC method. Several system sizes were

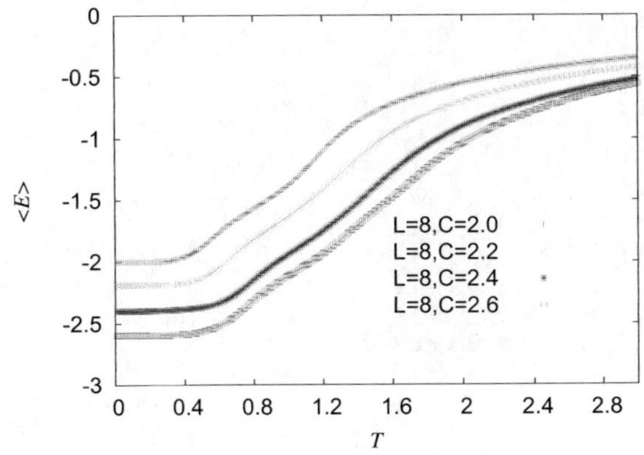

Figure 3. (Color online) Temperature dependence of energy of system size $L = 8$ for several bond concentrations. The unit of energy is dimensionless J/k_B.

simulated, namely $L = 8, 10, 12$ and 14, with $N = L^2$ is the total number of sites. As a periodic boundary condition is imposed, each lattice site, for the pure case, has equal number of nearest neighbors. For the rewired lattices, number of nearest neighbours (direct connections) of each site varies from 4 to 7. The lattice structure and the trivial ground state (GS) configuration of the system can be used to check the accuracy of calculation. For example, the density of states (DOS) for the case of pure lattice is normalized using the condition $\sum_E g(E) = 6^N$, and the degeneracy of GS energy, $g(E_{\text{GS}}) = 6$. Fig. 2: (a) shows the energy DOS of pure system for several system sizes and (b) the DOS for $L = 8$ for various values of C. Here, the energy is represented in units of J, and the GS energy is given by $-CNJ$.

According to the standard relation written in 5, the ensemble average of energy $\langle E \rangle$ can be extracted from the DOS, which is the main data obtained from the extensive simulation using WL algorithm. Combined with the second order moment of energy, $\langle E^2 \rangle$, we can define specific heat as follows

$$C_v(T) = \left[\frac{1}{NkT^2} (\langle E^2 \rangle - \langle E \rangle^2) \right]_{av} \tag{6}$$

where the energy E is in unit of J. All temperatures are expressed in unit of J/k where k is the Boltzmann constant. The square bracket symbolizes avaraging with respect to a number of diluted lattice realizations. This is a standard procedure in dealing with disorder system, where each realization of a diluted lattice with size L has diferent configuration of bond dilution. Therefore, for a reliable statistics, we average over several dilution configurations for each specified size, i.e., for $L = 8$, we take 10 realizations.

We plot the temperature dependence of energy E as shown in Fig. 3. There is a systematic increase of the absolute value of energy as concentration increases. This is due to the fact that the number of pair interactions increase, resulted from the presence of extra bonds, where each bond costs energy. The shift may signify an increase of critical temperature T_c for any existing phase transition. In diluted cases, both for the 2D Ising and clock model, T_c decreases as the number of depleted bonds increase. The difference in DOS (DDOS) is another physical quantity which can be directly extracted from DOS. This quantity can be used to analyze the order of phase transition, for example to probe the order of phase transition of q-state Potts model as value of q varies [14]. As KT transition is a special type of second order phase transition, it

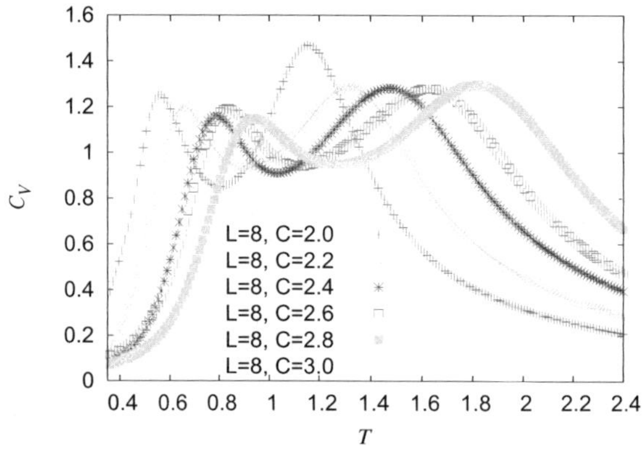

Figure 4. (Color online) Temperature dependence of specific heat of system size L=8 with various concentartion of bonds, C. The peak of the specific heat shifts to the right as bond concentration C increases. The error bar is in the order of the size of the symbol.

is interesting to study the characteristics of DDOS for q-state clock model; by which one can distinguish between the KT and the usual second order phase transition. This topic will be considered as prospective subject and be reported the result elsewhere.

A KT phase is a topological ordering in the form of vortex-antivortex pairs [6, 7]. For the continuous planar spin model (XY model), these pairs unbind at high temperature KT transition which separates the paramagnetic phase and the KT phase. This corresponds to the T_2 transition of the clock models which have another transition at lower temperature T_1. The presence of peaks in the temperature dependence of specific heat could generally signify the existence of phase transitions. As shown in Fig. 4, the specific heat plot possesses two peaks, which correspond to two KT temperature transitions, T_1 and T_2, each is asscociated with the abscissa of peak of specific heat as $L \to \infty$. For pure case, i.e., the lattice with $C = 2.0$, T_1 and T_2, are roughly estimated around 0.55(2) and 1.15(2), respectively, where number in bracket is the error bar of the last digit. The complete estimate of our critical temperatures are tabulated in Table 1. For the pure case, our rough estimates are comparable to the results reported by several previous studies [15, 16, 17], although significantly larger (for T_2 estimate) as we presently treat quite small system sizes. The temperature difference of the two peaks tends to be larger for smaller system size. For example, the estimates by Challa and Landau [15] and by Tomita and Okabe [16] are respectively $(T_1 = 0.68(2), T_2 = 0.92(1))$ and $(T_1 = 0.7014(11), T_2 = 0.9008(6))$. Nonetheless, our results have indicated the elegance of the WL algorithm applied to discrete model with multiple integer of energy.

It is to be noticed that the physical quantity extracted from the WL calculation can be presented in any temperature scale without having to perform extra MC simulation, which is different from that of obtained from such microcanonical ensemble methods such as the Metropolis algorithm. Therefore, using WL algorithm, one can obtain critical temperatures with very fine scale, which is more precise as long as the calculated DOS is truly convergent. As explained above, the KT transition temperatures presently reported are rough estimates as they are extracted from the specific heat plot. For a more precise estimate of critical temperature, one should perform finite size scaling of an appropriate order parameter, which is the magnetization for a ferromagnetic system having a TLRO. However, as the 2D XY model and its discrete counterpart q-state clock model have no TLRO (if $q > 4$), the magnetization is not a suitable

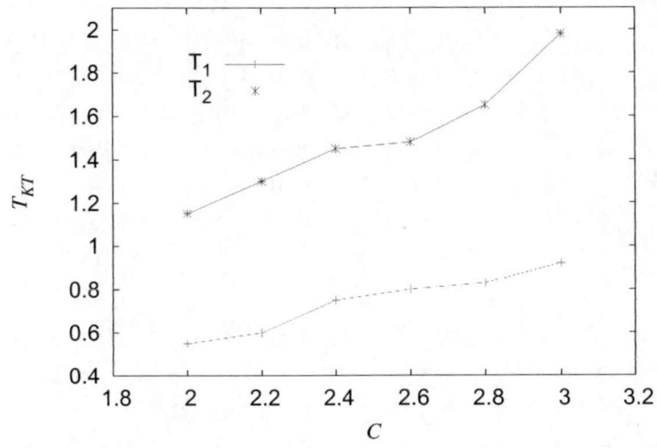

Figure 5. (Color online) Phase diagram of rewired six-state clock model. The QLRO is in between of the two solid lines. The true magnetic order is below the red line, while the paramagnetic phase is above the blue line.

Table 1. The estimate of KT temperatures T_1 and T_2 for each value of C. The number in bracket is the error bar of the last digit.

ANOB (C)	T_1	T_2
2.0	0.55(2)	1.15(2)
2.2	0.60(2)	1.30(2)
2.4	0.75(2)	1.45(2)
2.6	0.80(1)	1.48(2)
2.8	0.83(2)	1.65(2)
3.0	0.92(2)	1.98(2)

order parameter to detect the existence of KT phase (QLRO). A more precise quantity to detect the existence and analyze the KT transition is the correlation ratio, as used in previous studies [16, 17, 18]. Our calculation on correlation ratio is still in progress and the analysis of its results will be reported elsewhere.

4. Summary and Conclusion

This paper reports the study of the six-state clock model on rewired square lattices with various average number of bonds (ANOB). The lattices are obtained by randomly adding an extra bond to each site of the original square lattice, lead the rewired lattices to have fractional ANOB, symbolized as C, varied from 2.0 to 3.0. ANOB is analogous to coordination number for regular lattices. An added bond is constrained to connect one of the diagonal neighbors to the site. Rewiring the square lattice is associated with an increase in lattice dimension, which affects the existing QLRO (KT phase). Due to the increasing average number of bonds, spins become more cooperative; which may lead the system to possess a true long range order at certain value of C, in particular if we allow the rewiring not limited to nearest neighbors.

Monte Carlo method with Wang-Landau algorithm is used to study the physical properties of the system. This algorithm is considered to be very powerful in dealing with models with multiple integer of energy, such as Ising and six-state clock model. From the DOS, the energy and its second moment were extracted then used to calculate the specific heat. A clear indication of the effect of randomness on the existing KT phase is observed, which shifts the KT phase to higher temperature side. This result suggest that randomness in the form of irregular connectivity of lattices, which is weak type of randomness, does not change the universality of the system Therefore, an interesting aspect to probe is the interplay between randomness rendered by rewiring procedure and the partially frustrated state, which is the case for the anti-ferromagetic system not for FM. In addition, to find the lower dimension (smallest ANOB) and the higher dimension (largest ANOB) for the existence of KT phase on rewired lattices is still desirable. It will be our next topic to probe, for example searching for spin glass phase of AF six-state clock model on rewired lattices.

Acknowledgment
The author wishes to thank A. G. Williams, A. W. Sandvik and Bangsawang BJ for fruitful discussions. The computation of this work was carried out using parallel computing facility in the Physics Department, Hasanuddin University and the HPC facility of The Indonesian Institute of Science. The work is financially supported by KLN Research Grant, FY 2018 from Hasanuddin University.

References
[1] P. Ehrenfest and T. Ehrenfest, Phys. Z. **8**, 311 (1907).
[2] L. E. Reichl, *A Modern Course in Statistical Physics*, 2nd. John Wiley and Sons, (2009).
[3] A. B. Harris, J. Phys. C 7, 1671, (1974).
[4] N. A. Ridha, M. F. Mustamin and T. Surungan, J. Phys.: Conf. Ser. 979, 012086 (2018).
[5] T. Surungan, B.J. Bansawang , M. Yusuf, J. Phys.: Conf. Ser. 1011, 012079 (2018).
[6] J. M. Kosterlitz and D. Thouless, J. Phys. C: Solid State Phys. **6**, 1181 (1973).
[7] J. M. Kosterlitz, J. Phys. C: Solid State Phys. **7**, 1046 (1974).
[8] N. D. Mermin and H. Wagner, Phys. Rev. Lett. **17**, 1133 (1966); P.C. Hohenberg, Phys. Rev. **158**, 383 (1967).
[9] J. V. José, L. P. Kadanoff, S. Kirkpatrick, and D. R. Nelson, Phys. Rev. B **16**, 1217 (1977).
[10] S. A. Leonel *et al.*, Phys. Rev. B67, 104426 (2003).
[11] B. Berche *et al.*, Eur. Phys. J. B36, 91 (2003).
[12] T. Surungan, Y. Okabe, Phys. Rev B71, 184438, (2005).
[13] F. Wang and D.P. Landau, Phys. Rev. Lett. 86, 2050 (2001); Phys. Rev. E 64, 056101 (2001).
[14] T. Surungan, Y. Komura and Y. Okabe, AIP Conference Proceedings 1617, 79 (2014).
[15] M.S.S. Challa and D.P. Landau, Phys. Rev. B **33**, 437 (1986).
[16] Y. Tomita and Y. Okabe, Phys. Rev. B **66**, 180401(R) (2002).
[17] T. Surungan et al., Cond-mat-archive: 1901.03936, (2019).
[18] T. Surungan, Y. Okabe, and Y. Tomita, J. Phys. A: Math. Gen. **37**, 4219 (2004);
[19] T. Surungan, Bansawang BJ, D. Tahir, AIP Conf. Proc. 1719, 030006 (2016).

IOP Conf. Series: Journal of Physics: Conf. Series **1290** (2019) 012029 doi:10.1088/1742-6596/1290/1/012029

Large Deviations of Convex Hulls of the "True" Self-Avoiding Random Walk

Hendrik Schawe, Alexander K Hartmann

Institut für Physik, Universität Oldenburg, 26111 Oldenburg, Germany

E-mail: `hendrik.schawe@uni-oldenburg.de`

Abstract. We study the distribution of the area and perimeter of the convex hull of the "true" self-avoiding random walk in a plane. Using a Markov chain Monte Carlo sampling method, we obtain the distributions also in their far tails, down to probabilities like 10^{-800}. This enables us to test previous conjectures regarding the scaling of the distribution and the large-deviation rate function Φ. In previous studies, e.g., for standard random walks, the whole distribution was governed by the Flory exponent ν. We confirm this in the present study by considering expected logarithmic corrections. On the other hand, the behavior of the rate function deviates from the expected form. For this exception we give a qualitative reasoning.

1. Introduction

The random walk is a very simple model for diffusive processes with Brownian motion [1] as the prime example. Though its applications range from financial models [2] over online search engines [3] to the very sampling algorithm used in this study [4]. Its simplest variation lives on a lattice and takes steps on random adjacent sites at each timestep, which is exceptionally well researched [5]. With the further constraint that no site may be visited twice, such that the walk is *self-avoiding*, it becomes a simple model for polymers [6]. Interestingly, depending on the exact protocol how the self-avoidance is achieved, they can also be used to study the perimeter of, e.g., critical percolation clusters [7] or spanning trees [8, 9]

The distance of a random walk from its starting point is the most prominent and simple measurable quantity. Nevertheless, here we go beyond this by considering the convex hull of all T sites visited by the random walk, i.e., the smallest convex polygon containing all these sites. It can be seen as a measure of the general shape of the random walk, without exposing all details of the walk. Thus, the area A or perimeter L of the convex hull can then be used to characterize the random walk in a very simple way. This method is also used, for example, to describe the home ranges of animals [10, 11, 12], spread of animal epidemics [13] or classification of different phases using the trajectory of intermittent stochastic processes [14]. For standard random walks its mean perimeter [15] and mean area [16] in the large T limit are known exactly since a long time. More recently different approaches generalized these results to multiple random walks [17, 18] and arbitrary dimensions [19]. Even more recently the mean perimeter and area for finite (but large) walk lengths T were computed explicitly [20] if the random walk is discrete-time with jumps from an arbitrary distribution. If the distribution of the jump length is Gaussian, even an exact combinatorial formula for the mean volume in arbitrary dimensions is known [21]. For higher moments however, there is only one analytic result for the special

IOP Conf. Series: Journal of Physics: Conf. Series **1290** (2019) 012029 doi:10.1088/1742-6596/1290/1/012029

case of Brownian bridges [22], i.e., closed walks with Gaussian jumps. When asking for more, i.e., for the full distributions, no exact analytical results are available. This motivated the numerical study of the full distributions—or at least large parts of the support—using large-deviation sampling techniques to sample even far into the tails of standard random walks [23] and multiple random walks [24], also in higher dimensions [25]. These numerical studies are rather expensive, since they usually require Markov chain Monte Carlo simulations, allowing one to measure the distribution in regions where the probabilities are as small as 10^{-100}.

Since self-avoiding walks are considerably more difficult to treat analytically than standard random walks, there are no analytical results about the properties of their convex hulls yet. Therefore, the authors of this contribution very recently published a numerical study of the full distribution of perimeter and area of three different types of self-avoiding random walks [26], notably the classical *self-avoiding walk* (SAW) and the *smart kinetic self-avoiding walk* (SKSAW). While the SAW is combinatorial in nature and describes the set of all self-avoiding configurations with equal probability, the SKSAW is a growth process, which assigns more weight to some configurations. In [26] we also give an estimate for the functional form of the rate function Φ describing the far right tail behavior of the distribution, i.e., $P(S) \approx \mathrm{e}^{-T\Phi(S)}$. It was found to depend only on the dimension d and the scaling exponent ν. For two-dimensional random walks these scaling exponents are often known exactly through Schramm-Loewner evolution [27, 28, 29, 30].

In this study we test this prediction for the *"true" self-avoiding walk* (TSAW), which has a free parameter β governing how strictly self-avoiding the walk is. Introduced in [31] the TSAW was a counter model to the SAW, especially it should demonstrate that the behavior of the combinatorial SAW is very different from more natural growing random walks which avoid themselves. Indeed, in two dimensions, where the end-to-end distance r of a T step SAW scales as $r \propto T^{\nu}$ with $\nu = 3/4$, the TSAW will scale as

$$r \propto T^{\nu} (\ln T)^{\alpha} \tag{1}$$

with $\nu = 1/2$ [32] and a correction α, which is not known rigorously, but estimated as $\alpha = 1/4$ [33]. Here we show, for large-enough values of β, that in contrast to previous work [26] the rate function Φ is not simply determined by the value of ν, since the growth process of the TSAW in the large-area region of the tail is indistinguishable from the SKSAW growth process, although they have different values of the scaling exponent ν determined by the behavior of the high-probability part of their distributions.

2. Models and Methods

This section will introduce the TSAW model and the sampling method in enough detail to reproduce the results of this study. For more fundamental methods, like the construction of the convex hull, we will sketch the main ideas.

2.1. Large Deviation Sampling Scheme

To obtain good statistic in the far tail, it is not sufficient to perform naive simple sampling, since configurations of probability P would need about $1/P$ samples to occur at all. It is therefore not feasible to explore the distributions down to the tails of $P < 10^{-100}$ with simple sampling. Instead we use an importance sampling scheme to generate more samples in the low probability tails. Thus, we generate Markov chains consisting of sequences of TSAWs and use the well known Metropolis algorithm [34] with a Boltzmann sampling weight. For this purpose, we identify the quantity S we are interested in—here the area A but it could be any measurable quantity in principle—with the energy occurring in the Boltzmann factor and introduce an artificial "temperature" Θ. Since the TSAW is a growth process, it is not trivial to come up with a local change move within the Markov chain, i.e., it is difficult to change a configuration by a small amount while preserving the correct statistics. Therefore our Markov chain is not directly a chain of configurations of TSAW but rather a chain of random number vectors $\boldsymbol{\xi}_i$. Each vector $\boldsymbol{\xi}_i$ determines a configuration of a TSAW by performing the growth process and using for each of the T decisions a random number from $\boldsymbol{\xi}_i$. This approach is sketched in Fig. 1 and extremely general since it can be applied to any model [35]. A change move is a simple change of one entry of $\boldsymbol{\xi}_i$.

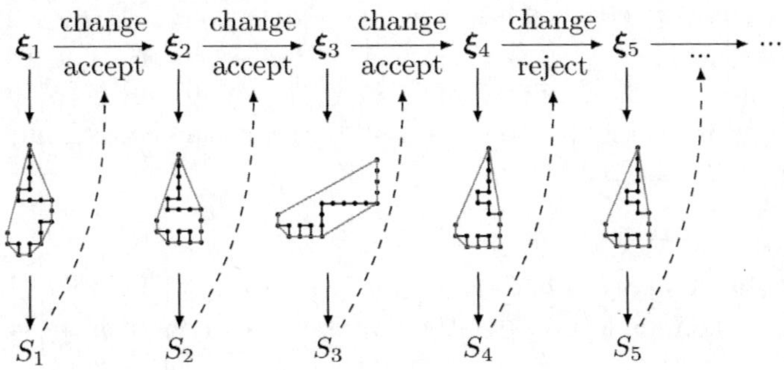

Figure 1. Sketch of the Markov chain of random number vectors $\boldsymbol{\xi}_i$. The change move is performed on the $\boldsymbol{\xi}_i$ and a new TSAW is generated from scratch, its energy difference to the previous configuration is used to accept or reject the change.

Following the Metropolis algorithm, we propose a new $\boldsymbol{\xi}'$ by replacing a random entry with a new random number $\xi \in U[0,1)$, generate a new TSAW configuration from scratch using the random numbers $\boldsymbol{\xi}'$ and calculating its energy S', i.e., its area. The proposed configuration is then accepted, i.e., $\boldsymbol{\xi}_{i+1} = \boldsymbol{\xi}'$, or rejected, i.e., $\boldsymbol{\xi}_{i+1} = \boldsymbol{\xi}_i$, depending on the temperature and energy difference with respect to the previous configuration with probability $p_{\rm acc} = {\rm e}^{-\Delta S/\Theta}$, where $\Delta S = S' - S_i$ is the energy difference caused by the change. Replacing a random entry by a new entry is clearly ergodic, since any possible $\boldsymbol{\xi}_i$ can be generated this way. Since we use the classical Metropolis acceptance probability, detailed balance is also given. This Markov process will therefore yield configurations $\boldsymbol{\xi}$ according to their equilibrium distribution $Q_\Theta(\boldsymbol{\xi}) = \frac{1}{Z(\Theta)} Q(\boldsymbol{\xi})\,{\rm e}^{-S(\boldsymbol{\xi})/\Theta}$, where $Q(\boldsymbol{\xi})$ is the natural, unbiased distribution of configurations and $Z(\Theta)$ the corresponding partition function. For small temperatures this will lead to small energies, i.e., smaller than typical perimeters or areas. For large temperatures typical configurations will be generated and for negative temperatures larger than usual energies dominate. Since this Metropolis algorithm will generate instances following a Boltzmann distribution we can easily undo this bias, i.e., we can derive the actual distribution $P(S)$ from the biased, temperature dependent distributions $P_\Theta(S)$ as

$$P_\Theta(S) = \sum_{\{\boldsymbol{\xi}|S(\boldsymbol{\xi})=S\}} Q_\Theta(\boldsymbol{\xi}) \tag{2}$$

$$= \sum_{\{\boldsymbol{\xi}|S(\boldsymbol{\xi})=S\}} \frac{\exp(-S/\Theta)}{Z(\Theta)} Q(\boldsymbol{\xi}) \tag{3}$$

$$= \frac{\exp(-S/\Theta)}{Z(\Theta)} P(S). \tag{4}$$

The unknown $Z(\Theta)$ can be numerically determined by enforcing the continuity of the distribution. Therefore we need to simulate the system at many different temperatures Θ, such

IOP Conf. Series: Journal of Physics: Conf. Series **1290** (2019) 012029 doi:10.1088/1742-6596/1290/1/012029

that all histograms $P_\Theta(S)$ overlap with adjacent temperatures. $Z(\Theta)$ can now be calculated in overlapping regions, which should coincide for continuity, i.e.,

$$e^{S/\Theta_i} Z(\Theta_i) P_{\Theta_i}(S) = e^{S/\Theta_{i+1}} Z(\Theta_{i+1}) P_{\Theta_{i+1}}(S) \tag{5}$$

$$\Rightarrow \quad \frac{Z(\Theta_i)}{Z(\Theta_{i+1})} = \exp\left(S/\Theta_{i+1} - S/\Theta_i\right) \frac{P_{\Theta_{i+1}}(S)}{P_{\Theta_i}(S)}. \tag{6}$$

This relation fixes all ratios of consecutive $Z(\Theta)$. The absolute value can be fixed by the normalization of the whole distribution.

This method is applicable to a wide range of models, and already successfully applied to obtain, e.g., the distributions over a large range for the score of sequence alignments [36, 37, 38], work distributions for non-equilibrium systems [35], properties of Erdős Rényi random graphs [39, 40, 41], and in particular to obtain statistics of the convex hulls of a wide range of types of random walks [23, 24, 26].

2.2. "True" Self-Avoiding Walk

The "true" self-avoiding Walk (TSAW) is a random walk model, in which the walker tries to avoid itself, but self-avoidance is not strictly imposed. To construct a TSAW realization one

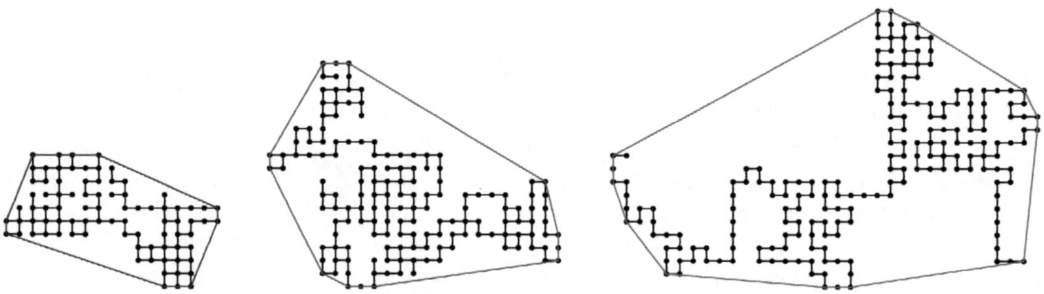

(a) $\beta = 0$ (b) $\beta = 1$ (c) $\beta = 100$

Figure 2. Examples of typical TSAW realizations at different values of the avoidance parameter β with their convex hulls. Each walk has $T = 200$ steps. Larger values of β lead to larger extended walks characterized by larger areas of their convex hulls.

grows a standard random walk on a lattice and records the number of visits n_i to each site i. For each step the probability to step on a neighboring site i is weighted with the number of times that site was already visited

$$p_i = \frac{\exp\left(-\beta n_i\right)}{\sum_{j \in \mathcal{N}} \exp\left(-\beta n_j\right)}, \tag{7}$$

where the sum over all current neighbors $\mathcal{N}$ is for normalization. The free parameter β governs the strength of the avoidance. Large values of β lead to stronger avoidance, negative values of β lead to attraction and $\beta = 0$ is the special case of the standard random walk. For a selection of β values typical examples are visualized in Fig. 2. The TSAW is not to be confused with the classical self-avoiding walk (SAW), which describes the ensemble of all configurations which do not intersect themselves each weighted the same. In Fig. 3 two partial decision trees are displayed which visualize the fundamental differences in the weights of the configurations. Even in the $\beta \to \infty$ limit ($Z_1 = 3$, $Z_2 = 2$) the weights differ. In particular its upper critical dimension is $d = 2$ [31], which means that the exponent ν, which characterizes the scaling of the end-to-end distance $r \propto T^\nu$, is $\nu = 1/2$ with logarithmic corrections, i.e., $r \propto T^\nu (\ln T)^\alpha$, where $\alpha = 1/4$ [33] is conjectured.

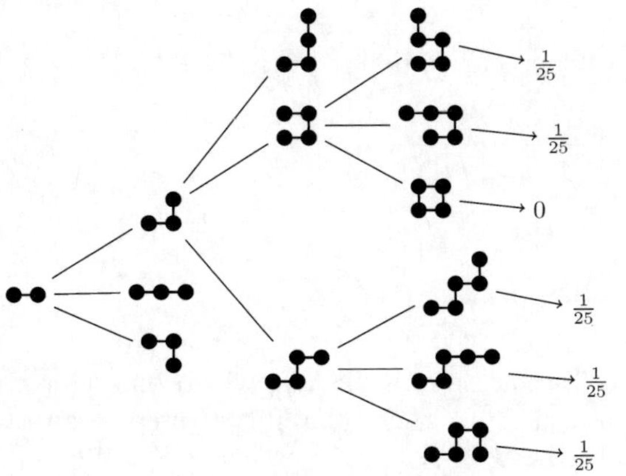

(a) SAW

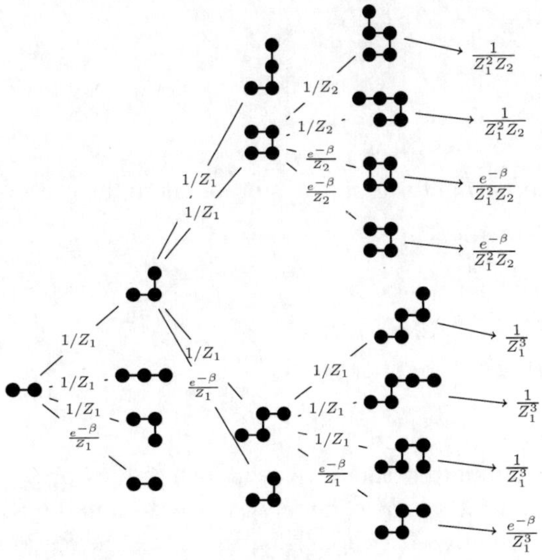

(b) TSAW

Figure 3. Partial decision trees for SAW and TSAW of walks up to length $T = 5$. On the right side of each tree the weight of the configuration is displayed. While the weights for the SAW are by definition uniform for every valid configuration, the TSAW not only allows self-intersection, but also has different weights depending on the history of the walk. Note that $Z_1 = 3 + \exp(-\beta)$ and $Z_2 = 2 + 2\exp(-\beta)$.

2.3. Convex Hulls

The convex hull of a set of points $\mathcal{P}$ in the plane is the smallest convex polygon enclosing every point $p \in \mathcal{P}$ and hence also every line between any pair of points. Some examples of convex hulls are visualized in Fig. 2. The construction of a convex hull of a planar point set is a solved problem, in the sense that an optimal algorithm exists [42, 43] with result-dependent run time $\mathcal{O}(T \log h)$, where T is the number of points $|\mathcal{P}|$ and h is the number of vertices of the resulting convex hull. In practice, however, suboptimal but simpler and for point sets as small as in this study ($T \approx 10^6$) faster algorithms are used. Especially for planar point sets one can exploit the fact that a polygon can be defined by the order of its vertices, instead by a list of its facets. The Graham scan [44] algorithm is based on this fact. After shifting the coordinate origin into the center of the point set, it sorts the points according to their polar coordinate. Then starting at an arbitrary point all points are filtered out which are oriented clockwise with respect to the the previous and next (not-filtered out) points. Iterating this over a full revolution, leaves only the points which constitute the vertices of the convex hull. This algorithm is dominated by the time to sort the points, which can be done in $\mathcal{O}(T \log T)$. Here, we use Andrew's monotone chain algorithm [45], which is a variation of the Graham scan sorting the points lexicographically, which is slightly faster, instead of by polar angle. Note that this type of algorithm does not generalize to 3 or higher dimensions. For those cases a different algorithm, like quickhull [46] has to be used. Before applying the exact algorithm, we reduce the size of the point set with Akl's elimination heuristic [47], which removes all points inside, in our implementation, a octagon of extreme points. Of a few tested polygons the octagon showed the best performance in the instances we typically encounter in this study.

To calculate the area A of a convex polygon, where the coordinates are sorted counterclockwise, one can sum the areas of the trapezoids extending perpendicular to the x-

axis

$$A = \frac{1}{2}\sum_{i=0}^{h-1}(y_i + y_{i+1})(x_i - x_{i+1}). \tag{8}$$

The perimeter L is the sum of the line segments of consecutive points of the hull

$$L = \sum_{i=0}^{h-1}\sqrt{(x_i - x_{i+1})^2 + (y_i + y_{i+1})^2}, \tag{9}$$

with $x_h \equiv x_0$ and $y_h \equiv y_0$.

3. Results

We simulated the TSAW at two values of β. The limit case of a TSAW, which only steps on itself, if it has no other choice, was simulated at $\beta = 100$. Since the probability to step on already visited sites is exponential in β, this corresponds to the $\beta \to \infty$ case. Further, we simulated at $\beta = 1$, to capture also the case, which does sometimes voluntarily step on itself.

First, we will look at the behavior of the mean of the perimeter and area. Here, we used simple sampling for walk lengths in the range $T \in \{2^k | 10 \leq k \leq 23\}$. Each value is averaged over 10^6 TSAWs. Naturally, the mean of geometric volumes scale with their intrinsic dimension d_i and a typical length scale r, e.g., the end-to-end distance, as r^{d_i}. Using the scaling of r from Eq. (1), we expect the mean values of the perimeter $\langle L \rangle$ ($d_i = 1$ in $d = 2$) and the area $\langle A \rangle$ ($d_i = 2$) to scale as

$$S \propto T^{d_i \nu} \ln(T)^{d_i \alpha} \tag{10}$$

for large values of T. We can even calculate the asymptotic prefactors μ^∞ by extrapolating the scaled values for finite sizes $\mu_L = \langle L \rangle T^{-1/2} \ln(T)^{-\alpha}$ and $\mu_A = \langle A \rangle T^{-1} \ln(T)^{-2\alpha}$ to their asymptotic values μ_L^∞ and μ_A^∞. For the extrapolation, which is shown in Fig. 4(a), we use a simple power law with offset $\mu = \mu^\infty - aT^{-b}$, which were already used for this purpose in [23, 24]. The asymptotic values are listed in table 1. As expected the values for the TSAW are larger for larger β. To our knowledge, there are no analytical calculations for these asymptotic values to which we could compare to. The given error estimates are only statistical and do not include the systematic error introduced by the ad-hoc fit function. Nevertheless the convergence of the values is very well visible, confirming $\nu = 1/2$ and $\alpha = 1/4$ to be very good estimates.

Direct fits of the form Eq. (10) yield values in good agreement with the expected exponents for the end-to-end distance r at $\beta = 1$, but most other data sets lead to fits overestimating α and slightly underestimating ν. A possible, at least partial, explanation for this is be that the relation $L(r)$ is not perfectly linear for the sizes we obtained data for.

We now focus on the main result, on the distribution $P(A)$ of the convex-hull area. These results were obtained using the large-deviation Markov-chain simulations. We had to perform simulations for different "temperatures" ranges for each system size and parameter β. For example $T = 128$ at $\beta = 1$ needed seven temperatures for the right tail $\theta \in [-40, -9]$ and three more for the left $\theta \in [7, 40]$. For larger system sizes more temperatures are usually needed. For the $\beta = 1$ case at $T = 2048$ we used 32 temperatures $\theta \in [-3200, -105]$ to obtain the right tail. For the $\beta = 100$ cases we could use similar values for the temperatures. Equilibration was ensured as described in [26]. In Figure 4(b) we compare distributions of different random walk types with the result for the TSAW at different values of β. By using the large-deviation algorithm, we were able to obtain this distribution over hundreds of decades in probability, down to values as small as $P(A) \sim 10^{-800}$ for the largest value of T. Notice that while SKSAW and

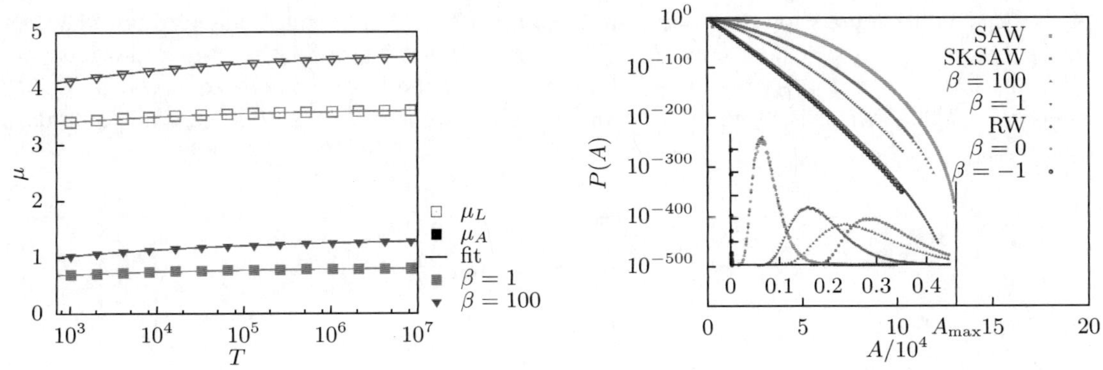

(a) Asymptotic mean values

(b) Comparison of different types of random walks

Figure 4. (a) Extrapolation of the asymptotic mean values of the perimeter and area of the convex hull. (b) Distribution of all scrutinized walk types with $T = 1024$ steps. The inset shows the peak region. Note that the standard random walk (RW) and the $\beta = 0$ TSAW coincide. Also the far right tail of the $\beta = 100$ TSAW and SKSAW coincide, but not the main region. The vertical line shows the maximum area constructable with 1024 steps, which has an area $A_{\max}$ about $512^2/2 \approx 1.3 \times 10^5$. The distributions are thus not sampled over their whole support, but a large region.

	μ_L^∞	μ_A^∞
TSAW $\beta = 0$ (exact)	3.5449...	0.7854...
TSAW $\beta = 1$	3.636(2)	0.820(1)
TSAW $\beta = 100$	4.641(3)	1.339(3)

Table 1. Asymptotic mean values of the area and perimeter scaled by Eq. (10). The values are obtained by the fit shown in Fig. 4(a). The given error estimates are only statistical and do not include the systematic error introduced by the ad-hoc fit function. The exact values are from [19] and converted to a square lattice as described in [26].

TSAW with high values of β show the same behavior in the far tail, where the walks are so stretched out such that trapping does not play a role anymore. In the main region however, they are clearly distinct, as is expected due to their different scaling exponent ν. Further, the parameter β can apparently be used to interpolate the tail behavior between the standard random walk case and the SKSAW case.

Since we have obtained large parts of the distribution, it would be interesting if the whole distribution scales the same as the mean values (cf. Eq. (10)). For other types of walks, the distribution of perimeter and area could indeed be scaled [23, 24, 25, 26] across their full support only knowing ν, as

$$P(S) = T^{-d\nu} \widetilde{P}\left(ST^{-d\nu}\right). \tag{11}$$

For the TSAW, this collapse, when considering the logarithmic corrections as visualized in Fig. 5, exhibits an apparent drift towards a limiting shape. Nevertheless, severe finite size effects are visible, especially in the tails but also in the main region. Despite far larger system sizes T considered, here the main region collapse is worse than for other kinds of self-avoiding walks as shown in [26]. The stronger finite size effect may be caused by the fact that all walks start on an empty lattice. This means for our case that the first steps of the walk behave differently from the last steps of the walk, when many sites are occupied. Although for the limit of large system sizes T, the latter should determine the behavior. A possible improvement to simulate TSAWs is suggested in [33], which is to simulate a much longer walk with $t \gg T$ steps and look at the last T steps.

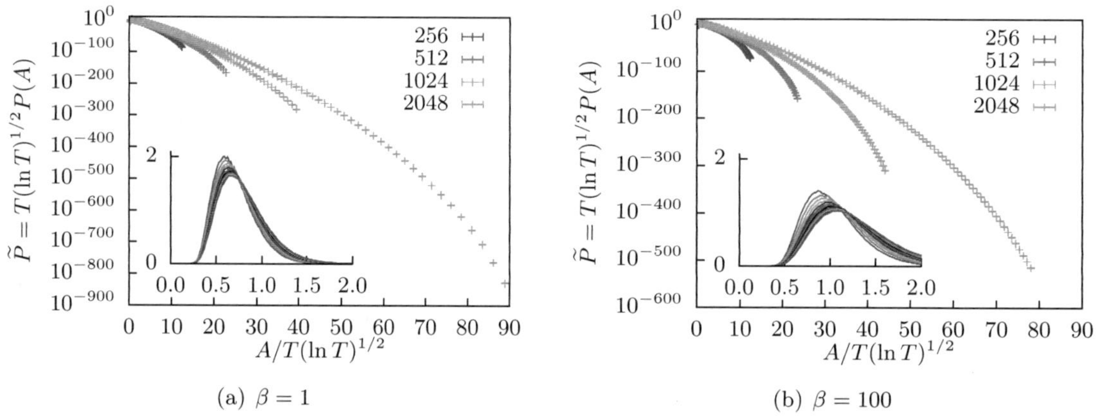

Figure 5. Distributions of the area of the "true" self-avoiding random walk scaled according to Eq. (11) plus logarithmic corrections for different values of β and lengths T. The insets show the main region for 14 values of $T \in \{2^k | 10 \leq k \leq 23\}$ obtained by simple sampling.

The rate function Φ is defined if the distribution obeys the *large deviation principle.* This means that the distribution, for large values of T, should decay exponentially in the length T scaled by the rate function as

$$P_T(s) \approx \mathrm{e}^{-T\Phi(s)} . \tag{12}$$

Usually the parameter s is between 0 and 1. We achieve this by dividing our measured area by the maximum area, i.e., by measuring $s = \frac{A}{A_{\mathrm{max}}}$. In two dimensions the walk configuration with maximum area is L-shaped with arms of equal length (for odd T) and therefore $A_{\mathrm{max}} = \frac{1}{2}\left(\frac{T+1}{2}\right)^2 \approx \frac{T^2}{8}$.

Similar to [26] we assume the rate function to be a power law

$$\Phi(s) \propto s^{\kappa}, \tag{13}$$

which seems to agree reasonably well with our data, since the double logarithmic plot Fig. 6 shows that the rate function appears as a straight line in the intermediate tail. The far tail is dominated by finite-size effects caused by the lattice structure, which leads to a "bending up" of the rate function. For small values of s, in the high-probability region, the rate function does not have any relevance. Assuming that the rate function is a power law Eq. (13) and scaling of the form Eq. (11) is possible, with d_i being the intrinsic dimension of the observable, e.g., $d_i = 2$ for the area, we can derive a value for the power law exponent of the rate function κ. Using the definition of the rate function Eq. (12)

$$\mathrm{e}^{-T\Phi(ST^{-d_i})} \approx T^{-d_i\nu}\widetilde{P}\left(ST^{-d_i\nu}\right) \tag{14}$$

should hold in the right tail. The $T^{-d_i\nu}$ term can be ignored next to the exponential, also the logarithmic correction is subdominant and would not allow to add any insight. Apparently the right hand side is a function of $ST^{-d_i\nu}$, such that the left hand side also has to be a function of $ST^{-d_i\nu}$. This is the case for [26]

$$\kappa = \frac{1}{d_i(1-\nu)}. \tag{15}$$

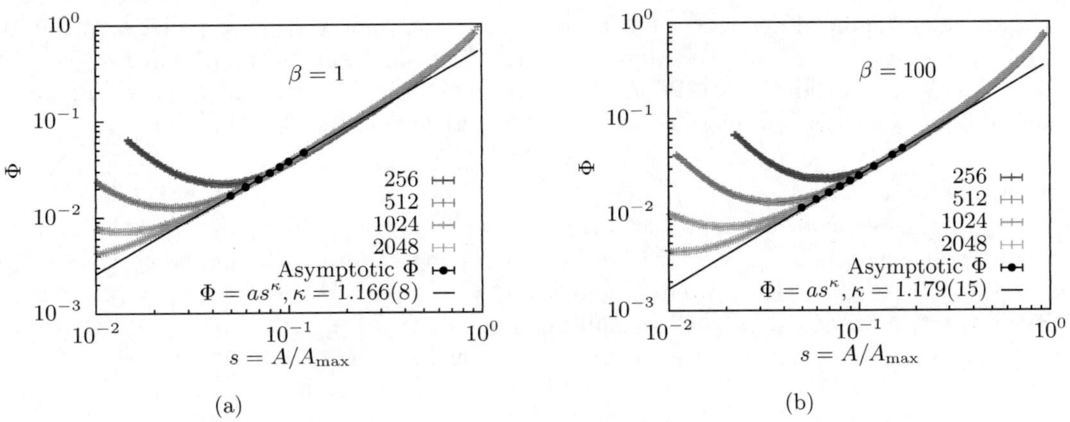

Figure 6. Rate function Φ for the area with fits to the assumed power-law form. The fit is performed over a range, where the finite-size influence of the lattice should be small, but the large T behavior can be extrapolated. Finite-size effects seem more pronounced for larger values of β.

To test whether the results for the rate function in case of the TSAW obeys this relation, we estimate the value of κ from our data. Since we have data for various values of the walk length T, we first extrapolate our data point-wise to large T. For this purpose we fit a power law with offset, where the parameters depend on the value of s:

$$\Phi(s,T) = a(s)T^{b(s)} + \Phi_\infty(s). \tag{16}$$

This results in the value of interest $\Phi(s) \equiv \Phi_\infty(s)$, the parameters $a(s)$ and $b(s)$ are only auxiliary quantities. We perform this extrapolation in a region which is far away from the finite-size effects of the far tail. In this range of medium values of s the extrapolation according to Eq. (16) works robustly. Since the bins of the logarithmic histograms we use do not have the same borders for every system size T, we have a-priory not access to the same value of s for different values of T. Thus, we use cubic splines to interpolate such that we obtain results for the same value of s for all walk lengths. We found cubic splines to be sufficient since the bins are rather dense such that systematic errors introduced by the interpolation should be small. The final values $\Phi(s)$ obtained from the extrapolated values to the assumed form Eq. (13) are shown in Fig. 6 as symbols. Next, we fit power laws to this data. The values for κ obtained are within errorbars consistent with $\kappa = 7/6$ which is the expected value for the SKSAW ($\nu = 4/7$) and incompatible with the expected value $\kappa = 1$ of $\nu = 1/2$ walks. This behavior is nevertheless plausible since in the tail (large-area) region, structures, which enable self trapping, i.e., loops, are rare since they do lead to smaller areas of the convex hull than straight regions. Therefore the influence of trappings should diminish in the large area tail, which is the main difference in the behavior of SKSAW and TSAW. Without trappings the TSAW in the $\beta \to \infty$ limit is functionally identical to the SKSAW. Apparently already $\beta = 1$ is large enough to produce this behavior. Therefore it is natural that the large-area tail behaves the same as the SKSAW. On the other hand, to possibly see a range where the rate function behaves like a power law with $\kappa = 1$ according to $\nu = 1/2$, one would have to go to much larger system sizes, because one would have to obtain data to the right of the peak, but for very small values of $s = A/A_{\max} \ll 10^{-2}$, where trappings still do play a role. In particular the analysis might be hampered by the presence of the logarithmic correction to the mean end-to-end distance.

This means that the TSAW is more complex in comparison to some other types of self-avoiding walks for which it was possible to predict the tail behavior from the same exponent ν which predicts the mean behavior. The other types of random walk were under scrutiny in [26], namely the self-avoiding walk, the loop-erased random walk and the smart kinetic self-avoiding walk (SKSAW). Instead for TSAW in the large deviation region a different scaling exponent seems to hold, which is very close to the scaling exponent of the SKSAW.

4. Conclusions

We studied the behavior of the distribution of the area of the convex hull of the "true" self-avoiding walk, especially in the large deviation regime of larger than typical areas. With a sophisticated large-deviation sampling algorithm, we obtained the distribution over a large part of its support down to probabilities smaller than 10^{-800} for a typical avoidance parameter of $\beta = 1$ and a large avoidance $\beta = 100$. The distributions seem to approach a limiting scaling form when rescaled by the behavior of the mean, but with much stronger finite-size effects as compared to other types of random walks, which were previously studied.

Using this data we calculated the rate functions. The rate function seem also to behave qualitatively similar in comparison to other types of self-avoiding walks studied earlier [26] in that they seem to be well defined and well approximated by a power law. In contrast to other types of random walks, this power law can apparently not be derived from the scaling exponent of the mean values ν. Instead it seems that a second exponent governs the scaling behavior of the tail for the TSAW, which is close to $4/7$, the scaling exponent of the smart kinetic self-avoiding walk. This is plausible since the large-area region should be dominated by configurations in which no trappings are possible, which is the major difference between these types.

Finally, we also provided estimates for the relevant scale factors of the mean of area and perimeter of the convex hulls of TSAWs. They might be accessible to analytic calculations in the future.

For future numerical work it would be interesting to look for further types of random walks, which show similar effects of distinct scaling exponents for different parts of the distribution but do not show the strong logarithmic corrections to scaling, and would therefore be easier to analyze. On the other hand, it would be very exciting if one was able to obtain data in the range where the rate function exhibits the exponent $\kappa(\nu = 1/2) = 1$, with the need to simulate really large system sizes, but closer to the typical behavior.

Acknowledgments

This work was supported by the German Science Foundation (DFG) through the grant HA 3169/8-1. We also thank the GWDG (Göttingen) for providing computational resources.

References

[1] Einstein A 1906 *Annalen der Physik* **324** 371–381 ISSN 1521-3889
[2] Fama E F 1965 *Financial Analysts Journal* **21** 55–59
[3] Page L, Brin S, Motwani R and Winograd T 1999 The pagerank citation ranking: Bringing order to the web. Technical Report 1999-66 Stanford InfoLab
[4] Newman M and Barkema G 1999 *Monte Carlo Methods in Statistical Physics* (Clarendon Press) ISBN 9780198517979
[5] Hughes B D 1996 *Random walks and random environments* (Clarendon Press Oxford)

[6] Madras N and Slade G 2013 *The Self-Avoiding Walk* (New York, NY: Springer New York) ISBN 978-1-4614-6025-1
[7] Weinrib A and Trugman S A 1985 *Phys. Rev. B* **31**(5) 2993–2997
[8] Manna S S, Dhar D and Majumdar S N 1992 *Phys. Rev. A* **46**(8) R4471–R4474
[9] Majumdar S N 1992 *Phys. Rev. Lett.* **68**(15) 2329–2331
[10] Mohr C O 1947 *American Midland Naturalist* **37** pp. 223–249
[11] Worton B J 1987 *Ecol. Model.* **38** 277–298 ISSN 0304-3800
[12] Boyle S A, Lourenco W C, da Silva L R and Smith A T 2009 *Folia Primatol.* **80** 33–42
[13] Dumonteil E, Majumdar S N, Rosso A and Zoia A 2013 *Proceedings of the National Academy of Sciences* **110** 4239–4244
[14] Lanoiselée Y and Grebenkov D S 2017 *Phys. Rev. E* **96**(2) 022144
[15] Letac G and Takács L 1980 *The American Mathematical Monthly* **87** 142–142 ISSN 00029890, 19300972
[16] Letac G 1993 *Journal of Theoretical Probability* **6** 385–387 ISSN 1572-9230
[17] Randon-Furling J, Majumdar S N and Comtet A 2009 *Phys. Rev. Lett.* **103**(14) 140602
[18] Majumdar S N, Comtet A and Randon-Furling J 2010 *Journal of Statistical Physics* **138** 955–1009 ISSN 1572-9613
[19] Eldan R 2014 *Electron. J. Probab.* **19** no. 45, 1–34 ISSN 1083-6489
[20] Grebenkov D S, Lanoiselée Y and Majumdar S N 2017 *Journal of Statistical Mechanics: Theory and Experiment* **2017** 103203
[21] Kabluchko Z and Zaporozhets D 2016 *Transactions of the American Mathematical Society* **368** 8873–8899
[22] Goldman A 1996 *Probability Theory and Related Fields* **105** 57–83 ISSN 1432-2064
[23] Claussen G, Hartmann A K and Majumdar S N 2015 *Phys. Rev. E* **91**(5) 052104
[24] Dewenter T, Claussen G, Hartmann A K and Majumdar S N 2016 *Phys. Rev. E* **94**(5) 052120
[25] Schawe H, Hartmann A K and Majumdar S N 2017 *Phys. Rev. E* **96**(6) 062101
[26] Schawe H, Hartmann A K and Majumdar S N 2018 *Phys. Rev. E* **97**(6) 062159
[27] Cardy J 2005 *Annals of Physics* **318** 81 – 118 ISSN 0003-4916 special Issue
[28] Lawler G F, Schramm O and Werner W 2002 *arXiv preprint math/0204277*
[29] Lawler G F, Schramm O and Werner W 2011 *Conformal Invariance Of Planar Loop-Erased Random Walks and Uniform Spanning Trees* (New York, NY: Springer New York) pp 931–987 ISBN 978-1-4419-9675-6
[30] Kennedy T 2015 *Journal of Statistical Physics* **160** 302–320 ISSN 1572-9613
[31] Amit D J, Parisi G and Peliti L 1983 *Phys. Rev. B* **27**(3) 1635–1645
[32] Pietronero L 1983 *Phys. Rev. B* **27**(9) 5887–5889
[33] Grassberger P 2017 *Phys. Rev. Lett.* **119**(14) 140601
[34] Metropolis N, Rosenbluth A W, Rosenbluth M N, Teller A H and Teller E 1953 *The Journal of Chemical Physics* **21** 1087–1092
[35] Hartmann A K 2014 *Phys. Rev. E* **89**(5) 052103
[36] Hartmann A K 2002 *Phys. Rev. E* **65**(5) 056102
[37] Wolfsheimer S, Burghardt B and Hartmann A K 2007 *Algorithms for Molecular Biology* **2** 9 ISSN 1748-7188
[38] Fieth P and Hartmann A K 2016 *Phys. Rev. E* **94**(2) 022127
[39] Engel A, Monasson R and Hartmann A K 2004 *Journal of Statistical Physics* **117** 387–426 ISSN 1572-9613
[40] Hartmann A K 2011 *The European Physical Journal B* **84** 627–634 ISSN 1434-6036
[41] Hartmann A K and Mézard M 2018 *Phys. Rev. E* **97**(3) 032128
[42] Kirkpatrick D and Seidel R 1986 *SIAM Journal on Computing* **15** 287–299
[43] Chan T M 1996 *Discrete & Computational Geometry* **16** 361–368 ISSN 1432-0444
[44] Graham R 1972 *Information Processing Letters* **1** 132–133
[45] Andrew A 1979 *Information Processing Letters* **9** 216 – 219 ISSN 0020-0190
[46] Barber C B, Dobkin D P and Huhdanpaa H 1996 *ACM Trans. Math. Softw.* **22** 469–483 URL `http://www.qhull.org`
[47] Akl S G and Toussaint G T 1978 *Information Processing Letters* **7** 219 – 222 ISSN 0020-0190

IOP Conf. Series: Journal of Physics: Conf. Series **1290** (2019) 012030 doi:10.1088/1742-6596/1290/1/012030

Tunneling Dynamics of interacting bosons in a quantum seesaw potential

Sunayana Dutta[1,*], Budhaditya Chatterjee[2], Pankaj Kumar Mishra[1], Axel U. J. Lode[3,4], Marios C. Tsatsos[5], and Saurabh Basu[1]

[1] Department of Physics, Indian Institute of Technology Guwahati, Assam, India
[2] Department of Physics, Indian Institute of Technology Kanpur, India
[3] Wolfgang Pauli Institute c/o Faculty of Mathematics, University of Vienna, Oskar-Morgenstern Platz 1, 1090 Vienna, Austria
[4] Vienna Center for Quantum Science and Technology, Atominstitut, TU Wien, Stadionallee 2, 1020 Vienna, Austria
[5] São Carlos Institute of Physics, University of São Paulo, P.O. Box 369, 13560-970 São Carlos, São Paulo, Brazil.

E-mail: * sunayanadutta@iitg.ac.in

Abstract. We study the tunneling dynamics of $N = 10$ one-dimensional interacting bosons confined in a temporally driven double well potential that imitates a quantum seesaw and how we can manipulate these dynamics by changing the drive of the seesaw potential. We emulate the seesaw with a driven double well potential and consider two driving protocols: an harmonic constant-frequency drive and a chirped drive with linearly increasing frequency. We consider the time-dependent many-body Schrödinger equation of a repulsively interacting quasi-one-dimensional few-boson system. We solve it by using the multiconfigurational time-dependent Hartree method for bosons (MCTDHB) as implemented in the MCTDH-X software. For an harmonic drive and at small values of the driving amplitude, the dynamics of the particles become very slow rendering a stationary-like state. In a phase-space picture the population imbalance between the wells follows a trajectory which is restricted to a comparatively small region of space. For an harmonic drive at intermediate amplitudes, the dynamics become periodic in nature, implying that the bosons populate each of wells periodically. At comparatively large amplitudes of the harmonic drive, the dynamics show features of chaos in phase-space representation. For the chirped drive with a driving frequency increasing linearly in time, the imbalance of the atoms in the seesaw, however, has a temporal evolution that is faster for certain frequency ranges. The tunneling dynamics in such cases, for small amplitudes, show the appearance of quasi-periodicity with simultaneously present slow and fast oscillations. Increasing the amplitude of the chirped drive, we observe that the dynamics, although being periodic, become severely damped in their amplitude. Our study establishes that by tuning the temporal evolution of the quantum seesaw, a precise control of tunneling dynamics of the correlated bosons can be achieved. Since harmonic driving and chirp frequency modulation of the seesaw are experimentally achievable, our simulations can be experimentally realized in laboratories dealing with cold atomic gases.

1. Introduction

The experimental realization of periodically driven systems such as ultracold atoms in a periodically shaken optical lattice has received significant attention in the quantum physics community. One prototype of these systems is a kicked top or kicked rotor model where

IOP Conf. Series: Journal of Physics: Conf. Series **1290** (2019) 012030 doi:10.1088/1742-6596/1290/1/012030

periodic kicks are subjected to a particle moving on a ring [1]. Dynamical Anderson localization [2, 3], dynamical stabilization in the classical or quantum limit, and coherently controlled phase transitions from a superfluid to a Mott insulator [4] represent prominent features of such periodically driven systems.
A significant amount of literature exists in the field of driven Bose-Einstein condensates [5, 6]. In particular, a periodic drive can be used to control the tunneling and interaction parameters and, thereby, provide the possibility to bias the many-body states and quantum phases of Bose-Einstein condensates [7, 8, 9] in the context of optical lattices. Recently, the experimental realization of such systems led to the realization of many-body localization [10, 11] and other exotic physical phenomena.
The efficacious experimental realization of Bose-Einstein condensates in optical lattices has provided a suitable paradigm to study different exotic quantum phenomena of ultracold atoms [12]. Theoretically, the static and dynamic properties of Bose-Einstein condensates are conventionally modeled by the Gross-Pitaevskii mean-field theory (GPE). This theory is an excellent model for explaining non-correlated bosonic systems, albeit with the limitation that it cannot accurately describe correlation effects that may arise dynamically due to the presence of interactions. The usage of only one single-particle state (orbital) to describe a condensate prevents the GPE from explaining correlations which are intrinsically multi-orbital features. We name here the emergence of fragmentation [13, 14, 15, 16, 17, 18, 19, 20, 21, 22, 23, 24, 25, 26, 27, 28, 29, 30, 31] and the superfluid-to-Mott-insulator phase transition [27, 29, 31] as examples. Recently an essentially exact, numerically efficient and multi-orbital many-body theory was developed; the multiconfiguration time-dependent Hartree approach (MCTDH). The bosonic version of MCTDH, the MCTDH for bosons (MCTDHB) [32, 33, 34, 35] represents a systematic generalization of the (one-orbital) Gross-Pitaevskii mean-field theory.
In this work, we use the MCTDHB method to investigate the tunneling dynamics of bosons with contact interactions that are confined in a temporally driven double-well potential with a periodically oscillating spatial bias that imitates a *quantum seesaw*. Our main observable is the imbalance of the number of atoms in the left and the right well as a function of time. We consider two driving protocols: an harmonic, constant-frequency drive and a chirped drive with a frequency linearly increasing in time.
Our paper is structured as follows: In Sec. 2 we discuss the MCTDHB method, in Sec. 3 we discuss the system we study and its physics, and conclude in Sec. 4.

2. MCTDHB method

The dynamics of interacting many-body systems of indistinguishable bosonic particles are described by the time-dependent many-body Schrödinger equation. Usually, this problem is solved by the mean-field approximation [36, 37] or the Bose-Hubbard (BH) model [38, 39]. Here we go beyond these models and employ the fully variational multiconfigurational time-dependent Hartree for bosons (MCTDHB) approach [32, 33, 34, 35]. In the MCTDHB model, which we discuss in the following, there is no restriction on the number of orbitals, geometry, dimensionality, and interparticle interactions of the time-dependent many-boson problem.
The time-evolution of N correlated bosons is governed by the time-dependent many-particle Schrödinger equation (TDSE).
To scale the Hamiltonian and our calculations into dimensionless units, we divide it by $\hbar^2/(2mL^2)$, where m is the mass of the considered boson and L a convenient length scale.
The TDSE reads:

$$\hat{H}\Psi = i\frac{\partial \Psi}{\partial t}. \tag{1}$$

Here, the Hamiltonian $\hat{H}$ is

$$\hat{H}(x_1, x_2, ..., x_N) = \sum_{j=1}^{N} \hat{h}(x_j) + \sum_{k>j=1}^{N} \hat{W}(x_j - x_k), \tag{2}$$

where x_j is the position of the jth boson. The first term of Eq.(2), $\hat{h}$, is the one-body Hamiltonian

$$\hat{h}(x) = \hat{T}(x) + \hat{V}_{trap}(x, t), \tag{3}$$

where $\hat{T}(x)$ is the kinetic energy and $\hat{V}(x,t)$ is the external potential or the trapping potential. The second term in Eq. (2), $\hat{W}(x_j - x_k) = \lambda_0 \delta(x_j - x_k)$ is the two-body interaction of the bosonic gas. In this work, as is conventional for Bose-Einstein condensates [37], $\hat{W}$ is approximated as a contact interaction.

Here λ_0 is the interaction strength and is scaled with the total number of particles as, $\lambda_0 = \lambda/(N-1)$.

The use of time-adaptive optimized basis states is the key idea of the MCTDHB approach. In second quantization, the bosonic field operator corresponding to a set of M orthonormal, time-dependent functions (orbitals) $\{\phi_k(x,t)\}$ which annihilates a particle at position x can be written as,

$$\hat{\Psi}(x,t) = \sum_{k=1}^{M} \hat{a}_k \phi_k(x,t). \tag{4}$$

The bosonic creation and annihilation operators obey the usual commutation relations at any instant t. With the help of the bosonic creation operators, $\hat{a}_k^\dagger$, time-dependent permanents can be expressed as

$$|\vec{n}; t\rangle = \frac{[a_1^\dagger(t)]^{n_1}[a_2^\dagger(t)]^{n_2}...[a_M^\dagger(t)]^{n_M}}{\sqrt{n_1!n_2!n_3!...n_M!}}|0\rangle \tag{5}$$

where $\vec{n} = (n_1, n_2, n_3, ...n_M)$ are the occupations of the orbitals which conserve the total number of particles as $n_1 + n_2 + ... + n_M = N$. These time-dependent permanents are used as the basis set in MCTDHB.

The ansatz made for the many-boson wave function in MCTDHB is:

$$|\Psi(t)\rangle = \sum_{n} C_{\vec{n}}(t)|\vec{n}; t\rangle. \tag{6}$$

The summation in Eq. (6) runs over all possible configurations, that can be obtained by distributing N bosons in M orbitals. In the limit $N \to \infty$ the permanents $|\vec{n}; t\rangle$ span the N-bosons Hilbert space and the variational principle [40] guarantees that the solutions of the time-dependent many-body problem provided by the MCTDHB method are exact [41, 35].

To derive the MCTDHB equations we use the time-dependent variational principle (TDVP) [40] to determine the time evolution of the expansion coefficients $\{C_{\vec{n}}(t)\}$ and the orbitals $\{\phi_k(\mathbf{r},t)\}$ [33, 34].

In the following picture the MCTDHB equations are obtained by fixing the gauge freedom in the choice of the orbitals by the condition $\langle \frac{\partial \phi_j}{\partial t}|\phi_k\rangle = 0, \forall j,k \in [1,M]$.

The variation of the action with respect to the expansion coefficients $C_{\vec{n}}(t)$ yields the equations of motion for the expansion coefficients [33, 34],

$$\sum_{\vec{n}} \mathcal{H}_{\vec{n}\vec{n}'}(t) C_{\vec{n}} = i \frac{\partial C_{\vec{n}'(t)}}{\partial t} \tag{7}$$

IOP Conf. Series: Journal of Physics: Conf. Series **1290** (2019) 012030 doi:10.1088/1742-6596/1290/1/012030

This equation of motion is a first-order differential equation and the matrix $\mathcal{H}_{\vec{n}\vec{n}'}(t) = \langle \vec{n}; t|\hat{H}|\vec{n}'; t\rangle$ is time-dependent as the permanents $|\vec{n}; t\rangle$ and $\hat{H}$ itself are functions of time.
The variation of the action with respect to the orbitals $\{\phi_k(x,t)\}$ yields the equation of motion for the orbitals [33, 34]:

$$i|\dot{\phi}_j\rangle = \hat{\mathbf{P}}[\hat{h}|\phi_j\rangle + \sum_{k,s,q,l=1}^{M} \{\rho(t)\}^{-1}_{jk} \rho_{ksql} \hat{W}_{sl}|\phi_q\rangle] \tag{8}$$

where

$$\hat{\mathbf{P}} = 1 - \sum_{j=1}^{M} |\phi_j\rangle\langle\phi_j| \tag{9}$$

$\rho_{jk} = \langle\Psi|\hat{a}_j^\dagger \hat{a}_k|\Psi\rangle$ are the matrix elements of the one-body reduced density matrix (RDM) and $\rho_{ksql} = \langle\Psi|\hat{a}_k^\dagger \hat{a}_s^\dagger \hat{a}_l \hat{a}_q|\Psi\rangle$ are the matrix elements of the two-body RDM. The local interaction potentials read $\hat{W}_{sl} = \int dx' \phi_s^*(x',t) W(x,x') \phi_l(x',t)$. For the contact interactions that we consider in this paper, the $\hat{W}_{sl}$ are given by $\hat{W}_{sl} = \lambda_0 \phi_s^*(x,t)\phi_l(x,t)$.

3. Results

In this work, we consider $N = 10$ bosons in one spatial dimension interacting via contact interaction trapped in a driven double well potential $V_{trap}(x,t)$ that imitates a quantum seesaw:

$$V_{trap}(x,t) = V_0 \cos^2(x) - \alpha x[\cos(\omega t) + \cos(\beta t^2)]. \tag{10}$$

Here V_0 is the barrier height of the trapping potential and αx is the relative tilt between the two wells. The parameter α thus corresponds to the strength of the drive of our quantum seesaw. By tuning the frequency ω, we drive the tilt of the seesaw harmonically and by tuning the parameter β we drive the tilt of the seesaw with a chirp, i.e. a frequency βt that linearly increases in time. In our analysis below, we choose the interaction strength of the bosons $\lambda_0 = 0.05$, the barrier height $V_0 = 10$, and the strength of the driving force α between 0 to 8. Our numerical grid extends from $x_{min} = -\pi$ to $x_{max} = +\pi$.
We simulate the dynamics with the MCTDH-X software package [42, 43, 44] using $M = 3$ orbitals for all calculations.
As the initial state for our computations, we use the ground state of the interacting bosons in the potential $V_{trap}(x,t)$ with $\alpha \equiv 0$ in Eq. (10).
Fig. 1 shows snapshots of the temporally oscillating double well potential at various instants of time.
To get an intuitive idea about the dynamical behavior of the system in the driven seesaw-potential, we systematically investigate the time-evolution of the population imbalance between the wells,

$$n(t) = N - \int_0^\infty \rho(x,t) dx. \tag{11}$$

Here, N is the number of bosons and $\rho(x) = \langle\Psi|\hat{\Psi}^\dagger(x,t)\hat{\Psi}(x,t)|\Psi\rangle$ is the one-body density.
We visualize the imbalance $n(t)$ by plotting it as a function of time and by analyzing it in a "phase-space representation". To obtain this phase-space representation, we plot points $[n(t), n(t+\tau)]$ as a function of time t for a fixed τ. From these phase-space plots, we are able to distinguish different kinds of dynamics as function the driving protocol [α, ω, β in Eq. (10)] of our quantum seesaw.

IOP Conf. Series: Journal of Physics: Conf. Series **1290** (2019) 012030 doi:10.1088/1742-6596/1290/1/012030

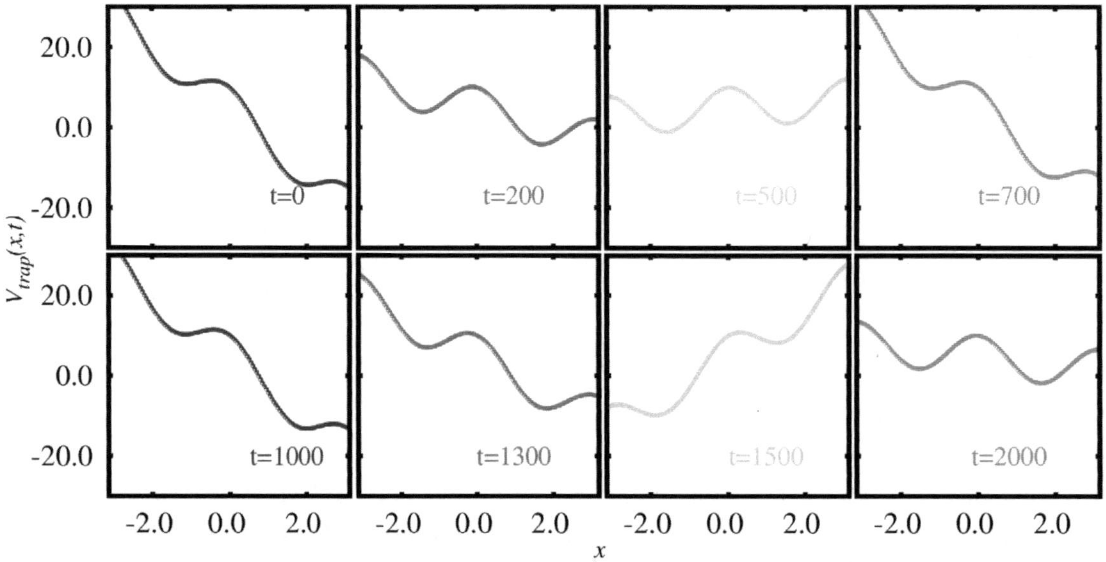

Figure 1. (Color online) Quantum seesaw potential as a function of time. This figure shows snapshots of the temporally driven double well in Eq. (10) potential for several times t for a chirped driven ($\omega = 0, \beta = 1, V_0 = 10$) tilt of amplitude $\alpha = 8$.

In the following, we exhibit our main result, the physics of the time-evolution of the population imbalance as a function of the force α that drives the seesaw. We showcase two different driving protocols for the seesaw, a harmonic drive ($\omega \neq 0$ and $\beta = 0$) and chirped drive ($\omega = 0$ and $\beta \neq 0$) in Subsections 3.1 and 3.2, respectively.

3.1. *Harmonic driving ($\omega \neq 0$ and $\beta = 0$)*

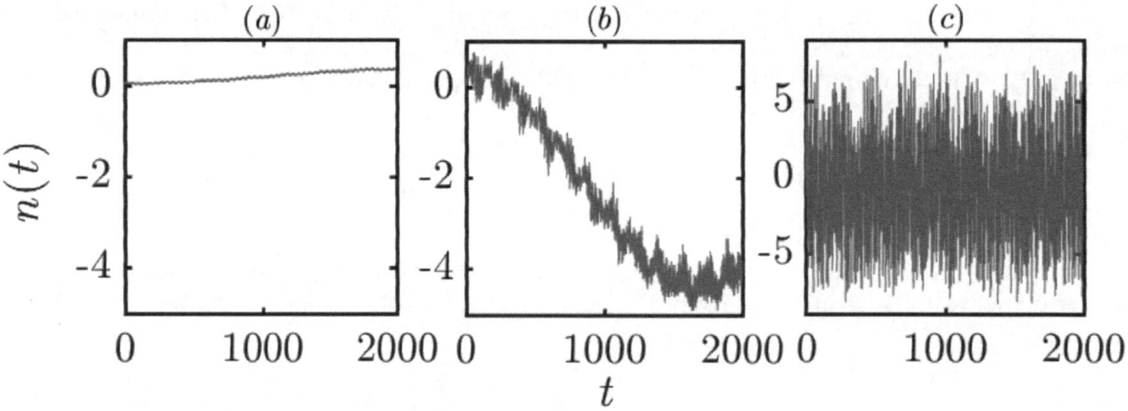

Figure 2. (Color online) Time-evolution of the population imbalance $n(t)$ for the harmonic drive ($\omega = 1$ and $\beta = 0$), for driving amplitudes $\alpha = 1$ (a), $\alpha = 3$ (b), and $\alpha = 8$ (c). See text for further discussion.

We drive our seesaw harmonically, i.e., with a constant frequency by choosing $\omega = 1$ and $\beta = 0$ in the trapping potential Eq. (10). At a small value of the driving amplitude of the seesaw,

$\alpha = 1$, we observe that the dynamics of the population imbalance of the bosons, $n(t)$, happen on a very long time-scale [Fig. 2(a)]. For larger amplitude driving, $\alpha = 3$, we find a periodic time-evolution of the imbalance $n(t)$ with a large period [see Fig. 2(b)]. For drives with very large amplitudes, $\alpha = 8$, the dynamics of the imbalance $n(t)$ start to feature fast oscillations [Fig. 2(c)]; the oscillations are much faster than those for intermediate and small amplitude drives depicted in [Fig. 2(a) and (b), respectively].

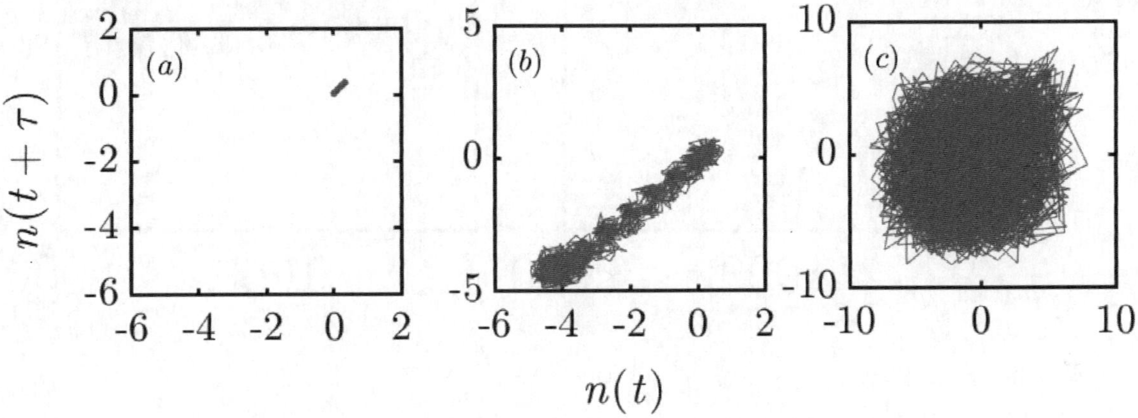

Figure 3. (Color online) Phase-space depiction of the time-evolution of the population imbalance $[n(t), n(t+\tau)]$ for the harmonic drive ($\omega = 1$ and $\beta = 0$), for driving amplitudes $\alpha = 1$ (a), $\alpha = 3$ (b), and $\alpha = 8$ (c). All panels show points $[n(t), n(t+\tau)]$ as a function of the time t for fixed $\tau = 50$. See text for further discussion.

We now analyze the phase-space representation of the population imbalance. We plot points $[n(t), n(t+\tau)]$ as a function of the time t for a fixed $\tau = 50$ in Fig. 3. For a small driving amplitude, $\alpha = 1$, the system behaves almost like a stationary state and very little of the available phase space is explored [Fig. 3(a)]. As the drive amplitude increases ($\alpha = 3, 8$), the trajectory of $[n(t), n(t+\tau)]$ explores increasing portions of the available phase space, Fig. 3(b) and (c). The $[n(t), n(t+\tau)]$ trajectory for the largest drive amplitude $\alpha = 8$ in Fig. 3(c) bears resemblance to trajectories expected for chaotic systems as it emphasizes that a large region of the phase space is visited multiple times by the system.

3.2. Chirping modulation ($\omega = 0$ and $\beta \neq 0$)

Now we consider a chirped drive of our quantum seesaw [$\omega = 0$ and $\beta = 1$ in Eq.(10)]. For small driving amplitudes ($\alpha = 1$), the tunneling dynamics in the chirped seesaw result in a quasi-periodic population imbalance: $n(t)$ shows oscillations that feature two frequencies – a small and a large one – see Fig. 4(a). When the amplitude of the drive is increased ($\alpha = 8$), the dynamics of $n(t)$ stay periodic with two frequencies; however, the amplitude of the oscillations in $n(t)$ is damped and the frequencies are changed in comparison to small driving amplitudes [compare Fig. 4(a) and (b)].

We now turn to the phase-space representation of the time-evolution of the population imbalance and plot $[n(t), n(t+\tau)]$ as a function of time t for $\tau = 50$ in Fig. 5. For small driving amplitudes [$\alpha = 1$, Fig. 5(a)] the phase-space representation shows an elliptical trajectory depicting the Rabi oscillations at a small frequency. The additional "curls" on this elliptical curve are the consequence of the second, larger, observed frequency in the oscillations of the imbalance [compare Fig. 4(a) and Fig. 5(a)]. For larger-amplitude chirped drives ($\alpha = 8$), we find a phase-space trajectory that is confined within a small volume of the available phase space, see

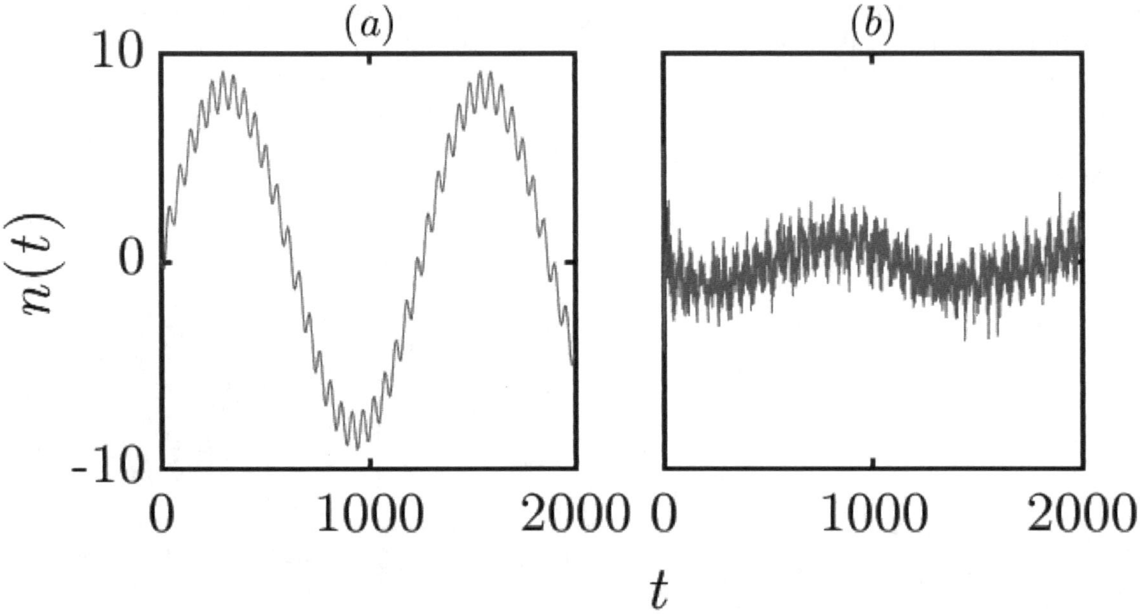

Figure 4. (Color online) Time-evolution of the population imbalance $n(t)$ for the chirped drive ($\omega = 0$ and $\beta = 1$), for driving amplitudes $\alpha = 1$ (a) and $\alpha = 8$ (b). See text for further discussion.

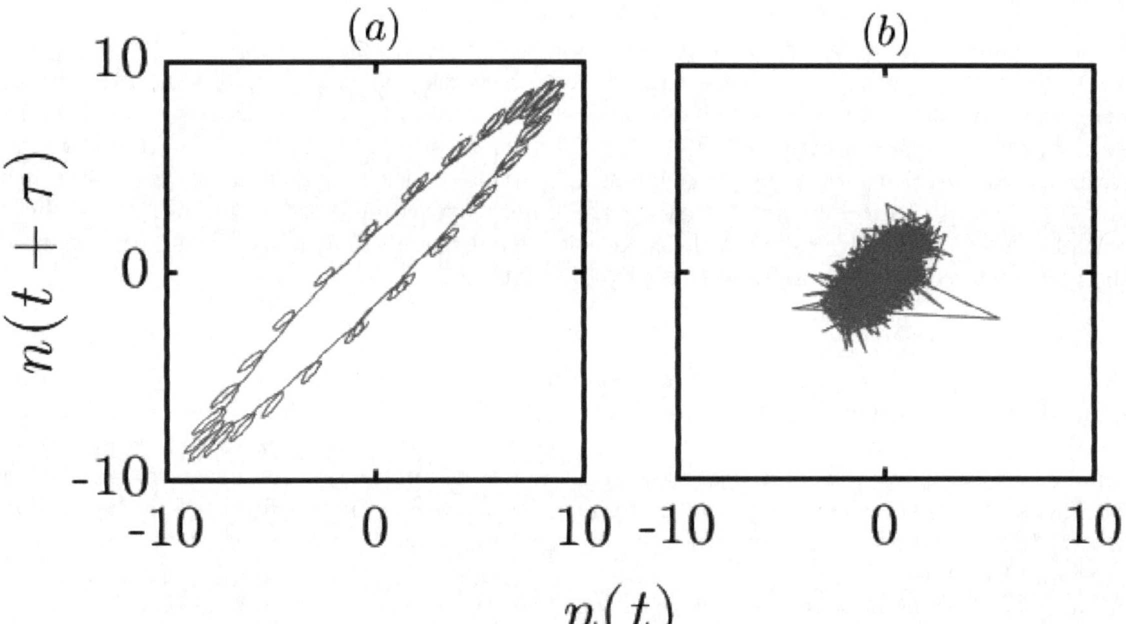

Figure 5. (Color online) Phase-space depiction of the time-evolution of the population imbalance $[n(t), n(t+\tau)]$ for the chirped drive of the seesaw ($\omega = 0$ and $\beta = 1$), for driving amplitudes $\alpha = 1$ (a) and $\alpha = 8$ (b). All panels show points $[n(t), n(t+\tau)]$ as a function of the time t for fixed $\tau = 50$. See text for further discussion.

IOP Conf. Series: Journal of Physics: Conf. Series **1290** (2019) 012030 doi:10.1088/1742-6596/1290/1/012030

Fig. 5(b). The observation of this confinement to a small volume of the available phase space is in mutual agreement with the smaller-amplitude oscillations of the imbalance [compare Fig. 4(b) and Fig. 5(b)].

4. Conclusions

We have investigated the tunneling dynamics of the quantum seesaw, a system of interacting bosons trapped in a temporally driven double-well potential.

We studied a harmonic drive of the seesaw with a constant frequency and a chirped drive of the seesaw with a frequency that increases linearly in time.

For the harmonic drive, at small values of the driving amplitude, the dynamics of the almost stationary system are very slow. The phase-space trajectory of the system is restricted to very limited region of the available phase space. For intermediate amplitudes of the drive, the dynamics become periodic in nature; the bosons populate different wells periodically in time. For large-amplitude drives, the periodicity disappears, the imbalance features fast oscillations, and – in phase-space representation – the dynamics resemble those of chaotic systems.

For a chirped drive of the quantum seesaw, we find quasi-periodic tunneling dynamics of the system: the time-evolution of the population imbalance features a small and a large frequency modulation with large and small amplitude, respectively. For chirped drives with a large amplitude, we observe similar two-frequency dynamics of the imbalance, but at a severely impeded amplitude with different frequencies as for small-amplitude chirped drives.

Our study thus establishes that by tuning the temporal evolution of the quantum seesaw, a precise control of tunneling dynamics of correlated bosons can be achieved. Since harmonic driving and chirped frequency modulation of the seesaw are experimentally achievable, our simulations can be experimentally realized in laboratories dealing with cold atomic gases.

As an outlook, it would be interesting to study the effects we found as a function of the interaction strength of the system. Furthermore, it would be highly interesting to understand how the observed different dynamical behaviors in the quantum seesaw are complemented by the many-body properties like the Glauber correlation functions [45] or the fragmentation [13, 14, 15, 16, 17, 18, 19, 20, 21, 22, 23, 24, 25, 26, 27, 28, 29, 30, 31] of the system.

References

[1] *Stochastic Behavior of a Quantum Pendulum under a Periodic Perturbation, in Stochastic Behavior in Classical and Quantum Hamiltonian Systems*, Lecture Notes in Physics, Vol. 93, G. Casati, B. V. Chirikov, F. M. Izraelev, and J. Ford, edited by Giulio Casati and Joseph Ford, Berlin Heidelberg; Springer (1979).

[2] F. Haake, M. Kus, and R. Scharf, Z. Phys. B **65**, 381 (1986); *Quantum Signatures of Chaos*, F. Haake, Berlin; Springer (2001); S. Fishman, D. R. Grempel, and R.E. Prange, Phys. Rev. Lett. **49**, 509 (1982).

[3] *The Transition to Chaos: Conservative Classical Systems and Quantum Manifestations*, L. E. Reichl, New York; Springer (2004).

[4] A. Eckardt, C. Weiss, and M. Holthaus, Phys. Rev. Lett. **95**, 260404 (2005).

[5] S. Longhi, J. Phys. B: At. Mol. Opt. Phys. **44**, 051001 (2011).

[6] K. Winkler, G. Thalhammer, F. Lang, R. Grimm, J. Hecker Denschlag, A. J. Daley, A. Kantian, H. P. Büchler, and P. Zoller, Nature **441**, 853 (2006).

[7] H. L. Haroutyunyan and G. Nienhuis, Phys. Rev. A **70**, 063603 (2004); G. Watanabe and H. Mäkelä, Phys. Rev. A **85**, 053624 (2012); W. Hai, C. Lee, G. Chong, and L. Shi, Phys. Rev. E **66**, 026202 (2002); G. L. Salmond, C. A. Holmes, and G. J. Milburn, Phys. Rev. A **65**, 033623 (2002).

[8] X. Luo, Q. Xie, and B. Wu, Phys. Rev. A **77**, 053601 (2008).

[9] J. Gong, L. Morales-Molina, and P. Hänggi, Phys. Rev. Lett. **103**, 133002 (2009).

[10] P. Ponte, Z. Papi, F. Huveneers, and D. A. Abanin, Phys. Rev. Lett. **114**, 140401 (2015); P. Bordia, H. Lschen, U. Schneider, M. Knap, and I. Bloch, Nature Physics **13**, 460 (2017).

[11] A. Eckardt, M. Holthaus, H. Lignier, A. Zenesini, D. Ciampini, O. Morsch, and E. Arimondo, Phys. Rev. A **79**, 013611 (2009).

[12] M. H. Anderson, J. R. Ensher, M. R. Matthews, C. E. Wiemann, and E. A. Cornell, Science **269**, 198 (1995).

[13] A. U. J. Lode, A. I. Streltsov, K. Sakmann, O. E. Alon, and L. S. Cederbaum *How does an interacting many-body system tunnel through a potential barrier to open space?*, Proc. Natl. Acad. Sci. USA **109**, 13521 (2012).
[14] I. Brezinova, A. U. J. Lode, A. I. Streltsov, O. E. Alon, L. S. Cederbaum, and J. Burgdörfer, *Wave chaos as signature for depletion of a Bose-Einstein condensate* Phys. Rev. A **86**, 013630 (2012).
[15] J. Grond, A. I. Streltsov, A. U. J. Lode, K. Sakmann, L. S. Cederbaum, and O. E. Alon, *Excitation spectra of many-body systems by linear response: General theory and applications to trapped condensates*, Phys. Rev. A **88**, 023606 (2013).
[16] I. Brezinova, A. U. J. Lode, A. I. Streltsov, L. S. Cederbaum, O. E. Alon, L. A. Collins, B. I. Schneider, and J. Burgdörfer *Elastic scattering of a Bose-Einstein condensate at a potential landscape*, J. Phys.: Conf. Ser. **488**, 012032 (2014).
[17] S. Klaiman, A. U. J. Lode, A. I. Streltsov, L. S. Cederbaum, and O. E. Alon, *Breaking the resilience of a two-dimensional Bose-Einstein condensate to fragmentation*, Phys. Rev. A **90**, 043620 (2014).
[18] M. C. Tsatsos and A. U. J. Lode, *Vortex nucleation through fragmentation in a stirred resonant Bose-Einstein condensate*, J. L. Temp. Phys. **181**, 171 (2015).
[19] A. U. J. Lode, B. Chakrabarti, and V. K. B. Kota, *Many-body entropies, correlations, and emergence of statistical relaxation in interaction quench dynamics of ultracold bosons*, Phys. Rev. A **92**, 033622 (2015).
[20] U. R. Fischer, A. U. J. Lode, and B. Chatterjee, *Condensate fragmentation as a sensitive measure of the quantum many-body behavior of bosons with long-range interactions*, Phys. Rev. A **91**, 063621 (2015).
[21] A. U. J. Lode and C. Bruder, *Dynamics of Hubbard Hamiltonians with the multiconfigurational time-dependent Hartree method for indistinguishable particles*, Phys. Rev. A **94**, 013616 (2016).
[22] S. E. Weiner, M. C. Tsatsos, L. S. Cederbaum, and A. U. J. Lode, *Phantom vortices: hidden angular momentum in ultracold dilute Bose-Einstein condensates*, Scientific Reports **7**, 40122 (2017).
[23] A. U. J. Lode and C. Bruder, *Fragmented Superradiance of a Bose-Einstein Condensate in an Optical Cavity*, Phys. Rev. Lett. **118**, 013603 (2017).
[24] M. C. Tsatsos, J. H. V. Nguyen, A. U. J. Lode, G. D. Telles, D. Luo, V. S. Bagnato, and R. G. Hulet, *Granulation in an atomic Bose-Einstein condensate*, arXiv:1707.04055 (2017).
[25] B. Chatterjee and A. U. J. Lode, *Order parameter and detection for crystallized dipolar bosons in lattices*, arXiv:1708.07409 (2017).
[26] P. Molignini, L. Papariello, A. U. J. Lode, R. Chitra, *Superlattice switching from parametric instabilities in a driven-dissipative BEC in a cavity*, arXiv:1710.02474 (2017).
[27] R. Roy, A. Gammal, M. C. Tsatsos, B. Chatterjee, B. Chakrabarti, and A. U. J. Lode, *Phases, many-body entropy measures and coherence of interacting bosons in optical lattices*, Phys. Rev. A **97**, 043625 (2018).
[28] A. U. J. Lode, F. S. Diorico, R. Wu, P. Molignini, L. Papariello, R. Lin, C. Lévêque, L. Exl, M. C. Tsatsos, R. Chitra, and N. J. Mauser, *Many-body physics in two-component Bose-Einstein condensates in a cavity: fragmented superradiance and polarization*, New J. Phys. **20**, 055006 (2018).
[29] S. Dutta, M. C. Tsatsos, S. Basu, and A. U. J. Lode, *Management of the Correlations of Ultracold Bosons in Triple Wells*, arXiv:1802.02407 (2018).
[30] S. Bera, B. Chakrabarti, A. Gammal, M. C. Tsatsos, M. L. Lekala, B. Chatterjee, C. Lévêque, and A. U. J. Lode, *Sorting Fermionization from Crystallization in Many-Boson Wavefunctions*, arXiv:1806.02539 (2018).
[31] B. Chatterjee, M. C. Tsatsos, and A. U. J. Lode, *Correlations of strongly interacting ultracold dipolar bosons in optical lattices*, arXiv:1806.05048 (2018).
[32] A. I. Streltsov, O. E. Alon, and L. S. Cederbaum, Phys. Rev. Lett. **99**, 030402 (2007).
[33] O. E. Alon, A. I. Streltsov, and L. S. Cederbaum, Phys. Rev. A **77**, 033613 (2008).
[34] K. Sakmann, *Many-body Schrödinger dynamics of Bose-Einstein condensates*, Springer Theses, (Springer, 2011).
[35] A. U. J. Lode, *Tunneling Dynamics in Open Ultracold Bosonic Systems*, Springer Theses, (Springer, Heidelberg, 2014).
[36] *Bose-Einstein Condensation in Dilute Gases*, C. J. Pethick and H. Smith, (Cambridge University Press, Cambridge UK, 2002).
[37] *Bose-Einstein Condensation*, L.P. Pitaevskii and S. Stringari, (Clarendon Press, Oxford, 2003).
[38] M. Greiner, O. Mandel, T. Esslinger, T. W. Hänsch, and I. Bloch, Nature **415**, 39 (2002).
[39] D. Jaksch, C. Bruder, J. I. Cirac, C. W. Gardiner, and P. Zoller, Phys. Rev. Lett. **81**, 3108 (1998).
[40] *Geometry of the time-dependent variational principle*, P. Kramer and M. Saracen, (Springer, Berlin, 1981).
[41] A. U. J. Lode, K. Sakmann, O. E. Alon, L. S. Cederbaum, and A. I. Streltsov, *Exact quantum dynamics of bosons with finite-range time-dependent interactions of harmonic type*, Phys. Rev. A **86**, 063606 (2012).
[42] A. U. J. Lode, Phys. Rev. A **93**, 063601 (2016).
[43] E. Fasshauer and A. U. J. Lode, Phys. Rev. A **93**, 033635 (2016).
[44] A.U.J. Lode, M.C. Tsatsos, E. Fasshauer, R. Lin, L. Papariello, P. Molignini, and C. Lévêque,

MCTDH-X: *The time-dependent multiconfigurational Hartree for indistinguishable particles software*, http://ultracold.org (2018).
[45] R. J. Glauber, Phys. Rev. **130**, 2529 (1963).

AUTHOR INDEX